Foundations of Economics

Foundations of Economics

A Christian View

SHAWN RITENOUR

WIPF & STOCK · Eugene, Oregon

FOUNDATIONS OF ECONOMICS
A Christian View

Wipf & Stock
A Division of Wipf and Stock Publishers
199 W. 8th Ave., Suite 3
Eugene, OR 97401
www.wipfandstock.com

ISBN 13: 978-1-55635-724-4

Manufactured in the U.S.A.

This book is dedicated to Michelle,
my virtuous wife.
You are my crown and far more precious than jewels.

Contents

Acknowledgments

M ANY PEOPLE ASSISTED ME in the completion of this book. I want to thank David Westrate and the Tithes and Offerings Foundation for providing a significant grant of financial assistance to help cover expenses. Grove City College graciously provided a sabbatical, allowing me to do much concentrated research and writing. Ronda Credille, Jeffrey Herbener, and David Whitlock read every word of earlier drafts, providing me with many helpful comments and suggestions. My student assistants, Jocelyn Bond and Robbie Long, provided invaluable help in research, editing, and graph composition.

It should be clear that my intellectual debts are from the Misesian tradition of Austrian economics and Reformed Protestant theology. My economic pedagogy relies heavily on Murray Rothbard's *Man, Economy, and State* and the intellectual foundation laid by Carl Menger and Eugen von Böhm-Bawerk and built into a full edifice by Ludwig von Mises. The Ludwig von Mises Institute has been a great blessing to me by providing material, intellectual, and moral support as I seek to integrate economic theory with a Christian view of man and nature. In lieu of a myriad of footnotes, I list specific sources I rely upon as suggested readings at the end of each chapter. The reader should think of these as a chapter by chapter bibliography. An exhaustive bibliography is also given at the end of the book.

Soli Deo Gloria.

Introduction

W HEN APPLYING TO GRADUATE school to study economics, I was
required to send letters of recommendation from three former
professors. My college was a small, Christian liberal arts college that only
had two economists on its faculty. In addition to soliciting letters from
them, I asked one of my religion professors to write the third. I had taken
two courses from this professor and knew him to appreciate at least my
academic potential. When I telephoned him, asking that he write me a letter of recommendation for graduate school, he inquired about my chosen
field of study. When I told him economics, he quickly replied, "Oh, studying the things of *this* world, I see."

While my former professor was only joking, a none-too-small number of Christians believe that economics is about how greedy people
satisfy their lust of the eye and pride of life. This view is incorrect for
two reasons. In the first place, God is interested in both the spiritual and
material aspects of his creation. Additionally, economics is a social science concerned with how people cooperate in the division of labor, not
propaganda for squeezing the blood of the exploited working class out of
the proletariat turnip.

God is interested in all of his creation, both spiritual and material.
We know this because of what God tells us about his own opinion and
care for the material world. God did, after all, create a material world. As
the Nicene Creed puts it, he made all things "visible and invisible," both
material and spiritual. God tells us that after he was finished with creation he declared his handiwork "very good" (Gen. 1:31). Additionally, the
apostle John tells us that one of the defining characteristics of a spirit that
comes from God is one who acknowledges that Jesus Christ has come in
the flesh (1 John 4:2). Jesus had to take on human flesh to become our
sacrifice and he ate and drank enough to be wrongly accused of gluttony
by the Pharisees. In light of these truths, it is hard to see how we can con-

clude that God is either dismissive of or actively opposed to the material part of his own creation.

It is easy to understand how people might dismiss economic subjects as too worldly if they read the Bible selectively. There are, after all, a number of passages in God's Word that warn us against loving the things of this world, loving money, and storing up our treasures here on earth. However, one should not conclude from these caveats and commands that the Bible advocates some sort of anti-material asceticism.

That God is not ascetic is evident throughout Scripture. In Proverbs 6:6–11, for example, the writer instructs us to learn from the ant. What is the ant doing that we are to use as an example? It is working to make material provision for itself in the winter by working and gathering its food in the summer. In Genesis 2 we find that before the fall God commands Adam to cultivate and keep the garden and gives him the fruit of the trees from which he may freely eat. Later in Proverbs 12:11, God affirms the principle that material sustenance is to be had by work. "He who tills his land will be satisfied with bread . . . ." In Genesis 24:35 we read that God blessed Abraham with "flocks and herds, silver and gold, male and female servants, and camels and donkeys." Needless to say, these are all material blessings.

In the book of Deuteronomy, God explains some principles relevant to tithing. He tells his people that they are to take a tenth of all their increase to the temple where a portion of it was to be joyfully eaten with the Levites and the poor. If people lived a long way from the temple, they could convert their commodities to be tithed into money. Then God commanded his people, "Spend the money for whatever you desire—oxen or sheep or wine or strong drink, whatever your appetite craves. And you shall eat there before the Lord your God and rejoice, you and your household." (Deut. 14:26). Clearly God is not Platonically anti-material.

Later in Deuteronomy, God pronounces a series of blessings and curses that will be meted out to his covenant people depending on whether they are obedient or disobedient to his law. In each series, the first blessings and curses are material in nature.

> And if you faithfully obey the voice of the Lord your God, being careful to do all his commandments that I command you today, the Lord your God will set you high above all the nations of the earth. And all these blessings shall come upon you and overtake you, if you obey the voice of the Lord your God. Blessed shall you be in the city, and blessed shall you be in the field. Blessed shall

be the fruit of your womb and the fruit of your ground and the fruit of your cattle, the increase of your herds and the young of your flock. Blessed shall be your basket and your kneading bowl. Blessed shall you be when you come in, and blessed shall you be when you go out (Deut. 28:1–5).

On a less happy note, God pronounces curses that are the exact opposite of the blessings:

But if you will not obey the voice of the Lord your God or be careful to do all his commandments and his statutes that I command you today, then all these curses shall come upon you and overtake you. Cursed shall you be in the city, and cursed shall you be in the field. Cursed shall be your basket and your kneading bowl. Cursed shall be the fruit of your womb and the fruit of your ground, the increase of your herds and the young of your flock (Deut. 28:15–18).

These verses sufficiently demonstrate that God is concerned with all of our existence, the material as well as the spiritual.

In the middle of his Sermon on the Mount (Matt. 6:19–34), Christ commands us not to be anxious about material considerations. He warns us not to seek security from earthly treasures, but to set our hearts on heavenly treasures. In verse 31, Jesus tells us explicitly not to worry about what we are to eat, drink, or wear. Instead we are to seek first his kingdom; God is to be our first priority. The Westminster Shorter Catechism tells us that the chief end of man is to "Glorify God and enjoy him forever." It does not say that our chief end is to glorify the Gap and to enjoy Pizza Hut forever. No, Jesus says to seek his kingdom *first* and *then* all these things shall be added. Notice that he does not say that food, drink, and clothing are worthless or something that we ought not own. His teaching is not some sort of Christian asceticism. Instead he is commanding us not to become anxious about these real material needs, because in doing so, we are failing to trust God to provide for us.

Economics is not only about money, stuff, or even resource allocation. Economics is really about helping man solve one of the greatest dilemmas confronting him; it is the dilemma that we discover after reading the first few chapters of Genesis. The first command that God gives man is found in Genesis 1. Here God tells Adam and Eve, "Be fruitful and multiply; fill the earth and subdue it; have dominion over the fish of the sea, and over the birds of the air, and over every living thing that moves on the earth" (Gen. 1:28). The above mandate involves a lot of activity focused on material concerns. It includes reproduction and building things. The Hebrew word translated as *dominion* implies more than merely to rule. It communicates the idea of dominion for use.[1] Consequently, the cultural mandate includes rearranging what we find in nature to suit our needs and to glorify God. Contrary to contemporary environmentalism, it includes ruling over every living thing and all of nature, not communing with nature as a co-equal. It includes taking God-given property and using it in the production of goods that benefit others and ourselves.

However, this mandate was made much more difficult as a result of what we read about in Genesis 3. In this chapter we have the record of the fall of man. As the result of Adam's disobedience to God's command not to eat the fruit of the tree of the knowledge of good and evil, God cursed all parties involved.

For the study of economics, it is important to remember that this curse is not only spiritual in nature. It is true that spiritual separation from God is the most devastating consequence of original sin. However, note that much of the curse was also material in nature. God tells Eve, "[I]n pain you shall bring forth children" (Gen. 3:16). He tells Adam, "[C]ursed is the ground for your sake; in toil you shall eat of it all the days of your life. Both thorns and thistles it shall bring forth for you, and you shall eat the herb of the field. In the sweat of thy face you shall eat bread . . ." (Gen. 3:17–19). As the result of the fall, man was banished from the Garden of Eden into a world of greatly magnified scarcity, where it became much more difficult to obtain one's daily bread. Nothing in Scripture, however, leads us to the conclusion that the fall and subsequent curse voided God's original command for man to have dominion over all creation. Indeed, God repeats the dominion mandate to Noah and his family (Gen. 9:1–7). What is true is that keeping this command has become much, much more difficult.

This leaves us with one of our greatest dilemmas: How do we have dominion over the creation in this world of scarcity without starving to death, killing one another, or both? It turns out that the discipline of economics provides great assistance as we attempt to answer this question. In the paragraphs that follow a number of propositions will be logically connected in order to reach an answer to the dilemma posited above. Some of these propositions may contain terms that are unfamiliar and some of

1. Hay, *A Christian Critique of Socialism,* 30.

the propositions themselves may seem debatable. The rest of the text will be spent demonstrating the truth of each of these propositions as well as many others, so be patient.

Readers who diligently read this text will discover that in order to escape starvation and a barbaric struggle for survival, we must take advantage of social cooperation through the division of labor. The only way to be fruitful and multiply our offspring without facing starvation or developing a base survival-of-the-fittest mentality is to reap the benefits of greater productivity resulting from the division of labor. In order for a division of labor to exist, people must be able to exchange. As more and more people have the ability to exchange what they produce for other things that they want, they are increasingly able to concentrate on producing those goods at which they are most efficient. However, there are also obstacles to overcome as the division of labor expands.

An economy based on the division of labor is a very complex economy. There are multitudinous natural resources, capital goods, and labor skills that can be combined in a myriad of ways to produce a vast array of goods. Given this complexity, it is not an easy task to coordinate all of the various production projects or to decide how to use our land, labor, and capital goods in the least wasteful fashion.

Readers later will find that a method of calculating profit and loss is necessary for any complex economy to be fruitful and to prosper. Profit is the positive difference between the benefit of an action and the total cost of that action. There are a number of different uses toward which land, labor, and capital goods can be directed. All of these alternatives must be compared to one another if people are to know how best to fulfill their desire for goods. Such a grand comparison only can be made if alternatives somehow can be compared to a common denominator for assessing how useful each quantity of land, labor, and capital are in the different projects in which they could be used. Such comparisons can be accomplished only through the use of free market prices denominated in terms of money. With the development of money as the medium of exchange, all exchange ratios for all goods are expressed in terms of money.

Consequently, producers have a way to compare the market price of their final product with the sum of the prices of all their inputs for every potential investment project. Market prices provide an objective way to appraise the relative wants of people in society. Diligent readers of this text will find that while all economic goods are valued subjectively by individuals, these subjective values of the buyers and sellers are manifested in the objective exchange ratios for all traded goods. Hence, while value is subjective and cannot be measured, the money prices are the objective manifestations of value that are determined by and based on personal subjective value scales. Entrepreneurs, therefore, are able to make wise decisions by using market prices as the basis for calculating expected profits and losses so that they are able to direct factors of production toward making those products that people want the most.

Note the use of the term *market prices* in the paragraph above. Readers of this text will discover that in order for entrepreneurial calculation to be appropriate, the prices they use to count potential profits and losses must be prices arrived at in the market. Socialism is doomed to failure because there is no way for a central planner to efficiently allocate factors of production toward making those goods most in demand. In a socialist economy there is no private ownership of the means of production. The state owns it all. There can be, consequently, no real exchange of the means of production. If there is no real exchange, there can be no real market prices that are manifestations of people's subjective value scales. Not only are there no market prices for consumers' goods, more importantly, there are no real market prices for producers' goods.

Without market prices of capital goods, there is no way for the central planner to compare the economic value of different materials and techniques that could be used to produce consumers' goods. Managers would be without the ability to calculate the expected profit or loss of various production projects. They would be left, as Ludwig von Mises put it, "groping about in the dark."[2] There is no way, then, for a central planner to rationally allocate resources which is the main reason why one sees so much absolute waste and subsequent poverty in a socialist economy.

Careful readers will discover that taking advantage of the division of labor requires the ability to exchange at prices voluntarily agreed upon by suppliers and demanders in a free market. Both of these requirements necessitate the existence of private property. People cannot exchange what they do not own. Additionally, only prices that result from people voluntarily trading their own property truly represent the relative value that they themselves place on different goods. So if we want to be fruitful and multiply and replenish the earth while we subdue it and if we want

---

2. Mises, *Human Action*, 696.

to be able to do this without killing one another, then we must maintain private property. Note, however, that the institution of private property is not simply a recommendation from economics. Private property is mandated by God as well.

Christians throughout history have recognized that when God commands us not to kill, not to steal, and not to covet, he is demanding that we recognize that there is a difference between what is mine and what is yours. Private property is not a mere human construct, but a divine right. Therefore, prosperity will generally be the result of obeying God on the issue of private property.

These conclusions and their further implications may strike some as rather radical, simplistic, plainly wrong, or even unchristian, but if one carefully reads and processes the following chapters, he should come to the same conclusions. Economic knowledge is gained by understanding that social regularities exist as a result of all people being made in the image of God. If we want to prosper by fulfilling God's mandate to have dominion in a civilized world of peace, we dare not dismiss what economics has to say.

1

The Biblical Foundations of Economics

PERHAPS THE QUESTION ADULTS ask each other more than any other when they first meet is, "What do you do?" When I respond, "I'm an economist," more often than not, I get one of two requests. I either receive a feeble plea to help someone balance their checkbook, or I am asked to supply the inside scoop on what the stock market is going to do in the near future. Both of these requests belie ignorance regarding what is economics.

The term *economics* is itself derived from the Greek word *oikonomikos* which originally meant *relating to household management*. To the ancient Greek, then, economics really did involve balancing the checkbook, or at least keeping track of the family budget, closer to what we used to call home economics.

However, the body of knowledge we today call economics is a social science, not household management theory. It is social in that it focuses on interpersonal action. It concerns itself with people . . . who need people.

Economics is a science. A science is "a systematic arrangement of the laws which God has established, so far as they have been discovered, of any department of human knowledge."[1] Economics is not just the regurgitation of singular, unrelated facts about how people behave. It is a body of thought that provides a number of principles that we can apply as we seek to answer important questions like, what will happen if we raise the minimum wage, decrease income taxes, put price controls on prescription drugs, or bail out insolvent mortgage companies.

A science involves the systematic arrangement of facts and truths. Science helps us make sense of the plethora of facts we experience, and by discovering the operation of general laws, science helps us arrive at true solutions to problems we face in the world. Note also that a science is not

1. Wayland, *The Elements of Political Economy*, 15.

about arranging a body of opinions or suggestions, but a body of truths so that we show the operation of real laws that are in force in the world in which we live. When I was in college, I many times heard my friends sympathetic to the welfare state bemoan that while communism is great in theory, it is a pity that it does not work in practice. To the contrary, if a certain economic policy is bad in practice, it usually means that it is bad in theory. Not only did communism fail miserably in practice, it did so precisely because it is a failure theoretically. Austrian economists Ludwig von Mises and Frederick Hayek demonstrated the theoretical failure of socialism as far back as the 1920s and 1930s, but few wanted to listen.[2] Although they had ears to hear, they heard not. It took seventy years of economic stagnation, culminating in the fall of the Soviet Union, for many scholars to take a second look at the arguments of Mises and Hayek. The object of economics is to discover economic principles that are true, so that policies developed in light of those principles will be suitable for achieving our goals.

WHAT IS TRUTH AND HOW DO WE KNOW IT?

Almost two thousand years ago, a certain Roman governor of Judea asked the question, "What is truth?" An observer standing there could have been excused for replying, "Truth is looking you right in the face." However, when most contemporary scientists (social or otherwise) ask the same question, they are usually not inquiring about the nature of Christ, but about the truths of the universe that Christ brought into being—whether scientists acknowledge the source of all truth or not.

So what is truth? The *Oxford English Dictionary* defines truth as "Conformity with fact; agreement with reality." A statement is true then, if it is factually correct and if it agrees with reality. It is possible for a statement to be factually correct and yet be untrue. For example, I can usually illicit a goodly amount of sympathy by telling the story of my brief life in telemarketing. I begin by explaining how at one point in my life, after having resigned from the United States Bureau of Labor Statistics, I moved back to Iowa and found a job working as a telemarketer with a firm in Omaha, Nebraska. They were in desperate need for workers because of an account that was fabulously successful. The firm was taking credit card

2. See the articles collected in Hayek, *Collectivist Economic Planning*. See also Mises, Ludwig von, *Socialism*.

applications over the telephone for a card issuer in the Midwest. There was a huge response that caught the telemarketing firm off guard. I responded to an advertisement to work telemarketing at a wage that was $2.50 above the norm for that type of work. Well, it did not take long for the popularity of the credit card to run its course and the calls slowed down and the firm was forced to let some people go. Because I was one of the most recent hires, I was one of the earliest to be laid off. I was given the news one week before Christmas. By stopping the story here, I can usually generate sighs of *Oh no!* as well as a general sympathy with my cause. It also tends to go away as I tell them, as Paul Harvey might say, the rest of the story. It turns out that my getting laid off a week before Christmas was no big deal, because I was living with my parents, so I did not have to worry about paying for housing or food. Suddenly it becomes clear that what some selected facts seemed to indicate regarding my plight, turns out to be not the whole truth. Indeed, people can communicate things that are not true, while only relating facts that are correct. Therefore we must understand truth to be something that is factually correct and that also corresponds to reality. It is rather easy to define what truth is; it is a bit harder to establish how we know truth.

The discipline that seeks to answer the question, "how do we know?" is a branch of philosophy called epistemology. The word *epistemology* is derived from Greek and means the study of knowledge. About the time I get to this topic in the course I teach, I usually get at least one student who asks me why we go over *this* topic. After all, this is a book about economics, not philosophy. True enough. However, as you make your way through this book, you will find that several principles, such as the law of marginal utility, the law of comparative advantage, or the law of demand, are presented as economic truths. It seems to make sense to spend a little time up front examining just how, exactly, we do know anything, so that we can be sure that the law of demand really is a *law* instead of some opinion of demand dreamt up by another ivory tower sophist. Besides, the word *epistemology* is good to toss into your vocabulary when your parents ask you what they are teaching you at college. This is especially good if it is their hard-earned money that is paying your tuition.

Throughout the history of philosophy there have been generally four schools of thought regarding how we know. The first theory of epistemology we will examine is *skepticism*. Actually, skepticism is more like the anti-epistemology. When people think of someone who is a skeptic, they usually think of someone who doubts something, such as someone doubting whether our politicians always tell the truth. A philosophical skeptic is a little different. The fundamental proposition of skepticism is that knowledge is impossible. You may have heard it put this way: there are no absolute truths. It is fairly easy to spot the problem with this theory of knowledge. The claim that knowledge is impossible implies that we at least know that knowledge is impossible. "There are no absolute truths," he said absolutely. Consequently, skepticism is internally inconsistent and, therefore, cannot possibly be true. If it is true then it is false.

Another closely related theory of knowledge is *relativism*. Relativists do not claim that truth does not exist, but that different people, different groups, or different cultures have different truths, all of them equally true. Many advocates of multiculturalism hold this view. Contemporary American philosopher Richard Rorty also argued that truth is relative to its cultural setting. In this sense, truth is what your peers let you get away with saying. Like skepticism, however, relativism falls quickly flat on its own propositions. Notice, for example, that the claim that truth is relative to different people, groups, or cultures, is a universal claim. It is always asserted absolutely, as if it applies to everyone. The relativist asserts the proposition that for all people, groups, and cultures, it is true that all people, groups, and cultures have their different truths that are all equally true. But if this proposition truly applies to everyone, everywhere, in every culture, it contradicts what is being asserted. If this proposition is indeed true for all people, groups, and cultures, then at least one truth is not relative. Again, if relativism is true, then it is false.

A third theory that attempts to explain how we know truth is *empiricism*. At some point in your education, you may have heard something referred to as empirical. If knowledge is empirical, that means it is based on experience. Empiricism claims that knowledge is only the result of experience. Note that it does not argue that we learn *some* things by experience, but that *all* knowledge is the result of experience. British philosophers John Locke and David Hume were in this camp, as well as a lot of modern scientists. Empiricists are rather like intellectual Joe Fridays imploring, "Just the facts, ma'am."

Empiricists have argued that at birth our minds are merely blank slates upon which our experience impresses ideas, and these ideas are the source of all of our knowledge. These ideas are the result of sense experience and inner experience. Ideas of sense experience are what one would expect: sensations like red, hot, spicy, and fuzzy. We know the bacon has burned by smelling the odor of rancid charred pork. Physical scientists use their sight to gain knowledge when looking at their various measurement instruments. You can learn whether it is a hot and muggy day by going outside and feeling the moisture and heat in the air. People learn musical technique largely through hearing sounds coming from their voices or instruments.

People can also learn by their sense of taste. The truism that appearances can be deceiving was brought home to me very clear while a student in college. One day while making my way down the cafeteria food line, I spotted a delicious looking bowl of vanilla pudding waiting for me like gold at the end of the rainbow. Filled with gastronomical joy, I placed a bowl on my tray and thought of the culinary delight that awaited me as I consumed my salad, overdone green beans, and patty melt. Imagine my shock, then, as I passed that first dollop of the pale yellow pudding into my mouth, only to be made dreadfully aware that the pudding I had been longing for was not vanilla after all, but lemon! What a sour tang abused my taste buds and proceeded to travel about my mouth. You know, the kind that makes your teeth hurt. It was my sense of taste that brought home the fact that what looks like vanilla can actually be lemon. In other words, appearances can be deceiving.

Ideas of inner experience come as we reflect on our sense experience. We use our mind to undergo mental operations such as thinking, perceiving, and doubting. By using both ideas of sensation and ideas of reflection, it is argued, we derive ideas of relation. We can combine the ideas of spicy, red, and wet to form the idea of Louisiana hot sauce. We can consider Louisiana hot sauce, mustard, ketchup, horseradish, pickle relish, and Tabasco sauce to form the concept of condiment.

Empiricism is an improvement over skepticism and relativism in that it does not assert that there is no truth, or that different cultures have different truths that are all equally true. Empiricism does not immediately descend into self-contradiction. However, it does have its own problems that leave it wanting as a valid epistemology. Remember that empiricism does not argue that *some* knowledge is the result of experience. Empiricists argue that *all* knowledge that can be known is the result of experience. Ah, as Hamlet would say, there's the rub.

When an empiricist makes the claim that we can only know if something is true based on experience, one is tempted to ask him, *where is your data?* The only way empiricists could make such a universal claim, based on their own epistemology, would be if they had actually experienced all of the facts in the universe. No one has done that, so using their own epistemology, empiricists do not have enough knowledge to know whether their claim is true or not.

In fact, there are several truths that are not based on experience. Our knowledge of time, for example, is not the result of an image of time being impressed upon our minds. The concept of time does not stem from our experiences, but we use this concept to make sense of our experience. We understand experiences as occurring sooner and later because we use the already present concept of time to interpret our experience.

Neither are the truths of geometry and arithmetic based on our experience. Mathematicians do not go about measuring triangles to demonstrate the principles of Euclidean geometry. Even when teachers instruct children in principles of addition by taking two pennies and combining them with two more pennies to make four, such instruction relies on the children already possessing knowledge of numbers. He cannot count pennies from one to four until after he knows how to count numbers.

Additionally, the empiricist cannot logically accept the notion of causality. Causality can never be observed by our senses; only correlation can. For example, economics teaches that if the price of tea in China decreases, people will desire to buy a larger quantity of tea than they were willing to before the price drop. Empiricists cannot explain this cause and effect relationship because at most all they have to go on is data that tell them in the past higher prices for tea in China were correlated with decreases in the quantity bought. The cause and effect relationship known as the law of demand is not something that is known to be true through our sense experience, because causation is not understood through our senses. It is a logical concept that we use to make sense of our experience.

On top of that, there is no way for an empiricist to know whether the observed correlation will continue into the future even one day, let alone for all time. If all knowledge is based on experience, nothing can validly be said about the future, because we have not experienced it yet. Experience is always experience of the past.

While experience, by itself, can never provide us with universal principles—such as that of causation—that are true for all time, entire disciplines require such universal principles. Physics, mathematics, and economics, for example, all require necessary and universal judgments about the world that are always true. If I dropped a ball yesterday and it fell to the ground yesterday yet drop that same ball tomorrow and anything can happen, then the law of gravity is not really a natural law. Rather, it is merely a description of what has happened in history. In order to discover the sort of universal laws of physics and economics, empiricism does not suffice.

None of the above should be taken to mean that we can learn nothing from empirical studies. Indeed, there is much to be learned from empirical analysis in its proper place. This chapter is focused, however, not on usefulness, but on foundations. Although we can learn true knowledge by using the experience of history, we cannot base the truthfulness of truth only on experience.

The theory of epistemology that we have left then is that of some kind of *apriorism*. The root word *prior* gives us a hint regarding the meaning of apriorism. Prior means *before*. Apriorism argues that in order for us to gain knowledge, there is something that is logically prior to empirical facts. Apriorists recognize that our mind imposes unity on experiences by applying innate mental categories in classifying and judging our experience. Our minds are not blank slates scribbled upon by atomistic facts. Our mind makes sense of our experience by organizing and judging the facts we encounter.

The laws of logic by which we think and analyze our experience are not derived from the facts, but are instead used to make sense of the facts. For example, the truth that if A is greater than B and B is greater than C, then A is greater than C is not something derived from experience. This truth—the principle of transitivity—is a logical law with which our minds are equipped that allows us to make sense of the many facts that come our way every day.

The most famous of all apriorists is probably Immanuel Kant (1724–1804), a German philosopher. He believed that we do learn by experience, but only because we can make sense of it. He argued that we can only make sense of our experience because we possess a priori categories of thought, such as logic, unity, space, and causality. Further, he thought that these a priori categories that we use to make sense of our experience are subjective aptitudes. That is, Kant believed that everyone has personal categories of thought that are known by intuition. These subjective categories introduce order into our sensory data by the mind alone. However, Kant could not get past the supposition that these mental categories do not necessarily allow us to perceive truth, but only dictate how we must think about things. For instance, Kant would have to admit that, based on his theory, perhaps there really is no causality in the world, and that it only seems like there is because our minds are programmed to think that way. I cannot say that a drop in the price of tea in China caused an increase in the quantity of tea demanded in China, only that it is impossible for me to think otherwise. An order is impressed upon our experience, but the world as it really is remains unknowable.

A BETTER WAY

You may recognize that Kant's problem leads us back to relativism and skepticism. Your categories could be *true* for you and mine could be *true* for me and never the two necessarily have to meet. So our trek through various epistemologies seemingly has led us nowhere. Take heart. We need not despair. What if we do have mental categories but our categories are implanted in us by a Creator, and what if he has fashioned both our mind and the world so that they harmonize? This is the idea we get from God's revelation given in the Bible.

This approach is a biblical apriorism. It begins with God's revelation in both his creation and his written word. Christians believe first and foremost that God is. He exists. Additionally, as Francis Schaeffer aptly titled his important book, *He Is There and He Is Not Silent.* God has spoken to us by revealing in his word absolute truth. This truth is not necessarily exhaustive, but it is absolute.

What truths has God communicated to us in the Bible? God created and actively sustains the universe. The very first verse of the first book in the Bible teaches, "In the beginning, God created the heavens and the earth" (Gen. 1:1). God created all there is. This includes time, space, the facts of science, and the human mind.

However, God tells us not only that he created all there is, but that he also actively sustains all there is. In the letter to the Colossians Paul tells us, "For by him (Christ) all things were created, in heaven and on earth, visible and invisible, whether thrones or dominions or rulers or authorities—all things were created through him and for him. And he is before all things, and in him all things hold together" (Col. 1:16–17). The phrase *hold together* in the Greek implies that all things continue and cohere. In Hebrews 1:3 we are told that Christ upholds the universe by the word of his power. Not only did Christ create all there is, but he is right now, even as you read this text, holding everything together. The atoms that make up the pages of this book are actually moving and vibrating very fast. It is God's providence that keeps all of those atoms from exploding. God is personally involved with his creation; he is not some sort of divine watchmaker that spun the universe up and just let it run.

Additionally, we see that God created this universe with order and purpose. In Genesis 1:14–16 we read the account of God making the sun, moon, and stars. God explicitly states that these lights were made to serve specific purposes. In verse 14 God says that they were made to "be for signs and seasons, and for days and years." God causes the moon, stars, and planets to move in such a way to produce regular seasons, days, and years. In verse 15 he tells us that their purpose is to "be for lights in the expanse of the heavens to give light upon the earth." The things that God has created exist for a purpose and exhibit regularity. In Genesis 1:20–25 God tells us that he made everything to reproduce after its own kind. A horse does not give birth to a pig which gives birth to a monkey which gives birth to a rhinoceros. All living creatures were brought forth according to their kinds and reproduce accordingly.

The regularity with which God providentially upholds nature is so recognizable that God uses the constancy of natural laws as evidence that we can trust his covenantal word. Speaking of God's eternal covenant with David, this word of the Lord came to the prophet Jeremiah: "Thus says the Lord: If you can break my covenant with the day and my covenant with the night, so that day and night will not come at their appointed time, then also my covenant with David my servant may be broken, so that he shall not have a son to reign on his throne, and my covenant with the Levitical priests my ministers" (Jer. 33:20–21). Here we see that the regularity of day and night is God's covenant. It is God who ensures that day will follow night and night follow day. We also find that God treats the recurrence of day and night as part of a natural law that transcends humanity. Finally, God uses the certainty of the pattern of day and night as reason to believe that God will keep a son of David on the throne.

We find similar usage of natural regularities in the Psalms. Truthfulness and faithfulness of God's word is attested to by the stability and endurance of creation (Psalm 119:89–90). The unfailing process of nature serves as token of the certainty of the enduring rule of Christ in his kingdom (Psalm 72:5–7, 17). God's covenant is as certain as the regular appearance of the moon (Psalm 89:34–37).

So the Bible affirms there is a purpose and orderliness to creation. Without such orderliness, it would be impossible to have any sort of science. There could be no scientific laws. All would be chaos. Because God created a universe with order and purpose, however, we can undertake science. We can observe natural and social regularities. We can use concepts of cause and effect to derive scientific laws.

Why do we think we have the mental ability to discover scientific laws? The Christian doctrine of man can help us here. The Christian view of man begins with the fact that man is created by God in his image. We understand more about the nature of man, then, as we understand more about the image of God. Christians understand the *image of God* to mean God's likeness. In other words, man is God's replica on earth. God created man as "a reasonable and immortal soul endued with knowledge, righteousness, and true holiness, after his own image."[3] If man is God's replica, then we can learn about the nature of man by examining the attributes of God.

For the purposes of the subject of economics, it is sufficient to stress only a few characteristics of God as revealed in the Bible. One is that God thinks. In Isaiah 55:8–10 God tells us that his thoughts are not our thoughts, but are higher than man's thoughts. Now obviously God's thoughts could not be higher than our thoughts unless he actually has them. Therefore we know that God thinks. Note that this passage also affirms that *both* God and man think. In Jeremiah 29:11, God tells us that he has thoughts of peace toward his people.

Additionally God reveals to us that he is a rational being. When calling Judah back to himself, through the preaching of the prophet Isaiah we read , "Come now, let us reason together, says the Lord: though your sins are like scarlet, they shall be as white as snow; though they are red like crimson, they shall become like wool" (Isaiah 1:18). God wants us to engage in a reasoned discussion with him during which we consider the rational propositions that he makes about himself and us.

Christian doctrine also teaches that God is omniscient. He knows everything. If God knows everything, he must think thoughts and engage in cognition. As a being created in God's image, man possesses mental faculties he can use to know things. God fit us with minds that exhibit mental categories reflecting his image and with these mental categories,

---

3. Westminster Confession of Faith, Chapter IV, Section II.

we are able to perceive reality because the same Creator made the world so that it harmonizes with our mental categories.

Obviously, there are important differences between us and God. In the first place, while God is infinite, we are his finite creatures. Therefore, we do not have exhaustive knowledge. We can, however, know some knowledge and the knowledge that we are capable of knowing we can know with certainty. The apostle John, for example, wrote in his first letter, "I write these things to you who believe in the name of the Son of God that you may know that you have eternal life" (1 John 5:13).

We also suffer from the consequences of the fall. Man's mind has been corrupted since the fall, but we can still know things. We do not think perfectly, but we still think. Our world still reflects God's orderly being and is therefore coherent. Consequently, we are able to use our mind to investigate God's orderly creation to discover certain regularities in what God has made. We are not perfect and we make mistakes, including mental errors. We do not, however, always make mistakes. In his letter to Titus the apostle Paul approvingly quotes a Cretan poet who wrote "'Cretans are always liars, evil beasts, lazy gluttons.' This testimony is true" (Titus 1:12–13). Here Paul is affirming the general observation of an unregenerate writer.

THE CHRISTIAN VIEW OF MAN AS RATIONAL ACTOR

In economics, the object of our study is man. Therefore, the Christian view of man not only instructs us regarding the possibility of perceiving truth and pursuing scientific discovery. The Bible also provides information that helps guide us to the foundation of economic science.

The Bible characterizes man as a creature who engages in action, that is, purposeful behavior. We again see this by considering the doctrine of man being created in God's image. It has been explained above that God thinks and man also thinks.

There are other characteristics of God that, furthermore, indicate that man is a being who undertakes action. God does not only think. He plans. In Ephesians Paul says of believers that "even as he [God] chose us in him before the foundation of the world, that we should be holy and blameless before him. In love he predestined us for adoption through Jesus Christ, according to the purpose of his will, to the praise of his glorious grace, with which he has blessed us in the Beloved" (Eph. 1:4–6). Among other things, this tells us that God planned for the salvation of lost sinners before the foundation of the world. God plans.

Scripture likewise affirms that, not only does God think, but he acts as well. Within only the first four verses of Genesis, we learn that God created (v. 1), spoke (v. 3), and separated (v. 4). All of these are actions. The Bible also describes the actions of God in the framework of choice. Isaiah 14:1 is only one verse among many that explains that God chose Israel as his people from all possible alternative races. As already noted, Genesis 1:14–17 reveals that God specifically acts with a purpose. God created the sun, moon, and stars with specific ends in mind.

Because God thinks and acts with purpose and because man is made in the image of God, it is reasonable to conclude that man is able to think and act with purpose. It can be inferred, then, that a very important part of the image of God is reason: the ability to think rationally in terms of cause and effect.

Furthermore, the Bible explicitly characterizes man as one who reasons. Throughout the Bible, God deals with man in a rational manner. As we have already mentioned, when calling to man that we should follow him, God implores in Isaiah 1:18, "Come now, and let us reason together." God appeals to our reason and expects us to make the reasonable choice based on the facts as God reveals them. In Matthew we find the instance when Jesus was warning against the doctrine of the Pharisees and Sadducees. He couched his warning in figurative language, warning the twelve against the "leaven" of the religious leaders. When the disciples did not understand and were puzzled over Jesus' figurative language, Jesus says to them "O ye of little faith, why reason ye among yourselves?" (Matt. 16:8 KJV). The Greek word translated as *reason, dialogizomai,* from which we get our English word *dialogue,* means *to reckon thoroughly* and is rendered elsewhere in Scripture as *consider, muse,* and *think.* Jesus, then, indicates that man is a being who uses his mind to rationally contemplate questions.

From the Biblical doctrine of creation then, we arrive at a number of conclusions relevant for economics. One is that we can discover natural and social regularities. Because God's creation is orderly and because God created us in his image with the ability to think and know, we have the ability to perceive creation and its regularities. Natural and physical regularities we refer to as natural and physical laws—such as the law of gravity. However, because part of bearing the image of God includes acting with a purpose, we can discover social regularities, some of which are economic laws concerning human action—such as the law of marginal utility and the law of demand.

We do not, therefore, believe the teachings of economics because many people in Western Civilization have believed them, although they have. We do not believe the teachings of economics because experience verifies the truth of economics, although it does. We do not even believe the teachings of economics because all humans have rational minds which allow us to understand that humans act purposefully, although we all do have minds fit for rational thought. We believe the truths of economics because God has created us in his image with the ability to know and perceive truth and one of these truths communicated to us in his creation and his Word is that, like God, we act with a purpose.

As mentioned, economics is a social science and, therefore, studies how humans engage in social (i.e. interpersonal) activity. All of this interpersonal activity is not merely behavior. Being made in God's image, humans act with purpose. Therefore, all sound economics begins with the axiom that *humans act*. It is this doorway that opens up into the wonderful world of economics.

SUGGESTIONS FOR FURTHER READING

Clark, Gordon H. *A Christian View of Men and Things*, Chapter VII, "Epistemology," 285–325. An excellent and thorough survey of the various schools of epistemological thought from a Christian perspective. Clark concludes that a consistent epistemology must begin by presupposing the truth of the Bible as God's Word. This chapter's discussion of epistemology draws heavily on Clark.

———. *Thales to Dewey*. This is Clark's textbook history of philosophy focusing on questions relating to the issue of knowledge. An excellent place to begin if one is interested in the history of philosophy from a Christian point of view.

Jaki, Stanley L., *The Savior of Science*. An impressive explanation of the importance of Christian faith and doctrine for the development and progress of science in Western Europe during the High Middle Ages.

North, Gary, *The Dominion Covenant*, 1–26. Chapters 1 and 2 of this book provide a good explanation of the importance of the doctrine of creation for all science in general and economics in particular.

2

General Principles of Human Action

A S STATED IN THE previous chapter, economics is a social science. Economics is not a bunch of unrelated facts mined from years and years of data spewed out by academics and government statisticians alike. A science involves systematically arranging facts and deriving general laws. Isaac Watts, the hymn writer, pastor, and logician, described science as "a whole body of regular or methodical observations or propositions . . . concerning any subject of speculation."[1] In order for any science to yield valid conclusions, it must proceed using a valid method. Our next question to consider is *what is the proper method of economics?*

There have been three general approaches to economics as a social science throughout history. One approach is to follow the method used in the biological sciences. A number of economists have argued that the economy is like a living organism that is constantly adapting to change over time. Consequently, to discover economic laws, economists should do empirical studies.

Ideally, this would involve performing controlled experiments like those done by biologists and botanists. In the 1990s, for example, NASA was conducting a number of studies designed to find out which plants would grow well in space. NASA is interested in eventually forming space colonies in the future on places like Mars. One of the requirements for journeying to such faraway places is food. It makes no sense to send people to colonize Mars if they die of starvation before they get there. On the other hand, including enough food on board a shuttle to feed all of the passengers for a trip as long as the one to Mars would make the vessel so heavy that the chance of getting it beyond the earth's atmosphere is rather slim. A suggested solution is taking the seeds of plants that can be

---

1. Watts, *Logic*, 173.

used to grow food instead of the food itself. That way food can be grown as needed without requiring so much weight.

A question that presented itself to NASA researchers was which type of food would be the best to grow in space. Which plant could grow relatively quickly and provide the most essential nutrients with the least amount of weight? One of their winners, it turns out, is the sweet potato. NASA set out studying the growth patterns of sweet potatoes, investigating what is the optimal amount of nitrogen to add to the soil to provide the best, most nutrient-packed, fastest growing sweet potato.

How do you think they attempted to discover this? Well, you might think they simply varied different levels of nitrogen added to different sweet potato plant beds and measured their growth rates and nutrient levels, and you'd be right. However, this is not the whole story. Many other factors also play a role in sweet potato development. Soil quality, the amount of water each plant gets, the amount of sunlight exposure, and the type of sweet potato seed that is planted all will affect growth and quality. Consequently, in order to isolate *only* the effects of different nitrogen levels on sweet potato growth, NASA researchers had to perform controlled experiments. That is, they had had to keep everything besides the nitrogen levels the same. They planted the same seeds in the same soil and gave them the same water and light, and then varied only their nitrogen levels to decide which amount was best for maximum sweet potato production.

Now, that works well for sweet potatoes, because the NASA biologists can control for all of the other variables. Let's consider economics, however. Remember that economics is a social science. Its object of study is people engaging in interpersonal action. Where do people actually interact with one another? Everywhere there are people, of course. That means that the whole world would have to be our test tube or Petri dish. How easy would it be for an economist to control for every variable in the real world? Not very. In fact, it is impossible.

Human action does not occur in a rarified laboratory in a controlled environment. There is no way for us to observe how people react to a change in one variable, say their income, if all other things were held constant, because those other things are never constant. The only economic data we possess is the result of actions that have already taken place, and every action is always the result of a composite of unique factors. Action is always undertaken by particular people with particular values at particular times in particular circumstances.

For example, you may think that a lot of people go to the movies these days. You would be right. In 2007 there were approximately twenty-seven million movie tickets sold in the United States each week. You might be surprised to know, however, that twenty-seven million per week is not even close to how many tickets were sold in an average week in the 1940s. Motion picture theater attendance in the United States peaked in 1948 when on average 90 million tickets were sold every week for an average price of 36 cents per ticket. Because of inflation (about which you will learn more later) 36 cents could buy a lot more in 1948 than it can now. In 2007 people had to pay about $3.10 to buy what in 1948 cost 36 cents. If we lowered all movie ticket prices today to $3.10 would movie attendance increase to 90 million per week? It is doubtful. There are so many more outlets for video entertainment today compared to 1948. Television is, of course widespread. We have DVD players, pay-per-view satellite networks, and internet sites that provide movies via streaming video. All of these allow us to watch films in the privacy of our own homes. There are a myriad of video games we can play as well. The point is that the people who bought movie tickets in 1948 were people who lived at a particular time, with particular incomes, with particular tastes, and particular alternatives for video entertainment. These particulars can never be exactly duplicated and can never be held constant. One cannot do economics the same way botanists study growth rates of sweet potatoes.

Another problem with the biological approach, which is based on empiricism, is that the data used in the attempt to discover economic laws is always data of past economic activity. You will remember from the last chapter that this exclusive focus on experience is a major deficiency with the empirical approach to epistemology. Such a focus is also very problematic for economics.

Because economics is the study of the actions of living, thinking human beings, we cannot assume that human behavior that was observed under past conditions will be repeated exactly. Conditions are always changing. Even if we are able to reproduce exactly those conditions, however, we still could not assume that people would do exactly the same thing. This is because, unlike sweet potatoes that cannot refuse to grow if placed in the correct environment, humans are not merely biological organisms running on instinct and hormones. Humans are creatures made in the image of God with the ability to think and act. We have the ability to choose to eat pizza today or not to eat pizza today. We can choose to eat pepperoni and green olive pizza or hamburger and mushroom. Humans are not merely acted upon, but are themselves actors who are catalysts for change according to their own wills.

Because humans cannot be counted on to do exactly the same thing as they did yesterday, there is no way to establish universal economic laws based on the biological/empirical approach to economics. Even if past experience tells us that people's demand for cappuccino increased whenever its price decreased for the past 100 years, the data, by itself, cannot guarantee that this observed negative relationship between price and quantity demanded will remain today, tomorrow, and forever. Therefore, although the study of economic history can be enlightening in its own right, such empirical methods will not lead us to universal economic truth.

Another approach to economics has followed the model of Newtonian physics. This approach expresses all economic relationships in mathematical terms and applies differential calculus in an attempt to solve problems. This approach has become very popular, but has weaknesses every bit as problematic as the empirical approach.

One of the weaknesses of the mathematical approach is that it violates the principle of *Occam's razor*. Occam's razor is the principle that says the simplest explanation out of a group of explanations with equal explanatory power is the best. If, for example, I want to scientifically analyze human relationships I may, as some do, decide that the best way to start is with the following: Suppose that H = f(F) such that= $\frac{\partial H}{\partial F} > 0$. What does this mean? The reader does not really have a clue until the scientist reveals that H is happiness and F is the number of friends one has. The scientist is saying that happiness is a function of the number of friends one has, but that is not all. Those familiar with calculus will recognize that the scientist is positing that the functional relationship between happiness and the number of one's friends is such that the first derivative of one's happiness with respect to the number of friends he has is positive. Note two things about the above example: the mathematical equation does not communicate the meaning of the scientist until he explains what H and F represent, and after all is said and done, it would have been a lot simpler to simply state that a person's happiness is positively related to the number of friends he has.

A mathematical economist could respond by arguing that, while using mathematical symbols and functions does require some interpretation, such usage actually has more explanatory power than English, because using math allows us to learn things that cannot be learned or learned as easily without it. This brings us to the true Achilles' heel of mathematical economics. A fatal weakness with the mathematical approach to economics is that a mathematical function implies a constant quantitative relationship between variables. If, for instance, we were to propose that the relationship between the number of friends we have and our happiness is such that $H = 2 + 3F$, this tells us that *every* time we gain a new friend, our happiness increases by 3.[2] If the above happiness function is true, then a friend never causes our happiness to increase by 2 or 4 or any other number but 3. Of course, this makes little sense when applied to friendship and happiness.

Quantitative constancy also does not apply to economic concerns either. There are never quantitatively constant relationships between economic variables. Economists have discovered that, other things equal, the price of a good and the quantity of that good people will buy are inversely related. When the price of an ice cream sundae decreases, the quantity of sundaes demanded increases. However, we can never say exactly how much the quantity demanded will increase, because that will always depend on the actions of people. As circumstances continually change, it is likely that their consumption of ice cream will change. More importantly, because people have volition, even in identical circumstances it is possible for the same person to act differently. Human action is not merely the outcome of external forces working on the person. It turns out that in human action, everything is variable. There are no constant mathematical relationships between any variables. These weaknesses indicate that mathematical economics after the pattern of physics is inadequate for discovering economic truth as well.

If the empirical and mathematical approaches are not the best way to do economics, what is? The approach that is the most realistic and meaningful method of discovering economics is verbal logical deduction. Beginning with the axiom that *humans act*, we can use verbal statements to logically deduce principles of economics that—as long as we do not make any mistakes in our logic—are themselves true. True, not just at one point in time, but true for all time. This method has been called the *praxeological* method (praxeology meaning the study of human action) and is the method most identified with the Austrian School of economics[3]

---

2. Let us not concern ourselves just now with the problem of just how we are supposed to measure happiness.

3. The Austrian School of Economics began in the 1870s with the work of Carl Menger.

SOME FIRST PRINCIPLES OF HUMAN ACTION

As Maria von Trapp (another Austrian, by the way) once sang, "Let's start at the very beginning. It's a very good place to start." Years ago, many writers on economics, such as Nassau Senior, John Stuart Mill, John Baptiste Say, and Francis Wayland characterized economics as the science of the nature, production, and distribution of wealth. Later, writers viewed it more as the study of exchange. While economics certainly does touch on both of these topics, it is broader than both. Economics is the study of how people seek to make their lives better through human action. Exchange—trading one thing for another—is action. Two youngsters trading a Manny Ramirez baseball card for an Alex Rodriguez, a middle-aged woman paying forty-five dollars for two sacks of groceries, and a college student giving twenty dollars to his buddy to get a used copy of a textbook like this one all have one thing in common. They are all engaging in human action. Production is action. Eating a plate full of buttermilk pancakes for breakfast instead of a bowl of Life cereal is action. Therefore we begin our study of economics by discovering some of the first principles of human action. If we want to begin our journey down the yellow brick road of economic knowledge from a spot called human action, it might be a good idea to know just what is meant by the term human action. Human action is most succinctly defined as purposeful behavior. A more detailed definition of human action is *applying means according to ideas to achieve ends.*

There are three very important words in this definition. One of them is *end*. We say that human action is goal-oriented in that it always seeks to attain an end. An end is our purpose for acting. They don't call human action purposeful behavior for nothing. If a ten year-old boy trades away his Manny Ramirez baseball card in order to receive an Alex Rodriguez, it is because his goal, his end, is to obtain the Rodriquez card.

Another important word in the definition of human action is *means*. Means are the things we use to achieve our ends. These means are part of the

The Mengerian tradition was carried on by economists Eugen von Böhm-Bawerk, Ludwig von Mises, and Murray Rothbard. Other influential economists connected to this tradition include F. A. Hayek and Israel Kirzner. This tradition is referred to as the Austrian School, because the founding members were citizens of Austria. Following World War II, the Austrian tradition moved to the United States and today adherents to the Austrian School can be found throughout the world. Students interested in the Austrian tradition in economics are encouraged to explore the website of the Ludwig won Mises Institute at www.mises.org.

created order. The means we use can be found both in us and in the world around us. Part of our environment consists of things we cannot control, such as sunlight and air. These things are called general conditions and are not means used in action in the sense that we alter them in order to achieve our ends. Means are things that we can appropriate for our purposes. When you brush your teeth (assuming of course that you are not opposed to good oral hygiene), your toothbrush, toothpaste, water, and hands are all things you can control in your effort to achieve your end of cleaning your pearly whites. In other words, they are means for action.

While we will address this point in much more detail later, it should be noted at the outset that human action necessarily implies the concept of ownership. If you are to attain your end by brushing your teeth, then you must have a toothbrush, toothpaste, water, and hands that are yours to do with as you please. All of these things must be at your disposal which means that they must be your property. Hence, the very concept of human action implies the existence of property.

The third very important word in our definition of human action is *idea*. An idea is the thought about the way that we can achieve our end by applying our means. It is the notion that, *if* you use a toothbrush, toothpaste, water, and hands, you can, indeed, clean your teeth so that they too resemble a flock of sheep (see Song of Solomon 6:6).

Our definition of human action as applying means according to an idea to achieve an end immediately presents us with a number of implications about human action. One such implication is that things do not act. Things do not have ends, means, or ideas. A rock does not have an end it is trying to achieve. A tree does not have means at its disposal that it directs toward an end. A lake does not think. It does not have any ideas. Consequently, things do not act; humans do. Another implication of human action is that action is conscious behavior, not reflex. Human action is purposeful behavior that is the result of ideas we have linking means and ends. Not all human behavior is human action. Action is conscious behavior. We do some things subconsciously—such as breathing—but this is not action.

Suppose you are in the kitchen making dinner for your significant other. (This is a big winner especially for you men.) Further suppose that your sweetheart enters the kitchen, naturally distracting you from the point at hand, and as you lean over to share romantic conversation, you set your hand on a burner turned to medium-high heat. What do you do? Do you begin thinking to yourself, "Hey, this is hot. In fact, this is *very* hot. In fact, it is so hot, that I think that if I leave my hand in its present position, I will burn it. In fact, if I do not remove it from the burner soon, it will be burned rather badly. Yes, I shall remove it." Do you move your hand away from the stove burner only after such ratiocination? Of course not! The nanosecond you touch the burner, you jerk your hand away without any need for deliberation. Thank God that he created us with sub-conscious physical reflexes. It saves us a lot of pain. However, such reflexive behavior is not what we mean by human action. Human action is purposeful, the result of conscious decisions made by human beings as the result of ideas linking means and ends.

Another thing we learn about action by contemplating its definition is that it can only be undertaken by individual persons. Only individuals can have ends and can act to attain them. Societies or groups of people have no independent existence outside of their individual members. They do not have autonomous minds that are able to think ideas, choose ends, or employ means that are separate from the ideas, ends, and means of their individual members.

Some years ago, the Southern Baptist Convention passed a resolution encouraging the members of its churches to consider whether they should continue to patronize the Disney Corporation in light of certain content in recent Disney films, as well as Disney's policy of extending health benefits to domestic partners of homosexual employees. Predictably, media outlets proclaimed, "Baptists boycott Disney!" To what extent was it actually true that Baptists were boycotting Disney? If only three Baptists decided not to buy any more Disney DVDs for their children, would it be true that the Baptists boycotted Disney? Not exactly. It is only true to the extent that individual Baptists stopped buying from Disney. A group or society only does things that are done by its members. All action begins with *individual* humans using means according to ideas to achieve their ends.

© United Feature Syndicate, Inc.

Still more about action can be learned from our definition. Unachieved ends will not be enough to produce actions. Despite the rock and roll personality Meatloaf's assertion that two out of three ain't bad, having two out of the three things necessary for action will not produce action. This should not surprise us. Proverbs 13 tells us, "The soul of the sluggard desireth, and hath nothing: but the soul of the diligent shall be made fat" (Prov. 13:4 KJV). Merely because someone has an end (the sluggard desires something) this does not mean that he will necessarily act in such a way to achieve it. In fact, the verse says that while the sluggard has an end that he desires, he has nothing.

In addition to what we find in the book of Proverbs, it is only logical that if human action is applying means according to ideas in order to achieve ends, we must have means, ideas, and ends in order for action to occur. Suppose at the end of a stressful day, you return to your room looking for some mental peace and refreshment. Suppose further that you have the idea that if you could only put on your favorite recording of Beethoven's *Pastoral* Symphony, his Sixth Symphony, you could receive peace and refreshment. However, if you have left your recording at your parent's house, you will not act to achieve your end by listening to Beethoven's *Pastoral*. If you have only an end and an idea, action will not occur. Alternatively, if you have your favorite recording of Beethoven's Sixth and you know that it will provide you with mental peace and refreshment, but you do not at that time want mental peace and refreshment, then action will not occur. If you have means and an idea, but no end, you will not act. Finally, if you have a copy of Beethoven's Sixth Symphony performed by the Columbia Symphony Orchestra conducted by Bruno Walter and you do indeed wish to receive peace and refreshment after a long day, but you have no idea that listening to Beethoven will provide the peace and refreshment you crave, you again will not put on the Beethoven. If you have the means and an end, but no idea linking the two, action will not occur. For a person to act, he must have an end, and means, and ideas about how to use his means to achieve his ends.

It should also be noted that the word action does not necessarily imply a lot of activity. Doing things that are considered passive are still actions. If one has been sitting for an hour in a chair reading and then decides to take a nap for the next hour, taking a nap is an action, because it is purposeful behavior, even though napping is a rather passive endeavor.

What are some other first principles regarding human action that we should note? One existential fact is that action always takes place in time. Time is a flux during which every action takes place. Some actions, such as producing a crop of corn, can take a relatively long time—several months in fact. Other actions, such as snapping a photograph, take such a short period of time that they seem almost timeless. However, even if the time involved is only a nanosecond, one nanosecond is a period in time.

Additionally, action is always future-oriented. We always act in the present in order to reap an end in the future. The end can be reached in the near future or more distant future. You may want to achieve that pause that refreshes, so that you take a swig of Coca-Cola. Swallowing some Coke takes maybe one second, so that you will receive refreshment one second into the future. On the other hand, perhaps your goal is to graduate with a bachelor's degree, Lord willing. From the time you begin your higher education, your goal—graduation—will be reached four years into the future (or longer for some).

Because we are finite creatures, scarcity is one of the existential facts of life. Something is scarce if the desire for it is greater than the quantity freely available in nature. It is important to note that scarcity is not the result of the fall. Even in the Garden of Eden Adam's body was finite. Consequently, he could only be at one place at one time. Adam was commanded to tend the Garden. Even though he had not sinned yet, so that death had not yet become a constraint to his life span, he still could not do everything at the same time. He had to choose to do some things sooner and other things later. In acting, people must economize their labor and their time. The same is true for other means.

The scarcity of time, however, is not the same as the scarcity of other means. We cannot appropriate a unit of time the same way we can appropriate other means. The temporal succession of time cannot be reversed. One cannot save up time like he can save money or apples for future consumption. One cannot decide to save an hour by living only twenty-three hours one day in early September, so that he can have twenty-five hours the day before the final exam, giving him an extra hour to study. Neither can a person produce more hours in a day or years in his life.

The scarcity of means requires choices regarding what we want to do most. If I spend three hours on a Sunday evening in February watching the Super Bowl, I cannot at the same time worship the Lord in a Sunday evening service. If your friend Joe spends his time producing barbecue beef sandwiches for sale, he cannot at the same time produce Pez dispensers. A person that works all day in a meat packing plant cannot at the same time lounge around the house reading or watching television. Because we cannot do everything we want during the same period of time, we must choose to satisfy some ends and leave others unfulfilled. Consequently we must prioritize our ends. We must decide which are more important and which are less important.

All means are scarce, because there are more ends they can be used to satisfy than there are means themselves. All means must therefore be economized. *Economizing* can be defined as using our means to serve our most desired ends. Our means can be put toward satisfying more than one end, so I must choose between which ends I want the most. Your notebook paper can be used for either taking notes or making paper airplanes. Butter can either be used to make the world's greatest breakfast, buttermilk pancakes, or the world's greatest dessert, flourless chocolate cake.[4] Butter used in making pancakes cannot also be used in the flourless chocolate cake. Therefore, we must choose which we want more, the pancakes or the flourless chocolate cake. We must rank our ends.

VALUE IS SUBJECTIVE

Another fact of human action is that people rank ends by a process of subjective evaluation. By *subjective* economists mean personal. People rank ends according to how much they value them. Take the case of Helena Sophia for instance. Helena has a stick of butter that she can use to achieve a number of ends. She can use it to make buttermilk pancakes, flourless chocolate cake, or sauté a filet of orange roughy. How she chooses to use her butter will ultimately be determined by how she ranks her three ends. A person's ranking of ends can be called a value scale or preference ranking.

Helena will allocate the stick of butter toward that end she ranks most highly or what gives her the most utility. When economists use the word *utility*, they do not refer to the electric or water company, neither do they mean something that is necessarily *useful* in performing some task. For economists utility simply means satisfaction, that is, what Mick Jagger can get none of.

---

4. See "Light and Fluffy Pancakes," and "Ultimate Flourless Chocolate Cake," in The Editors of *Cook's Illustrated, The Best Recipe*, (Brookline, Massachusetts: Boston Common Press, 1999), pp. 397, 461.

Suppose that Helena ranks her ends as follows:

Helena's Value Scale

(First)   Flourless Chocolate Cake

(Second)   Buttermilk Pancakes

(Third)   Orange Roughy

She will allocate her stick of butter toward the making of a flourless chocolate cake because that is what she values most highly. Who is it that decides that flourless chocolate cake is Helena's most highly valued end for her butter? Helena does. It is Helena's own value scale that determines how the butter will be used by Helena. This is what economists mean when they say that value is subjective. People rank ends according to their *personal* preferences. In the context of economics, subjective simply means personal.

What of the buttermilk pancakes, however? She must forgo making them. She must do without. In other words, her decision to make the flourless chocolate cake requires that she not enjoy eating a batch of buttermilk pancakes. Economists refer to this *doing without* as a cost. In fact, they have a special name for it: *opportunity cost*. Opportunity cost is the value of the alternative that must be foregone as the result of choosing to achieve a certain end. For Helena, one flourless chocolate cake costs her a batch of buttermilk pancakes. Helena's opportunity cost of the flourless chocolate cake is the value of a batch of pancakes. You who are full-time college students give up the opportunity to work full-time and so for you, choosing to go to school brings with it the opportunity cost of the sacrificed income that you could have earned when working. This is one reason why high school dropout rates tend to decline during economic recessions. When the economy is in decline and businesses are not hiring as much, jobs are harder to come by and those that are available tend to pay lower wages. Hence, students give up less by foregoing such work and staying in school.

Who is it that decides what the opportunity cost for a particular action is? In Helena's case, who decided that the opportunity cost of the flourless chocolate cake was a batch of pancakes? Helena did, of course. It was her value scale that determined not only what is most valuable to her, but also her most highly valued alternative that must be sacrificed. Consequently, just as the value of the end achieved is subjective to the person doing the acting, it turns out that costs are subjective as well. In economics there is no such thing as objective costs, because costs are values of things forgone and value is subjective.

Much confusion has been generated by well-meaning folk who, from the discussion above, conclude that economists assume that everyone is selfish and greedy. These people rightly are concerned that economics provides a true view of man. Good economists, however, never assume that people are selfish. They merely understand that human action implies a value ranking that is subjective. Economics says nothing about what people actually value. Helena could desire to make a flourless chocolate cake in order to feed others before she is fed, in which case feeding others would be higher on her value scale. It is true that some people may be selfish in an evil, greedy sense. Some people do love money and as such, commit idolatry. Others, however, do not. When Mother Theresa received $190,000 that came with the Nobel Peace Prize, she did not blow it on a Ferrari, but spent it building a leprosarium. Was Mother Theresa selfish? I think we could all agree that she was not. Did she act in accordance to her own value scale? She used the money the way she thought it should be used, therefore the answer is yes. So when economists make the claim that people's actions are the result of subjective value, they are only claiming that such valuation is personal from the point of view of the actor, and not necessarily ascribing selfishness or charity.

Subjective value also implies, however, that the ranking of values is ordinal, not cardinal. Linguists will recognize that ordinal is a derivative of the word order. An ordinal ranking of ends ranks such ends as first, second, third, and so forth. A cardinal ranking would assign numbers to each end that assign a certain weight or measurement to the end. When we say that our temperature is 101°, it implies more than that it is higher than 100°. It implies a certain amount of degrees that can be measured.

Human value scales are always the product of an ordinal ranking, not a cardinal one. This is because there is no measure of value that can be added up. Imagine that, as often happens to me, you get invited over to dinner by a friend that is just crazy about the music of the German early baroque composer Heinrich Schütz (1585–1672). Eventually, of course, the conversation turns towards your host's various enthusiasms and finally touches on Schütz. You relate to your host that you also really like the music of the composer in question, especially his choral collection *The Psalms of David*. Then your host exclaims that he loves Schütz more than you do. You reply, "Well, I don't know." Your host quickly shoots back, "I do. I love Schütz more than you do. In fact I love Schütz *twice* as much as you do!" Now, if you knew economics, you would know that the best response would be, "What does *that* mean? Twice as much as what?" The point is that the preference one has for the music of a particular composer or any other good, for that matter, cannot be measured. The satisfaction you or I receive from Schütz's *The Psalms of David* is a mental phenomenon that cannot be measured. There is no degree of satisfaction like there are degrees Fahrenheit. There are no *utils* of utility. There are no *satees* of satisfaction.

The fact that value is subjective does not only mean that it cannot be measured and added. Because subjective value means personal value, value also is non-comparable between individuals. Only your host knows for sure the level of affection he has for Schütz. Only you know how much Schütz's music means to you. Because there is no objective unit of measurement for value, trying to compare values between people is like claiming that I am taller than my wife because she is 5´4" while my temperature is 98.6°. It makes absolutely no sense.

While the above conclusions regarding subjective value can be deduced from the fact that humans act purposefully by allocating scarce means to achieve their most highly valued end, we do not have to rely only on our reason to be sure that are conclusions are correct. God's Word, which is sure, also instructs us that humans act according to their own subjective value. Remember that humans are created in God's image, so we can learn about the nature of man by looking at God. One thing the Bible tells us about God is that he imparts value into his creation. God makes judgments regarding creation based on his holy nature. For instance, at the end of every day during the creation week, God made an evaluation and declared what he made to be good. After he made man, he said it was very good. Now, because it was God who was doing the evaluating and because God's judgments are always objectively true, God's valuation was true in an objective sense. However, it was also a subjective evaluation because God was making his evaluation according to his own criteria. Because we are created in God's image, we also make evaluations according to our own value scales. We are not God, so our judgments are not necessarily objectively true. In fact, they are only true to the extent that they agree with God's objective standards. However, as God's image bearer, we do subjectively evaluate our ends as we undertake action.

Additionally, the very vocabulary of the Bible characterizes man as a being who acts according to subjective evaluations. In the Old Testament, the Hebrew word that has been translated into English as *value* is *arak*. This word is used as *value* in Leviticus 27:8, 12. In this passage, God is instructing the priest regarding allowing people who rashly have made a vow to dedicate their lives to serving God in the temple to get out of their vow by paying a sacrifice. God then gives a specific price for redemption and then graciously makes allowances for those who cannot afford the price that God names. It is here that God tells the priest that he should set the redemption price depending on how much he *values* or *araks* the person wanting out of his vow. The word *arak* literally means to set in a row, arrange, or put in order. In Leviticus 27:23 *arak* is translated as *estimation*. Note that the Hebrew word for value literally refers to an ordinal ranking. In Psalm 119:127, 128, the same word is translated *esteem*.

A second Hebrew word used in the Old Testament in connection with human valuation is *mikcaw*. In Leviticus 27:23 it is translated in the Authorized Version as *worth*. According to Strong's *Exhaustive Concordance*, *mikcaw* literally means *enumeration* implying a valuation. Note again that an enumeration is an ordering—a subjective ranking of preference—not an objective assigning of cardinal values. It is clear that the Hebrew words pertaining to how humans value things refer to human subjective evaluation.

Such textual evidence is not limited to the Old Testament, but also is apparent in the New. This should not surprise those who understand that it is the same Holy Spirit who inspired all of the books in the Bible. The Greek word used in the New Testament in reference with human valuation is *timao* (pronounced tim-AH-o). This word is translated in the Authorized Version as both *valued* and *value*. Strong's *Exhaustive Concordance* tells us that its literal meaning is "to prize, i.e. fix a valuation upon; to revere."[5] The conception of someone prizing or revering an object or person clearly implies a subjective evaluation. So we see that in both the Old and New Testaments, God's Word describes us as people who make personal, subjective value judgments. Again, this does not mean, and sound economics does not teach, that moral values are subjec-

5. Strong, *A Greek Dictionary of the New Testament*, 72.

tive or relativistic. It simply means that humans value things according to their own personal value scales. The goal of Christians is to have our value scales line up in agreement with God's.

ECONOMIC GOODS

Action, as you will remember, is applying means according to ideas to achieve ends. In order to get a firm grasp of economic principles then, it is imperative to have a proper understanding of the concept of the means people use to achieve their ends. The means people use in action are called *economic goods*.

There are two categories of economic goods: consumer goods and higher order goods. *Consumer goods* are goods that are directly serviceable. Nothing has to be done to a consumer good to make it more fit for use. For example, a Twinkie is the quintessential consumption good. All that has to be done for it to serve its purpose is to be chewed and swallowed. Nothing has to be added to it. It does not need to be further baked or warmed up. It is fit to be eaten as soon as it comes out of its plastic wrapper. For reasons that will be explained below, consumer goods are also referred to as goods of the first order.

Twinkies, however, do not just appear out of nowhere. They do not spontaneously generate. They do not just show up on the ground for us every morning like manna from heaven, although there is some speculation that the manna left for the Israelites was of similar chemical constitution and flavor as Twinkies. Of course, some scholars also have speculated that the manna in question was actually secretions from scale insects.[6] Of course, one could then ask how such secretions came down from heaven.

In any event, for Twinkies to exist for our consumption, they have to first be made. The question presents itself: from what are Twinkies made? Hostess, the makers of Twinkies, must use other means to achieve their end of producing a Twinkie. These means are higher order or *producer goods*. These are goods that are not directly serviceable for consumption, but are transformed into directly serviceable goods in the future. They are indirectly serviceable.

What does Hostess use to produce a Twinkie that sits at your local grocery or convenience store? Well, they use yellow sponge cake, cream

---

6. I am not making up the bit about insect secretions. See the note on Exodus 16:14 in *The New Oxford Annotated Bible*.

filling, machinery, plastic wrappers, cardboard boxes, preservatives, and trucks. However, all of these things do not just spontaneously come together, do a little jig, and present themselves as a tantalizing snack food. No, it takes people to use the machinery that injects the cream filling into the cake, and people to drive the trucks and stock the shelves. Also, it should be pointed out that all of this Twinkie production does not take place floating in air, like some operation from the Jetsons television show. All production uses a bit of the old *terra firma*.

The producer goods can be divided into three general factors of production. One of the above-mentioned factors is *land*—the *terra firma*. Land includes, of course, the earth upon which production takes place, and also all the natural resources such as water and minerals that are in or on the land. Another factor of production that is distinct is labor. *Labor* is the human energy used in production. Labor is treated as a separate factor of production because people are obviously not land. Human effort is also not something that needs to be produced, but is readily available to the person who wants to use it.

The bulk of the other producer goods used to produce the Twinkie— sponge cake, cream filling, machinery, trucks, etc.—fall into a third category of producer goods. These producer goods are called capital goods. *Capital goods* are best defined as produced means of production. Sponge cake, cream filling, machinery, and trucks do not themselves spontaneously generate, neither are they merely ready to be appropriated in nature. Before Hostess can use sponge cake to make a Twinkie, the sponge cake itself must first be produced. The same thing is true for the cream filling, machinery, and trucks. Before they can be used in production, they themselves must first be produced.

THE STRUCTURE OF PRODUCTION

If one contemplates this process of Twinkie production, and meditates on the fact that a similar process occurs for every consumption good, it will not take long to realize that each consumption good is the result of a rather complex structure of production. The structure of production is the complex relationship between consumer goods and every producer good that had a hand in making it. The production structure is like a family tree, or genealogy, for a particular good. An example of the production structure for a Twinkie is illustrated in figure 2.1.

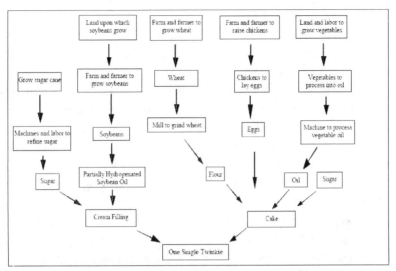

Figure 2.1: Simplified production structure for a Twinkie. In order to make one single Twinkie several higher goods produced at successively higher stages of production are needed. Not even included in this simplified diagram are any of the multiple processes and materials necessary to package Twinkies or to ship them to your local convenience store. As you can see, conceivably there could be hundreds of different materials and processes that go into the production of a single, seemingly simple product like a Twinkie.

Earlier consumer goods were identified as goods of the first order. This terminology comes from the fact that they are at the bottom of the production structure. In our example, the Twinkie is a good of the first order. All of the goods that are used to produce the Twinkie are higher up the structure of production by one stage. They, therefore, are called higher order goods.

Not all higher order goods are at the same stage of production. Some producer goods are at higher stages of production than others. We have already stated that land and labor do not need to be produced, so they are called original factors of production. However, all of the capital goods, by definition, need to be produced. Yellow sponge cake, for instance, does not just materialize out of nothing. It has to be made using other factors of production like flour, sugar, partially hydrogenated vegetable oil, preservatives, machinery, land and labor. These goods are at an even higher stage of production. Again the flour used to make the sponge cake does not spontaneously generate but must be produced from wheat, a grinder, land, labor, etc. These goods are at a still higher stage of production. We could trace the production structure for Twinkies back to the very beginning. Fortunately for our purposes, that is unnecessary. Suffice to say that the production structure for any consumer good produced in a modern capitalistic economy is highly-developed, complex, and far-reaching.

At this point it is sufficient to note three characteristics of the production structure that we want to keep in mind throughout our study of economics. One is that production effort moves down the structure of production, from higher order goods to lower order goods. Before a Twinkie can be produced, sponge cake, cream filling, and plastic wrappers must be made. Before sponge cake can be made, flour, sugar, and partially hydrogenated vegetable oil must be produced. Before flour can be produced . . . well, you get the picture. At every stage of production, producers must first have ownership of the necessary higher order goods before they can produce their lower order goods. Hence, the direction of effort runs down the structure of production. One implication of this fact is that, contrary to conventional wisdom, stimulating consumption will not actually make a society better off. Consumption does not just happen, but instead is the result of a lengthy structure of production. Without investment in the manufacture of producer goods, society will not be able to replenish its supply of consumer goods.

Another characteristic of the production structure that needs to be kept in mind is that the higher order goods are valued precisely because they can be used to produce a lower order good that is also valued. Because Hostess knows that consumers place a certain value on Twinkies, they themselves value yellow sponge cake and cream filling along with all the other factors of production used to make Twinkies. Value is transmitted up that structure of production.

The third characteristic is closely related to the second. Because value is transferred up the structure of production, by demanders, money income moves up the structure of production as well. Why does Hostess even produce Twinkies? Because the company's board of directors want to eat them all themselves? Because they like to watch them as they pile up in their warehouse? Because they just like that golden color? No, the very reason that Hostess makes Twinkies is that consumers will buy them with money. Money moves from the hands of Twinkie buyers to Twinkie producers. The makers of Twinkies in turn value certain factors of production such that they pay money to the owners of the ingredients, machinery, land, and labor, in return for the services of each of their factors. In this way monetary income moves up the structure of production. This fact will be important later as we examine how business firms attempt to maximize their monetary incomes and reap profits.

TIME

A final first principle of human action to further investigate is the fact that all action takes place in time. Thoughtful students will have already recognized that all of the production that goes into making one little Twinkie takes a lot of time, which brings us to the concept of the period of production. As indicated in figure 2.2, from the point of view of any action, time can be divided into three periods: the period before the action, the period of time during the action, and the period of time after the action.

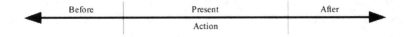

Before   Present   After

Action

Figure 2.2. Time from the perspective of human action.

If the action we are talking about is producing a good, the time it takes to do the action is called the period of production. It is the time period from the beginning of production to the time when the consumer good is available. The actual period of production will be unique to each good and is determined by the production process necessary to produce a given good. For example, the period of production for an ear of corn can be relatively long. Soil is tilled, then the seed is planted, and then around six months after the process began, the corn is harvested. For other goods the process could be even longer. Building skyscrapers often takes years to finish. On the other hand, it takes only a few minutes for McDonalds to assemble a Big Mac. Each good, then, has its own period of production.

Given that all action takes time, man prefers his end to be achieved sooner rather than later—the less waiting the better. This is because all people have what economists call time preference. *Time preference* refers to the fact that a person values a good in the present more than he values that same good at some point of time in the future. Someone who is to get paid one thousand dollars would rather be paid that amount today rather than have to wait six months to receive it.

Most people have watched at least one episode of the television game show *The Price Is Right* in their lifetime. If you have, you know that the goal of the game is ultimately to get yourself into the Showcase Showdown. The Showcase Showdown pits two contestants bidding on two different groups of prizes called showcases. The prizes usually include a luxurious vacation to someplace like the Bahamas, Italy, or Japan. It often also will include a bedroom or dining room furniture set. The first contestant in the Showdown has the option of bidding on the first showcase he sees or passing it on to the second contestant so that he can bid on the second showcase. The first contestant passes on the first showcase almost every time because the hopeful contestant is holding out for what? *A Brand New Car*!!!!! Imagine what would happen if, after the first contestant passes the first showcase so he can bid on the second and after listing several trivial prizes such as a year supply of Rice-A-Roni, the San Francisco Treat, the announcer blares that the contestant could win *a brand new car Ten Years From Now*!!!!! That last phrase sort of takes the luster off doesn't it? It is not nearly as exciting to receive a new car in ten years as it is to receive it now. Getting one in ten years is better than not getting one at all, but not as good as winning one today. People will value a car today more than receiving the same car ten years from now.

Not only does production take place in time, but the satisfaction received from consumption is temporally limited as well. A consumer good provides services to its user only for a certain period of time. This period of time during which a good can satisfy an end is called its duration of serviceableness. Different goods provide want satisfaction for different lengths of time. The more durable a consumer good is, the longer the duration of serviceableness. When considering purchasing a computer for example, a person considers not only the price of the computer, but also how long he thinks the computer will provide the quality of computing he desires. The more durable the good is the greater the amount of services it renders. However, as these services are spread over the duration of serviceableness, rather than provided all at once at the time of purchase, time preference is also, in this case, a consideration.

The fact that all action is undertaken in the present to achieve an end in the future also implies that all actions are speculative. All action is undertaken because we want our future state to be better than it would have been if we had not acted. What is the one thing we know for sure, however, about the future? That it is uncertain. When economists say that the future is uncertain they do not assert that people know nothing about the future. I know, for example, that I will never be even a third-string short stop for the Pittsburgh Pirates. This does not, however, keep the future from being uncertain. Uncertainty means that we do not know *everything* about the future. You do not know for sure what you will be doing ten years from now.

The future is uncertain for a number of reasons. One is that our actions are affected by the actions of other people. A young woman's decision to marry a particular man, for instance, is partially determined by how he treats her. Because each prospective suitor may decide one day to become afraid of commitment, aloof, and begin acting like an emotional jellyfish, it is quite possible that the man who looked like such a good potential life-partner ends up being a woeful mistake. The future is also uncertain because we do not have perfect knowledge of natural phenomena. People who contemplate building large, expensive houses on an ocean front do not know for sure when the next big hurricane will blow through, leaving their abode looking like the pile of twigs made famous in the story of the three little pigs.

Finally, the future is uncertain because we do not know everything that God is going to do. We do know *that* God knows. We also know some things that God will do in the future because he has revealed them to us. We know, for example, that Christ is returning, with glory, to claim his church and to judge the living and the dead. We do not know, however, exactly when this will occur. We also do not know exactly how God will choose to work out his eternal decree on a day-to-day basis.

Because the future is uncertain, all actions necessarily are speculative because they are based on our judgments of uncertain future events. When we act, we must predict what we think the actual outcome will be before we take the action. The results are never guaranteed. Because all action is speculative, it brings with it the possibility of success or failure. All action is undertaken because the person taking the action thinks that he will be successful. Virtually no one gets married hoping that the marriage will be a disaster ending in divorce. *Ex ante*, meaning before the action, a person has a reasonable expectation that action will successfully achieve his end.

Ex post, meaning after the action, a person can look back and assess whether his action was wise or foolish. When an action is successful, that is, when the expected end is actually achieved, the actor receives the satisfaction of having his end fulfilled. Some economists have referred to this as *psychic profit*. If Helena uses butter to make a flourless chocolate cake instead of pancakes and the flourless chocolate cake does provide her with all of the pleasure associated with a pure unadulterated chocolate experience as Helena was expecting, she can be said to profit because the benefit received from the cake was, after all, greater than what she would have received from the pancakes she did not make. In other words, her benefit was greater than her cost. Now when economists say that this profit is psychic, they do not mean that it has anything to do with fortunetelling, Shawn Spencer, or Dionne Warwick.[7] In this case, psychic simply means subjective. Perhaps it would be best to merely call it subjective profit.

Of course, an action could prove to be a mistake. People are not omniscient. Unlike God, they can err. Helena could make the chocolate cake and have it turn out not as good as she remembered it. She could be left with a bad taste in her mouth, pining for the pancakes she had to give up in making the cake. In this case Helena suffers a loss.

THE POWER OF THE RIGHT AXIOM

There is much knowledge packed into our beginning axiom that people act purposefully. Think of all we have derived so far. Action is purposeful behavior undertaken by individuals according to their subjective valuations. In acting, people apply scarce means to achieve their ends. All action takes time; we prefer to achieve our ends sooner rather than later, and because the future is uncertain, all of our action is speculative. We have discovered all this and so far have only touched the very tip of the iceberg.

It is time to move forward and derive our very first two laws of human action. One is a law regarding how people value goods. We already know that people value goods subjectively, but what about different units of the same good? What if someone has the opportunity of obtaining four of the same recordings of Beethoven's Fifth Symphony? Will that someone value each one the same? This is the question pondered next chapter.

The other is a law about the physical returns from applying varying amounts of a factor, holding other factors constant. Will the returns be constant? Will we reap exponential growth? These questions will likewise be answered next chapter.

---

7. Dionne Warwick had a successful singing career and then went on to be, among other things, the chief spokesperson for the Psychic Friends Network that used to have infomercials on television from time to time. The Network has since gone bankrupt, which causes one to wonder if their people were really psychic, couldn't they have predicted that they were going to fail?

SUGGESTIONS FOR FURTHER READING

Menger, Carl. *Principles of Economics*. 77–113. Menger's *Principles of Economics* is the fountainhead of the Austrian Tradition of Economics. While the entire book is worth reading, chapter 2 provides an especially important identification of economic goods as scarce objects. It also includes a vital discussion of the interrelationship between consumer goods and higher order goods.

Mises, *Human Action*, 1–118. The first six chapters of Mises' Magnum Opus provide the seminal explanation of the nature of human action and its implication for how we discover economic truth.

Rose, Tom. *Economics*, 96–100. This brief passage provides various economic insights gleaned from the Bible and documents that biblical language implies subjective valuation.

Rothbard, *Man, Economy, and State with Power and Market*. 1–21. The opening pages in Rothbard's economic treatise provide an outstanding explanation of the nature of and first principles implied by human action.

———. "The Mantle of Science," 3–23. Rothbard's essay is a definitive exposition and defense of praxeology, the economic method most appropriate for discovering economic truth in light of human action. Rothbard also provides an excellent critique of scientism in economics.

3

Laws of Human Action

THIS CHAPTER MARKS SOMETHING very exciting—the derivation of our very first laws of human action. We will apply several principles of human action that we have discovered thus far and derive the law of marginal utility and the law of returns. Both have profound implications for economics. First to be derived will be the law of marginal utility.

DIAMONDS OR WATER

The year 1776 was a banner year for notable events, of course the most famous being the signing of the Declaration of Independence in America. It was also the year, however, that perhaps the most famous economics text ever written was published. If you have only heard of one economics book in your life, it is probably this one. If you have only heard of one economist in your life, the chances are very good that he is the author of this book. Of course this economist is Adam Smith and his great book is *An Inquiry into the Nature and Causes of the Wealth of Nations*, or *The Wealth of Nations* as it is more commonly abbreviated.

The title of Smith's book pretty much states his purpose. He wanted to explain what wealth is and how nations become wealthy. While providing a number of insights along the way, such as the importance of the division of labor, of which we will have more to say later, Smith also made some rather large blunders, the worst of which was his value theory. Instead of recognizing that value was subjective from the point of view of the person acting, he argued that the value of an economic good is objective and equal to its cost of production—the cost of labor used to produce the good in particular. Smith failed to solve the paradox of value that he outlined at the end of book 1, chapter IV of his great work.

The paradox of value is often called the diamond-water paradox because of the two goods that Smith uses in his account. Before we tackle this particular paradox, however, we need to understand exactly what we mean by *paradox*. Contrary to what you might think, a paradox is not two university professors. A paradox occurs when two mutually exclusive statements or things seem to be both true. Smith's value paradox, that he did not solve, concerned the value of two different goods: diamonds and water. As Smith puts it, "Nothing is more useful than water: but it will purchase scarce any thing; scarce any thing can be had in exchange for it. A diamond, on the contrary, has scarce any value in use; but a very great quantity of other goods may frequently be had in exchange for it."[1]

The two, seemingly mutually exclusive truths of the paradox are as follows. Water is much more useful than diamonds. Yet, diamonds are much more valuable. Both are apparently true. Diamonds may be a girl's best friend, but she can do without diamonds for extended periods of time. If a girl tries to do without water for very long, she dies of thirst. We do not need diamonds in order to live. We do need water. In this sense, water is much more useful than diamonds.

Yet it is also true that the diamonds are much more valuable than water. If you were to go to a local snack bar and ask for some water, you could probably buy a bottle of filtered water for less than two dollars. You may even be able to get a cup of tap water for free. If you walked into a jewelry store, however and asked for a cup of diamonds, you could end up paying more than four years of college tuition. Why is something that is as frivolous to life as a diamond, so much more valuable than something as necessary to life as water? This is the value paradox that Smith could not answer. He failed to answer it because he did not accept that value is subjective and he did not think in terms of the marginal unit. We, however, are not shackled by such errors and will proceed to derive a law of human action that will unlock the paradox of value.

Before we proceed, however, we need to keep in mind three principles of human action. We learned in the previous chapter that every time a person ranks and chooses between various ends he tries to achieve, he is making a judgment of their subjective value to him. Second, when acting, the person is attempting to exchange a less satisfactory future state of affairs for a more satisfactory future state of affairs. In that sense, all action can be viewed as an exchange of one state for another.

1. Smith, Adam, *The Wealth of Nations*, 28.

The third principle requires some further elucidation. People value means according to the valuations of the ends they serve. Consider again the case of Hostess, the makers of Twinkies. Hostess does not value sponge cake and cream filling just because it seems like the right thing to do at the time. They value the factors it takes to produce Twinkies, because the Twinkies are valuable to them in earning money when sold. If it was revealed through scientific research that consumption of Twinkies leads to death by cancer within a year, people would, as Yogi Berra might put it, stay away from Twinkies in droves. Demand for Twinkies would plummet. No one would buy them and Hostess would have no incentive to produce them, because they could not sell them. Therefore, sponge cake, cream filling, and all of the other factors used to make Twinkies would not be valuable to Hostess.

THE LAW OF MARGINAL UTILITY

Keeping these three principles in mind, we can proceed to developing a law of human action that will solve the diamond-water paradox. This law is called the *law of marginal utility*. If we want to develop a law of something, it is helpful to understand both what we mean by economic law and what is that something. An economic law is an inexorable regularity of phenomena present in the field of human action. Just as there are regularities in biological and physical phenomena that man must submit to when he acts, there are, likewise, such regularities in social relations that God providentially upholds. Economics is about discovering and explaining these regularities.

The law of marginal utility is one of the most important of all economic laws. Marginal utility is most simply defined as the utility or satisfaction one receives from the marginal unit of a good. This, of course, raises the question: what is the marginal unit. We can help get at the answer by revisiting the value paradox. The solution to the diamond-water paradox lies in the fact that people are never asked to choose between all the diamonds in the world and all the water in the world. If one were indeed forced to make such a choice, it is probable that people would value water more than diamonds. However, no one is required to make such a choice. When choosing between spending money on water or diamonds, people choose between marginal units.

The *marginal unit* of any good is the unit that is relevant for action. Suppose you own five t-shirts and are contemplating buying another one.

The marginal shirt would be the sixth shirt you are considering purchasing. In this case marginal can be thought of as the last unit added, the sixth shirt. If you are thinking about giving away one of your five shirts to the Salvation Army the marginal unit is the one that you are giving up, that is, the first unit subtracted.

People value each unit of a good separately. When budgeting for food and gasoline, people do not choose between gasoline in general and butter in general, but between specific units of gasoline and butter. They decide to buy the next pound of butter instead of the next gallon of gasoline. People do not value time in general, but choose how to act within specific periods of time.

When economists speak of the marginal unit, they refer to the unit relevant for action, not necessarily just one of something. For example, when people purchase eggs, they often do not consider purchasing only one, but twelve at a time. In the case of eggs, the marginal unit is often a dozen. When purchasing shoes, the marginal unit is usually two, or a pair, so the marginal utility of shoes is the satisfaction received from the use of a pair of shoes.

With this working definition of marginal utility, we can proceed with deriving our first law of human action. The method we will use to do so is to make a conjecture and then use deductive logic to derive implications of human action. In so doing, we will discover how a person values different marginal units of a particular good. We must, however, remember an important condition. The law of marginal utility applies to the use of different units of the *same good*. Therefore, we must be speaking of marginal units of a good that are equally serviceable. In other words, no one unit is more suitable to serving a particular end than any other unit. For the units under consideration there are no quality differences.

If there were quality differences between different units, people would treat the units as different goods. When I visited my grandparents as a young boy, I found, instead of spreading margarine on their toast, biscuits and rolls, they often used real homemade butter. This butter was the pure, unadulterated, real thing and tasted delightful. It was made, however, without preservatives, which meant that it tended not to last as long as butter or margarine purchased at the grocery store. On more than one occasion I spread Grandpa's homemade butter on my biscuit only to find that it had "turned" as they say, meaning it had become sour like old milk will do. There are few things more nasty tasting than sour butter. No one wants their toast ruined by rancid butter, therefore, fresh butter and rancid butter are not perceived as the same good.

While deriving the law of marginal utility, it becomes apparent that what explains the value paradox is the relationship between the marginal unit of a good and the total quantity already owned by the actor. Often very profound principles can be discovered using rather simple examples, so we will begin by looking at the case of water. In this case the marginal unit would be a gallon.

There are numerous ends that water can serve. You will remember that in allocating scarce goods, people must rank their ends. Remember also that people value goods according to the ends that they serve. Suppose that Ben Boden has five ends that he can fulfill with a gallon of water, and that he ranks his ends as follows:

(First)	Drinking
(Second)	Cooking
(Third)	Washing
(Fourth)	Watering Plants
(Fifth)	Watercoloring

The first, most urgent end for which Boden uses water is drinking. He cannot live without it. Suppose that Boden fashions himself as a regular Iron Chef. Boden's second most highly valued end for water is cooking. The third most highly valued end for which he could use water is washing. Perhaps Boden's fourth most highly valued end for a gallon of water is to water his plants. If all of these ends were met, Boden would like to express his sensitive inner-artist, so his fifth most highly valued end for a gallon is to use it in water coloring.

With Ben Boden's ranking of ends in mind, we can consider how he will allocate different marginal units of water. Suppose for instance that Boden has only one gallon of water. Which end would he service with it? His most urgent one, of course. Which end is most urgent is decided by his own subjective preference ranking. Given Boden's preference ranking, he would use the first gallon of water for drinking. Now, how would he use his water if he had not one but two gallons? The first would be used for drinking, but the second would be used to satisfy the most highly valued end not yet satisfied, in this case his second most highly valued end, which is cooking.

Suppose that Boden has not two gallons but three. The third gallon now becomes the gallon on the edge, or the marginal gallon. What will he use this gallon for? His ends of drinking and cooking are already fulfilled by his first two gallons, so he will use the third gallon for washing. No doubt by now you have noticed a pattern. If Boden has four gallons of water, he will allocate the fourth one toward his fourth most highly valued end which is watering his plants, and if he has five gallons he will use the fifth gallon to do water coloring. Boden's allocation of different marginal units of water is shown below:

Marginal Unit		End
First Gallon	→	Drinking
Second Gallon	→	Cooking
Third Gallon	→	Washing
Fourth Gallon	→	Watering Plants
Fifth Gallon	→	Watercoloring

We now have all the information we need to derive the law of marginal utility. Remember that people value goods according to the value of the ends they serve. Boden values drinking more than cooking, so the first gallon of water, which serves a more highly valued end than the second gallon, will be valued more highly than the second gallon. We know that Boden values the third gallon he obtains less than the second, because it serves a less valued end. This pattern continues. Boden values the fourth gallon less than the third and he values the fifth gallon of water less than the fourth, because each successive gallon he has serves a less valued end. The result is that if Boden has more water, each successive marginal unit is valued less than the preceding marginal unit.

What happens if he must do without gallons that he already has? Suppose that after purchasing five gallons, one of them springs a leak on the way home from the store and he arrives home with four gallons and a puddle in the backseat of his car. He loses his fifth gallon, so his marginal unit is no longer the fifth but the fourth. Of those two gallons, which does he value higher? The fourth gallon serves a more highly valued end than the fifth, so it is more highly valued than the fifth. If he drops another gallon as he enters his house and it pours onto the floor, the marginal unit is no longer the fourth, but the third. The third gallon is more highly valued than the fourth, because it serves a more highly valued end. We see the pattern that as the stock of water decreases for Boden, the value he places on the marginal unit of water increases.

It does not matter which of the physical gallons of water is lost. When Boden buys five gallons, he does not write on their sides, first, second, third, etc. They are only identified as such as he uses them for each successive end. Suppose that the physical gallon of water that springs a leak in the car is the one he was prepared to use for cooking. Boden does not do without cooking, but merely reallocates the water so that he can achieve all of his most highly valued ends with the remaining four gallons. His marginal gallon becomes the fourth and he forgoes water coloring. In either case as the stock of Boden's water decreases, the value of the marginal unit increases.

From this example we are able to derive the *law of marginal utility*: there is an inverse relationship between the quantity of a good and the marginal utility of that good. If the quantity of water is larger, the utility received from the marginal unit is lower and if the quantity of water is smaller, the utility received from the marginal unit is higher.

It is this law of marginal utility that explains the diamond-water paradox. It is not that water in total is valued less than diamonds, but that the marginal unit of water is valued less than that marginal unit of diamonds. This is the case because water is relatively abundant compared to diamonds. The stock of water available is so large in more developed countries that we can get it whenever we want by merely turning on a tap or pressing a button on a drinking fountain. There is, as far as I am aware, no tap out of which diamonds run. The stock of diamonds is much smaller than water, so that the marginal unit of diamonds is much more valuable.

The above is true for what we could call normal conditions. If one was stranded in the desert, it is quite probable that the relationship between diamonds and water would be reversed. In a desert, water is not readily available. In fact, it is very scarce. Hence a small cup of water is seen as very valuable. You do not hear of people stumbling across the hot, endless sands gasping in a raspy voice, "Diamonds . . . diamonds . . . help me . . . please . . . someone give me some diamonds."

The law of marginal utility was a great breakthrough in economic science. The law was discovered independently by three different economists in three different countries: William Stanley Jevons in England, Carl Menger in Austria, and Leon Walras, who, though French by birth, worked for most of his career in Switzerland. All of these men published their findings in the early 1870s and changed the way economists thought about value so much that their discovery has been called The Marginal Revolution. We will find that not only does the law of marginal utility explain the diamond-water paradox, but it also undergirds the laws of demand and supply.

MORE THAN ONE GOOD

We have observed that if we have a larger amount of a good, the marginal utility of that good will be lower. However, none of us go through life seeking to obtain just one good. No one uses only water or only diamonds or only Twinkies or only recordings of *Bing Crosby's Greatest Hits* to achieve his ends. People obtain many different goods as means to fulfill their several different ends. The principles of marginal utility, however, apply to all goods, and to decisions we make when choosing between different goods.

Suppose, for example, that a Victoria Barkley of Barkley Ranch is considering building up her stock of horses and cattle. How will she do so? Will she buy all horses, all cows, or some combination of each? If she buys a combination, what will determine how many horses and how many cows that she will want to obtain? The answer lies in the respective marginal utilities of horses and cows. Barkley will value horses and cows according to her own personal value scale. Suppose that her preference ranking for both animals is as follows:

1. First Horse
2. First Cow
3. Second Cow
4. Second Horse
5. Third Cow
6. Fourth Cow
7. Third Horse

If Barkley can only obtain one animal, which will she choose? The one that gives her the most utility. In deciding what to do, the Barkley compares the marginal utility of horses with the marginal utility of cows. In this case, the first horse she could obtain gives her more utility than the first cow, so if she could procure only one animal it would be a horse. If she already has a horse and could obtain another animal, she chooses by comparing the value received from the second horse with that received from the first cow. In this case, the value of the first cow is greater than the value of the second horse, so she would choose a cow.

Note that the first and second cows are each individually valued more highly than the second horse, so that if she already had one horse and could obtain another two animals, they would be the first and second cows. In this way Barkley would increase her stock of animals by comparing the marginal utility of each animal.

What if Barkley decided she had to sell one animal? Let's say that now, for example, she owns three horses and four cows; which will she choose to give up? Again she compares the respective marginal utilities for horses and cows, and she will give up the one with the lowest marginal utility. In this case, of all the animals that she owns, her lowest valued is a horse, so she will let that one go.

A couple of things become clear from Victoria Barkley's example. One is that the law of marginal utility applies to all goods. It does not apply to only water, but to all goods that we use to achieve our ends. Notice that for both horses and cows, with larger stocks of each, the value of the marginal unit decreases, because each successive horse and cow serves a less valued end. Also notice that Barkley makes a direct comparison between the marginal utility of horses and cows. She does not have a hermetically sealed value compartment for horses and another for cows, but instead she ultimately places both horses and cows on a single value scale when choosing between the two.

When people choose between various means to achieve their ends, they rank all good on a unitary preference ranking. This is true for all action, and it follows directly from the fact that all action means choosing between alternatives. Action means achieving some ends and leaving others unfulfilled. It means choosing to obtain some goods and passing by others. Such choosing necessitates making judgments of preferences which means that all of the various alternatives are ranked on a single value scale that enables the actor to choose between them.

Some Christian economists disagree with this analysis. They argue that not all ends are the same and that people do not evaluate all ends and means in the same way. These economists argue that deciding which color shirt to buy is so far removed from other more spiritual actions such as helping to feed the poor, that it is pure folly to proceed as if a person merely places shirts and charity on the same value scale. When asserting that people do not evaluate all ends and means in the same way, they are not, however, arguing that people value everything the same. No one believes this. These economists, rather, claim that the method of evaluation for different ends and means are different, so it is a mistake for us to analyze human action as if all ends and means are ranked on a single preference ranking.

It is true that people do have different motives for different actions. Some decisions carry much more moral weight than others. The choice between a blue shirt and a red shirt is trivial compared to deciding whether or not to go into all the world to preach the gospel. However, this does not change the fact that when deciding between spending twenty dollars on a shirt or giving that same twenty dollars to a Christian charity that helps to feed and evangelize hungry people, the person must compare these two different alternatives. The person's choice will depend on which end, getting another shirt or helping to minister to more poor people, he prefers. The only way to choose between them is to compare them in value and the only way to compare the two is for both alternatives to be ranked on one value scale. Although one good being more mundane and one being more spiritual may affect where a person places each alternative on his value scale, this does not alter the reality that they are both ranked on a single scale.

THE LAW OF RETURNS

The second general law of human action that has many implications in economics is the law of returns, often called the law of diminishing returns. The *law of returns* states that if we vary the quantity of one factor that we use, while holding the quantity of other complementary factors constant, there is an optimal quantity of the variable factor to be used in order to achieve our end. This is the result of the physical limitations of finite goods.

The principle might most easily be illustrated by considering the agricultural industry. Consider the legendary Farmer Brown who decides that he wants to grow a nice crop of corn with the end being to maximize his yield per acre. We can consider each acre to be the factor that does not vary in quantity. An acre is an acre after all. Now, suppose that Farmer Brown knows that if he plants 100 seeds in one acre, that there is a good chance that he will see 100 plants grow up, each providing a certain yield. He reasons that if he plants even more seeds, then his output per acre would increase, and rightly so. However, if he continues to plant more and more seed corn, eventually he will reach a point where the additional yield he receives from planting each marginal 100 seeds is a smaller increase than that from the previous additional 100 seeds. What is more, if he continues to sow additional seed corn in units of 100, eventually the total yield per acre will decline because the seeds will begin to crowd each other out as the plants compete for nutrients and water. In which case, additional planting actually results in decreased yields.

The same thing holds true with fertilizer. At first, if Farmer Brown adds some fertilizer to a field that has received none previously, there will be a noticeable increase in yield per acre. If he continues to add more units of fertilizer, however, the subsequent increases to his yield will become smaller and smaller until, if he uses too much fertilizer, his yield will actually decrease because the fertilizer will begin to burn his crop, which is no way to make a living on the farm.

What this means in general is that as the quantity of the variable factor used increases, while the quantity of complementary factors remains constant, a point is reached where the Marginal Physical Product (MPP) of the variable factor begins to decrease. The marginal physical product of a factor is the change in physical output attributed to the use of the marginal unit of a variable factor, holding all other factors constant. It can be shown that in order to allocate different factors so that none are wasted, all should be used so that the MPP is decreasing, but not negative.[2]

Suppose that Brown's production function is such that his MPP schedule for seed corn per acre is as in Table 3.1.

2. See Rothbard, Murray *Man, Economy, and State*, 468–75.

Table 3.1

MPP of Seed Corn		
Seed (in thousands)	Yield (bushels per acre)	MPP
25	110	---
30	124	14
35	139	15
40	152	13
45	161	9
50	165	4

As Brown increases his seed corn from 25,000 per acre to 50,000 per acre his total yield per acre increases from 100 to 165. Note, however, that for each addition in seed corn, output increases by a smaller and smaller amount. The MPP curve for a factor of production can be derived by plotting different factor unit-MPP combinations as points on a graph. Doing so, as is done in figure 3.1, allows us to see clearly the law of returns in action.

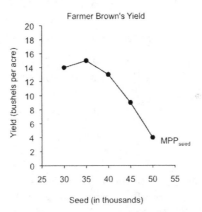

Figure 3.1. The marginal physical product for a variable factor.

The slope of the MPP curve is first positive as the quantity of seed corn planted increases, but after a quantity of 35 seeds per acre, the slope becomes negative.

TAKING STOCK

We have already travelled some distance in discovering principles of all human action. Before we proceed from general human action to the topic of economics proper, it would be a good idea to take stock of some of the most important principles of human action we have learned so far.

We have learned that human action is applying means according to ideas to achieve ends. Because means are scarce, people can only achieve some ends while others are left unfulfilled. This implies that humans must rank their ends and do so according to their own subjective value scales. Also all action takes time and people prefer to have their ends satisfied sooner rather than later because everyone has positive time preference. Additionally, people value means according to the value of the ends that they serve. Because each person uses whatever goods they have to serve their most highly valued ends, each successive unit of a good attained will be put toward satisfying a less valued end. In other words, there is an inverse relationship between the stock of a good and the marginal utility of that good. You will recognize this statement as the law of marginal utility. We discovered that in acting, each person ranks his many alternatives on a single value scale. Finally, we have found that there is always an optimal quantity of a variable factor to combine with fixed quantities of complementary factors. As we increase the quantity of the variable factor, at some point diminishing returns will set in. This is the law of returns. Now we are ready to proceed from considering first principles of human action in general, to action in exchange—the subject of the next chapter.

SUGGESTED READINGS

Menger, *Principles of Economics*, 114–48. This is a section from Menger's seminal book where he outlines the theory of subjective value and derives the law of marginal utility.

Mises, *Human Action*, 119–30. This is Mises' exposition of the law of marginal utility and the law of returns from chapter VII of his Magnum Opus.

Rothbard, *Man, Economy, and State*, 21–38. Rothbard provides an outstanding introduction to the law of marginal utility and the law of returns, both derived from human action.

Smith, *The Wealth of Nations*, 28. Here we find Smith's classic statement of the diamond-water paradox that he left unsolved.

4

Exchange, the Division of Labor, and Property Rights

THE PRINCIPLES OUTLINED so far in this book apply to all actions of individuals. This does not mean that all action involves only the individual actor. A large part of life is filled with social, or interpersonal, action.

There are indications from Scripture that humans were designed to interact with each other from the beginning. In the account of the creation of man, God says in Genesis "It is not good that the man should be alone; I will make him a helper comparable to him" (Gen. 2:18). From the very beginning of human history, man was not meant to be an individualistic hermit cut off from all other human contact. God tells us that the very reason he created Eve was to give a helper to Adam. Note that God does not say that he wanted to create a woman for Adam to give him companionship or to keep him from being lonely, although this is of course a benefit from our social relationships. Instead, God's primary reason as revealed in Scripture is so that Adam would have someone else to help him in the calling that God gave him. The Bible indicates that from the very beginning man benefited from cooperation in his attempt to achieve his ends. This general principle is affirmed in Ecclesiastes where the Preacher tells us "Two are better than one, because they have a good reward for their labor" (Eccl. 4:9).

The church itself is a great social undertaking. Hebrews 10:25 commands Christians not to forsake "the assembling of ourselves together, as is the manner of some, but exhorting one another . . . ." In 1 Corinthians 12 the Holy Spirit tells us through the writing of Paul that the church is made up of many different people with many different gifts and qualities that are all to function together in order for the church to achieve its divine purposes. God designed man as a creature who fulfils his calling through interpersonal action of a particular kind.

There are two types of interpersonal action. One of them is violence. If you see that your friend is wearing a pair of shoes that are stunning, and you decide you want a pair, you could go up to your friend and simply demand them. If he refuses, you could pop him one in the head and take them. Needless to say, this does not increase social cooperation or foster the growth of society.

The other type of interpersonal action is voluntary exchange. If you want your friend's shoes, you could offer to give him something that he values more than his shoes, so that you could trade with him. Alternatively, you could inquire where he bought them and, assuming the shoe store had an identical pair, you could offer to give the store manager something he values more than the shoes. In either of these cases, you receive the shoes through an act of voluntary exchange. Economics is not criminology, so this book will not focus its attention on acquisition through violence. Economics concerns itself with how people seek to achieve their ends via voluntary exchange.

PRINCIPLES OF EXCHANGE

As noted in earlier chapters, profound principles can be discovered through the use of simple cases. This is precisely the case with interpersonal exchange. We will begin by analyzing the simplest exchange possible: one person trading one unit of a good to another person for one unit of another good.

Suppose that there are two film buffs, Siskel and Ebert, both of whom want to see one of their favorite films at a local revival house, a theater that shows primarily older classics. Further suppose that the value scales for both Siskel and Ebert are as illustrated below:

Siskel	Ebert
(*Citizen Kane*)	(*Casablanca*)
Casablanca	*Citizen Kane*

We see from Siskel's value scale that he would prefer to see the film *Citizen Kane*, because it is higher on his value scale. However, also note that this good has parentheses around it. In this book, whenever a good has parenthesis around it, this serves as an indication that the person does not own that good. In Siskel's case, then, although he would prefer to watch *Citizen Kane*, he does not have a ticket to see it. Because the film

Casablanca does not have parentheses around it, we know that Siskel does have a ticket to see *Casablanca*. Now look at the case of Ebert. We can see from his value scale that he would prefer to watch *Casablanca*, but does not have a ticket to get in. He instead has a ticket to see *Citizen Kane*.

Is there any way through interpersonal action Siskel and Ebert can become more satisfied? Wise readers recognize that they can become better off by engaging in exchange. Siskel can give his ticket to see *Casablanca* to Ebert in return for Ebert's ticket to see *Citizen Kane*. If this trade is made, both Siskel and Ebert can achieve an end that they value more.

Necessary Conditions for Exchange

We can learn a number of principles from this simple example. The first is that there are three conditions that are necessary for exchange to take place. The first is that, in order for exchange to occur, each party of the exchange must value the goods being traded in *reverse order* from the other party. Sometimes economists refer to this reverse preference ranking as the *double coincidence of wants*. In any event, each potential trader must value the good the other person has more highly than what he himself possesses in order to trade to take place. Suppose, for example, that the value scales of Siskel and Ebert are not as documented above, but instead like this:

Siskel	Ebert
Casablanca	(*Casablanca*)
(*Citizen Kane*)	*Citizen Kane*

In this case, Siskel still has a ticket to see *Casablanca* and is without a ticket to see *Citizen Kane*. Ebert still has a ticket to see *Citizen Kane* but would rather see *Casablanca*. However, no trade will take place because *both* Siskel and Ebert value *Casablanca* more than *Citizen Kane*. Ebert would be quite happy to trade away his ticket to *Kane* in order to see *Casablanca*, but Siskel is not willing to give up his ticket, because he values that ticket more than the ticket that Ebert has. Just as it takes two to tango, it takes two to trade. There must be a double coincidence of wants for exchange to take place.

Another necessary condition is that each party must know of each other. Suppose that Siskel and Ebert value the goods in reverse order, but Siskel lived in his stylish penthouse in L.A. and Ebert inhabited a small apartment on the lower east side of Manhattan each with no knowledge that the other was even alive. In this case, they could obviously not be able to trade their tickets, much less co-star on a syndicated television show in the 1980s. Therefore in order to make an exchange, both parties must know of each other.

To say that both parties must know of each other does not mean that they must be personally acquainted. A great number of exchanges take place between people that have never met before. When people buy goods at a Wal-mart, a convenience store, or a gas station, for example, it is not unusual for them not to even recognize the cashiers taking their money. Internet markets like EBay see a multitude of exchanges take place every day between people who will never meet face-to-face. This, of course, is perceived as risky by a number of people, which is why websites like EBay often allow customers to rate sellers regarding quality and reliability. In any event, in order to engage in exchange, the people involved do not have to know each other personally. They only need to know that there is someone out there who values the goods in question in reverse order compared to them.

A final condition for exchange to take place is ownership. Suppose that Siskel and Ebert valued both goods in reverse order and know of each other's existence. They are, consequently, about ready to make their trade. They both reach out their tickets with their right hand and are preparing to grasp the other's with their left when someone brandishing a forty-four magnum, the most powerful handgun in the world, steps in-between, begins slapping their hands shouting, "Stop that!" When they try the trade again, the man with the gun cocks it and makes it clear that if Siskel and Ebert commence to exchange, it will be the last trade they ever make.

In this case, neither Siskel nor Ebert ultimately owns his ticket. From the economic point of view, ownership is defined as ultimate control. If Siskel and Ebert owned their tickets, they could use them to get into their film. They could throw them away. They could scribble notes on the back of them. However, ownership also includes the ability to transfer owner-ship. If a person can use his property in some ways, but cannot trade it to someone, he does not have full ownership. The issue of ownership will be examined in depth in a later chapter. For now it is sufficient to remember that ownership is a necessary condition for exchange. You cannot trade what you do not own.

So, in order for exchange to take place, each party must rank the goods in question in reverse order, each person involved must know of each other's existence, and each must own the goods they are trading away. There are a number of important implications that we can derive from the conditions necessary for exchange.

Implications of Voluntary Exchange

The first implication is that trade is mutually beneficial. If reverse preference rankings by both trading partners are necessary for exchange to occur, then it is clear that both parties benefit from the exchange. In the above example, who was better off as a result of the exchange of movie tickets? Both Siskel and Ebert were able to achieve an end that was higher on their own value scales. Voluntary exchange is never a case of one person winning while the other loses. Both perceive himself to be better off as a result of the exchange. If either party would expect himself to be worse off, he would not have made the trade in the first place. To the contrary, if the exchange is voluntary, we know that both are able to obtain something they value more.

Sometimes people find it hard to accept the principle that trade is mutually beneficial. For example, a lot of people were dismayed that, immediately following the terrorist attacks of September 11, 2001, several gas stations were selling gas for five dollars per gallon. There were complaints to the government and a toll-free hotline for people to call reporting those stations that were suspected of so-called price gouging. Were the owners of these gas stations benefiting at the expense of those buying gas? One might be tempted to think so, but forming a correct answer to this question requires that we think very clearly about the nature of exchange. We have already seen that if an exchange is voluntary, then *both* parties must think that such a trade will be beneficial. If gas stations are, in fact, able to sell gasoline to buyers at a price of five dollars per gallon in a voluntary exchange, this must mean that there are buyers who think that the gallon of gas they receive is worth more to them than the five dollars they give up. It is hard to see how the sellers of the gas are doing something that harms the buyers.

There was one story of a man and wife who owned their own station, and soon after the terrorist attacks, saw long lines forming at their pumps and they became concerned that they would be running out of gas soon. They decided that in order to ration the gasoline to those who really wanted it, they would raise their price to five dollars a gallon. The owner walked out to the curb where the line was still quite long and explained to customers about the price increase. Still the line stayed in place as customers were quite willing to pay the higher price. Seeing that the first price increase did little to reduce the quantity of gas demanded, the owner increased the price further to seven dollars a gallon. There were still buyers willing to buy gas.

After the panic buying stopped a few days afterwards, this gas station owner and others like him were shamed by the press and threatened by the government, so they offered to give money back to consumers who paid the higher prices. The station owner said later that he realized that he should not have raised prices like he did, but that, to prevent running out of gasoline, he should have closed down altogether. In other words, the solution that would make the press and the government most happy is that, instead of selling gas to those who were willing to pay for it, no one should have received any. This solution, of course, does not take into account the fact that if an exchange is voluntary, then one party does not benefit at the expense of the other. On the contrary, both parties are left better off than they would be without the trade.

It should be noted that these judgments regarding the benefits of trade are made by the people trading before the exchange is made. These evaluations are *ex ante*. It is always true that one of the parties could be in error. I once shelled out five dollars to see a matinee of the wretched movie *Titanic* early during its release in 1997. I had heard good things about it from members of the press and even some friends. I willingly gave away five dollars in order to receive my ticket to see the movie, which it turns out would have been more accurately named *Tramp Steamer*. After all was said and done, I felt like I had paid five dollars for someone to insult me for over two hours. This does not mean that the movie theatre unjustly took advantage of me. No one forced me to believe the press or buy the ticket. I simply made an error and reaped a subjective loss for my mistake.[1]

Another implication following from the fact that trade only occurs if both parties value the goods in reverse order is that a trade is never an exchange of equal values. In any voluntary exchange both parties must value the good they receive *more* than the good that they give up. Siskel

---

1. For a much better experience, both dramatic and historical, see the 1958 film adaptation of Walter Lord's *A Night to Remember*.

and Ebert each valued the movie ticket he received more than the one he gave up. In neither case did the value of the *Citizen Kane* ticket equal the value of the *Casablanca* ticket.

If the goods were valued equally by each party then no exchange would take place because there is no way for either party to increase their state of satisfaction from an exchange, unless they just find it thrilling to trade tickets back and forth. Carl Menger explained why this is very unlikely in his *Principles of Economics*. Menger was taking to task Adam Smith for attributing the universal practice of trade to some "propensity to truck, barter, and exchange." Menger's point is that people trade, not because of some mystical force, but because they rationally find it mutually beneficial. In one of the more humorous paragraphs written by an economist in the last half of the nineteenth century, Menger explains:

> Suppose that two neighboring farmers each have a great abundance of the same kind of barley after a good harvest, and that there are no barriers to an actual exchange of quantities of barley between them. In this case, the two farmers could give free rein to their propensity to trade, and could exchange 100 bushels or any other quantity of barley back and forth between themselves. Although there is no reason why they should desist from trading in this case if the exchange of goods, by itself, affords pleasure to the participants, I believe nothing is more certain than that these two individuals will forgo trade altogether. If they should nevertheless engage in this sort of exchange, they would be in danger, precisely because of their enjoyment of trade under such circumstances, of being regarded as insane by other economizing individuals.[2]

Exchange is not the result of the pleasure we get from trading the same things back and forth again and again. People trade because each party values the good received more than the good given up. This implies that exchange is never a trade of equal values.

Another implication of voluntary exchange is that the commonly dreaded middleman serves a valuable economic function. This is implied from the necessity of both parties knowing of each other. One of the first things people who are moving do when looking for a house to buy in their new location is go to a realty office. Why don't they cut out the middleman and thereby avoid his four to seven per cent commission? The reason is that realty agents specialize in making house buyers and sellers aware of

---

2. Menger, *Principles of Economics*, 175–76.

each other. An alternative to going to the realty office is to drive around in search of signs that read *for sale by owner*. Prospective buyers could, alternatively, stop at every house that looks appealing and ask the owner if he would be willing to sell. It could take a very long time for a buyer to find someone willing to sell a house that is right for the buyer and in a price range the buyer could afford.

All of this looking around costs time, wear and tear on the car, and gasoline. Economists call such costs *search costs*. In order to engage in exchange, buyers and sellers must search out trading partners with preferences reversed from theirs. Middlemen reduce search costs, and as such, provide a productive service to both the buyer and seller for which they are willing to pay. The revenue reaped by middlemen is not merely money siphoned off from the *real* buyers and sellers. It is income earned by making it easier for buyers and sellers to trade. We know that they provide a productive service as long as they can make a profit as the result of voluntary exchange.

Examples of such helpful middlemen abound. We already have seen the value they add in the real estate market. Additionally, one way banks make their money is by bringing borrowers and lenders together. Suppose that our friend buying a house wants to borrow money to pay for the house. If he cannot go to a bank, his alternative is to take out a want ad requesting that someone lend him tens of thousands of dollars. He could also, of course, go door-to-door trying to find a lender. He does not have to do either of these if he can simply go to a bank. The same holds true for the saver who wishes to lend. Where does the bank get the money to lend to the homeowner? One source is from other people who save. These savers could go knocking on doors seeing if there is someone who would like to borrow their money to be paid back with interest in the future. This searching would cost time and effort. Banks, consequently, serve a valuable economic function in bringing borrowers and lenders together. A banker serves as a middleman.

Advertising generally serves much the same purpose. Instead of personally going to every store to check who has the best price for peaches this week, a shopper can look at the store ads in their local newspaper and cut down on search costs. This makes it easier for buyers to identify those merchants who have the goods of a certain quality for sale at a certain price. Therefore, the advertiser also performs a valuable economic service as a middleman.

Reputation also reduces search costs. Some producers and merchants have a reputation for providing good products or service as promised while others develop a reputation for selling shoddy merchandise and giving poor service. Potential buyers can reduce their options by considering only those sellers with good reputations. In relying to some extent on reputation, they are able to reduce the time it would take to evaluate the products of every seller.

All of these methods of communicating information tend to reduce losses from exchange by making *ex ante* expectations more in line with reality, and hence, *ex post* realizations. As buyers discover whose information they can trust and use this information in their decision making, they are less likely to make errors and more likely to trade for goods that turn out to perform as advertised. They are able, therefore, to make actions that more frequently successfully satisfy their ends.

MORE THAN ONE

So far we have learned the conditions necessary for exchange, that trade is mutually beneficial, that trade is never for equal values, and that people who specialize in reducing search costs perform a valuable service. All of this economic knowledge was gleaned from an example of trade for one unit of each good.

What about trade when there is more than one unit of each good? How many units of each good will people trade if they have more than one unit to potentially offer? All? None? Some? It turns out that the answer is not blowing in the wind, but is determined by marginal utility. Take, for example, the case of Groucho who has a stock of mangoes that he owns. He likes mangoes as much, or perhaps more, than the next guy. But after awhile he gets tired of eating only mangoes and, besides, he realizes that he needs more nutrients supplied to his body than only the simple carbohydrates provided by the mangoes. He needs and wants protein. In particular he wants beef! Not some tofu or quiche. He wants what he perceives to be a manly meal. He wants to sink his teeth into a medium-well thick cut rib eye. Alas, he has none.

On the other hand, there is a fellow named Harpo who does have a stock of beef that he consumes quite liberally. But, like Groucho, Harpo is moving closer and closer to gastronomical ennui from eating the same thing over and over. He decides that for variety's sake he would like some fruit. Fruit would be a nice palette topper and would also provide him with carbohydrates of which his beef is bereft. If Groucho and Harpo come to know of each other's situations, then the opportunity for mutually beneficial trade exists. The question we set out to answer, however, is how many units of each good will be traded between parties?

We have already seen that people will be willing to exchange as long as they value what they receive more than what they give up. This same principle applies for exchanges of more than one unit of each good. Recall that the value received from the marginal unit of a good is called marginal utility. Consequently, we arrive at the conclusion that people are willing to trade as long as the marginal utility of the good they are receiving is greater than the marginal utility of the good they are giving up.

Suppose that Groucho and Harpo have value scales such that each ranks different units of beef and mangoes as follows:

Groucho	Harpo
5th bu. Mangoes	(1st bu. Mangoes)
(1st lb. Beef)	5th lb. Beef
4th bu. Mangoes	(2nd bu. Mangoes)
3rd bu. Mangoes	4th lb. Beef
(2nd lb. Beef)	(3rd bu. Mangoes)
(3rd lb. Beef)	3rd lb. Beef
2nd bu. Mangoes	(4th bu. Mangoes)
(4th lb. Beef)	2nd lb. Beef
1st bu. Mangoes	(5th bu. Mangoes)
(5th lb. Beef)	1st lb. Beef

The respective value scales show the relative level of value Groucho and Harpo ranks each successive bushel of mangoes and pounds of beef that could be either accumulated or traded away. Remember that preference rankings are ordinal, not cardinal, so these value levels are merely an ordinal ranking. Groucho's ranking shows the value he places on each pound of beef he gains from trade and the value he places on each bushel of mangoes he trades away. If Groucho accumulates beef as the result of trade, what happens to the value he places on each additional pound? If he obtains more beef, his stock of beef would be larger. If Groucho's stock of beef was larger, the value of each marginal pound would be lower, according to the law of marginal utility. In order to obtain a pound of beef, Groucho must give a bushel of mangoes to Harpo. If Groucho increased his stock of beef through trade, then, his stock of mangoes would have to decrease. Groucho's value scale reveals that if his stock of mangoes was lower, the value of each bushel would be higher, according to the law of marginal utility.

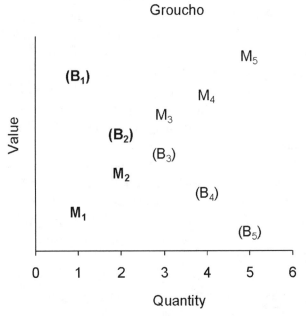

Figure 4.1. Ordinal ranking of beef and mangoes for Groucho. The marginal value for beef is greater than the marginal value for mangoes through two marginal pairs (highlighted in bold), but not for the third, fourth, and fifth. Therefore, Groucho's value scale indicates he is willing to trade two bushels of mangoes for two pounds of beef.

We now have enough information to figure out how many bushels of mangoes Groucho is willing to trade away for how many pounds of beef. Groucho will be willing to trade mangoes for beef as long as the marginal utility for beef is greater than that for mangoes. From Groucho's value graph, we see that he certainly values the first pound of beef he would receive more than the value of the first bushel of mangoes he would trade away. Consequently, he would be quite willing to trade away at least one bushel of mangoes for one pound of beef. Is that all, however? If we look at his graph again, we see that, even though the value Groucho places on the second pound of beef is lower than the value of the first, he still values the second pound of beef more than that second bushel of mangoes he would have to do without. Therefore, he is willing to trade at least two bushels of mangoes for two pounds of beef. Would he be willing to trade three for three? The answer is no. Examining his value graph reveals that the marginal utility for beef has declined enough and the marginal utility for mangoes has increased enough so that Groucho values the third bushel of mangoes he would have to give up more than the third pound of beef he would receive. No one, including Groucho, wants to give up something he values more for something he values less. Therefore the most that Groucho is willing to trade is two bushels of mangoes for two pounds of beef.

So how many mangoes for how much beef will actually be traded? Well, we do not know yet. There are two parties to every exchange. We will not know how many units of mangoes and beef will be traded until we look at Harpo's value graph shown in figure 4.2. Harpo's graph shows us the value he places on each bushel of mangoes he receives in exchange for each pound of beef he gives up. If Harpo accumulates mangoes, his stock of mangoes, of course, increases. Therefore, the value of each marginal bushel would be lower, according to the law of marginal utility. In order to receive a bushel of mangoes, Harpo must trade away a pound of beef. In order for Harpo to increase his stock of mangoes, then, his stock of beef would decline. If Harpo's stock of beef decreases, the value of the marginal pound is higher, according to the law of marginal utility.

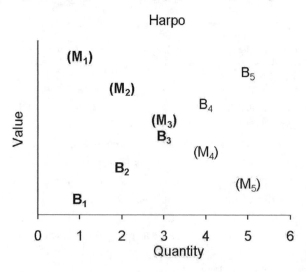

Figure 4.2. Ordinal ranking of beef and mangoes for Harpo. The marginal value for mangoes is greater than the marginal value for beef through three marginal pairs (highlighted in bold), but not for the fourth and fifth. Therefore, Harpo's value scale indicates he is willing to trade three pounds of beef for three bushels of mangoes.

Like Groucho, Harpo will be willing to trade beef for mangoes as long as the marginal utility of the good received is greater than the marginal utility of the good traded away. If we look at Harpo's value graph, it is clear that Harpo values the first bushel of mangoes he would receive more highly than the first pound of beef he would have to give up. So Harpo would be very willing to trade at least one pound of beef for one bushel of mangoes. Because his second bushel of mangoes would be used to serve a less valued end, Harpo values the second bushel of mangoes less highly than the first, but notice that he still values the second bushel more highly than the second pound of beef he would have to give up to get the mangoes. Consequently, we know that Harpo would be willing to trade at least two pounds of beef for two bushels of mangoes. What about three for three? Well, Harpo's value graph does indeed indicate that he values the third bushel of mangoes he would receive in exchange more highly than the value of the third pound of beef he would trade away. Harpo would not, however, be willing to trade a fourth pound of beef away to receive a fourth bushel of mangoes. Consequently, the most that Harpo would be willing to trade is three pounds of beef for three bushels of mangoes.

Now you know enough to answer our original question: How many mangoes for how much beef will actually be traded? One might be tempted to answer three bushels for three pounds because Harpo is obviously willing to make such an exchange. Remember, however, that exchange requires a *double* coincidence of wants. *Both* parties must value the traded units in reverse order. The marginal utility of the good received must be greater than the marginal utility of the good given up for *both* parties. In the case above, while Harpo is willing to trade a third pound of beef for a third bushel of mangoes, Groucho is not. Groucho values the second units of beef and mangoes the same as Harpo. The most Groucho is willing to trade is two for two. Consequently, the actual amount that will be traded is two bushels of mangoes for two pounds of beef.

THE POWER OF EXCHANGE

The possibility of exchange opens the door to great benefits for all of humanity. The opportunity for trade is such a boon because it allows us to produce for the market rather than only for ourselves. The key word in the preceding sentence is *market*. At the mention of this word, many will first think of the stock market: the shares of stock that are traded on Wall

Street. Indeed this reaction is so prevalent that investing in stocks is often referred to as *playing the market*, as if only this one market exists. However, there are also things called supermarkets, where groceries are bought and sold. You may have heard of my wife's personal favorite, the flea market, which is not, as far as I know, where fleas are bought and sold, but where people go to buy all sorts of different things that sellers no longer want. A flea market is like a more developed rummage sale or a less sophisticated Ebay, depending on your perspective. One seller at such a market quipped that he sold *junktiques*. He bought his merchandise as junk and sold it as antiques. There are several other markets you may have heard of, such as the grain market, the oil market, the money market, etc.

What is the one common characteristic that allows each of these to be called a market? Items are bought and sold. The market, then, is not a physical place, but is simply the vast network of interpersonal exchanges. EBay is a very large market that is not limited to any one physical location.

The possibility for exchange allows us to produce goods for the market. Imagine our situation if there was no exchange. Everything we consumed we would have to produce by ourselves. We would have to raise our own animals for our meat and milk. We would have to grow our own vegetables. We would have to make our own bread, which would require us to grind our own flour, which would require that we grow our own wheat. At the same time we attempted to do all this, we would have to build our own house if we wanted a roof over our head. We would have to make all of our own clothes, which would require that we make our own fabric, which would require that we grow our own cotton or obtain wool from sheep we raised ourselves. If we wanted to use an automobile to get around, we would have to make it ourselves, which means we would also have to spend time and resources making metal, vulcanized rubber, and glass for the body, tires and windows. We would have to also make all of the tools necessary to put the car together. We would not have the ability to trade for anything. It does not take very long to realize that if there were no trade, our standard of living would fall drastically. No one has the time to make *everything* listed above. All of our clothes, food, and shelter would be much cruder, lower in quality and much less plentiful. Many of the above goods we would simply have to do without. In such a world of self-sufficiency, many people would not be able to survive because there simply would not be enough food produced to feed them. Those who did survive would exist at a wretchedly squalid standard of living. In such a world, it would be very hard, if not impossible, to fulfill God's mandate to multiply, fill the earth, and subdue it. Exchange allows us the benefit of using higher quality and greater quantities of all goods. Why is this so?

It is because exchange opens the door to specialization and the division of labor. The phrase *division of labor* refers to different people specializing at producing different goods. If exchange takes place, each person must be relatively specialized in the ownership of the different goods being traded. Each has a different proportion of goods relative to his wants. In our example above, Groucho specialized in the ownership of mangoes while Harpo specialized in the ownership of beef. Such a specialization in ownership occurs usually because there is a specialization in production. The reason Groucho had so many mangoes and no beef is because Groucho produced mangoes and did not produce beef. Similarly Harpo produced beef and did not produce any mangoes.

Such specialization in production is due to the natural inequality of God's creation. The vast creation of God exhibits remarkable variety. He did not create just one sort of bird or fish or mammal. He created several different kinds of fowl, fish, and creeping things. He created many different kinds of geological formations and climates. Even man, who bears his very image, demonstrates great diversity. For example, he created human beings male and female. Men and women are not biological copies of each other. They each have different abilities and strengths. He created some weak and some strong, some tall and some short, some who are better at using their intellect and some better at using their hands. God even reminded Moses that he created some who could speak, hear, and see and others who could not (Exod. 4:11). God in his wisdom did not make everything the same. Everyone and everything is unique. From the point of view of human producers, this inequality of creation is manifested in three ways.

The suitability of natural resources at everyone's disposal is different. One reason why Groucho may specialize in the production of mangoes is that the area he lives in has soil well suited to growing mango trees but has no grassland appropriate for cattle grazing. Likewise, perhaps Harpo lives in plush grassland that does not support mango trees.

Another reason why people specialize in the production of different goods is that there are differences in their stock of capital goods. Groucho, for instance, might have inherited from his father machines designed for fruit picking, numerous fertilizers and pesticides to help mangoes grow well and free from insects, and a physical plant built specifically for packing mangoes. If such is the case, it would be economically foolish for Groucho to trash all of these capital goods and try to raise cattle instead. Likewise, Harpo may have inherited a large amount of grassland with a sizable feed lot, and butchering facilities; he would be foolish to walk away from these capital goods to try to grow mangoes.

Finally, God has created each of us with different skills and abilities. Groucho may be one of those people who seem to have a green thumb. He might also have an interest in growing mangoes and find it exhilarating to read and study how to be a better mango grower. Likewise, Harpo may simply be created with a knack for cattle husbandry. Often differences in skills are predicated by what we like. People seem more likely to develop better skills for the very tasks that they enjoy.

Because of differences in the suitability of natural resources, in given stocks of capital goods, and in labor skills, people are more suited for some lines of production than others. Therefore, they are more likely to specialize in making those goods at which they are relatively better at producing. This is only practical, however, if people are free to trade for other things they want, but cannot produce as well. It makes sense, for example, for Groucho to specialize in the production of mangoes, because he can trade mangoes to Harpo for beef. It makes sense for Harpo to specialize in the production of beef because he can trade beef for mangoes. If the opportunity for trade was not there, both Groucho and Harpo would have to produce both mangoes and beef themselves if they wanted to consume both.

People engage in exchange because they will be better off as a result and they can consume more than they could if they had to produce everything themselves. Participating in the division of labor allows the entire society to have more material blessings than they would without it. Remember, however, that it is the possibility of exchange that allows for the division of labor. Adam Smith got a number of things wrong, such as his already mentioned value theory, but one thing he got right was the title of the third chapter in *The Wealth of Nations*, "That the Division of Labour is Limited by the Extent of the Market."[3] This means that it will only make sense to specialize in the production of a good for which there is a market, or in other words for which there are enough people who are willing to pay enough so that it is economically wise to spend one's time making only that good.

3. Adam Smith, *The Wealth of Nations*, 17.

Let us return for a moment, to the case of Groucho and Harpo. Groucho and Harpo found it in their interest to specialize in the production of mangoes and beef respectively, because each could trade with the other. Now suppose that a third party, Chico, comes along. He wishes to join Groucho's and Harpo's society and desires to do so by specializing in the production of rocket science. What do you suppose would be the reaction of Groucho and Harpo? After they controlled their giggling, picked themselves up off of the grass, dirt, floor, or whatever else was down there, and recovered their power for speech, they would most likely reply, "Get away from here kid, you bother us." Why? What is the problem? Why would they react thus? Neither Groucho nor Harpo has any need for rocket science in their primitive conditions. Neither want rocket science enough to trade away mangoes or beef for it, so it would be foolish for Chico to specialize in producing rocket science if he wished to eat.

If, however, Chico had the ability to produce something that both Groucho and Harpo might be interested in—milk for example—then Chico could participate in the division of labor. Chico could specialize in the production of milk and produce some for himself and some for both Groucho and Harpo in exchange for mangoes and beef. In this way all three members of society are able to achieve more ends than if they all had to produce everything themselves. All three are now able to eat both mangoes and beef and drink milk. The addition of the milk was possible because Chico was able to specialize in its production, and it made sense for him to do so because he could trade milk to Groucho and Harpo for what they had to offer.

So we see that as the market extended beyond mangoes and beef to include the production and exchange of milk, the division of labor increased and the standard of living for everyone in our three-person society increased. This result provides a rudimentary theory of economic development and helps provide the beginning of an answer to the economic problem mentioned at the opening of this text: how do we fulfill God's mandate to have dominion over creation in a world of scarcity without killing one another or starving to death?

We are able to escape such a barbaric struggle for survival by taking advantage of the division of labor. As the extent of the market expands, the division of labor is able to develop as well. As the division of labor expands, the productivity of labor increases. This is because people will tend to specialize at producing those goods at which they are most efficient given their God-ordained differences. As each person's productivity increases, he will be able to consume more of whatever he produces and have even larger surpluses to trade for even more things that others are producing. As the productivity of each person increases, each person's wealth and standard of living increases as well. This permits a further extension of the market, which allows for further developing the division of labor, which reinforces the whole process. As Groucho, Harpo, and Chico increase their incomes and wealth, this creates the opportunity for other people to join their society and specialize in the production of other goods, such as clothing and shoes, so that each person has access to even more material blessings. It is in this way that exchange helps bring about economic growth and allows for an increasing number of people to have dominion over the earth, not by killing or robbing one another, but by peaceful, mutually beneficial voluntary exchange.

What is it that determines the exact pattern of the division of labor? What determines who will specialize in the production of what? By now you should be able at least to answer, "The different skills, resources, and capital stock with which God has endowed each of us." Can we go beyond that, however, to establish universal principles that are applicable generally to everyone everywhere? Let us see.

If two people are trading two goods and each person is better than the other at producing each of the goods traded, then obviously both parties can benefit via exchange and the division of labor. For example, suppose that a physician, Dr. Caligari, due to his natural intelligence and professional training, is better at providing medical services than Tully, the gardener. If Tully, due to his natural enthusiasm and experience with various flowers, bushes, and shrubbery is likewise a better gardener than physician, it is only wise for Dr. Caligari to spend all of his time doctoring, and pay Tully to tend his prized garden. If each person is better than the other at producing a good, this is known has having an *absolute advantage* in the production of that good. In the above case, Dr. Caligari has an absolute advantage in providing medical services and Tully has an absolute advantage in gardening. In a case where each party has an absolute advantage in producing something, it makes sense for each person to specialize in the production of that good at which they have an absolute advantage and trade to obtain the other good. By doing so, the standard of living of everyone involved increases due to the greater productivity of the division of labor.

However, what if, as is often the case, someone has an absolute advantage in producing everything? Is the division of labor and trade still beneficial? One might be tempted to respond in the negative because in the division of labor each person is wise to specialize in what they are better at producing. If a person is better at doing everything, then his potential trading partner is worse at everything and will have nothing to offer, or so it seems. As with all temptations, you would do well to avoid this one also.

As it is, even if one has an absolute advantage in the production of everything, that producer can still benefit via exchange and the division of labor. If one person has an absolute advantage in producing all goods in question, it will still benefit all parties to specialize in the production of that good at which they have a *comparative advantage*. The key word here is *comparative*. In this case the deciding factor is relative opportunity costs. Suppose that Dr. Caligari is both an excellent physician and gardener, so that he is better at both tasks than Tully. Dr. Caligari can treat one patient in an hour; Tully is completely unqualified to treat any. Additionally, it takes Dr. Caligari only one hour to cultivate his garden, while it takes Tully three hours to accomplish the same amount of cultivation. In this case Dr. Caligari has an absolute advantage in both lines of production.

However, both Dr. Caligari and Tully have a comparative advantage in different goods. Even though Dr. Caligari is better at both doctoring and gardening then Tully, it will still benefit him to spend all of his time doctoring and hire Tully to do the gardening. Suppose that the market rate for a doctor visit is $60 an hour and the market rate for gardening is $10 an hour. Every hour Dr. Caligari spends gardening instead of treating patients costs him $60 in foregone revenue. Each hour Dr. Caligari gardens himself might seem to save him ten dollars; after all if he gardened for only an hour, he would have the garden cultivated. However, do not forget that he would be foregoing the opportunity of receiving $60 of revenue he could earn from seeing patients. In other words, he would be out a total of $50. On the other hand, if Dr. Caligari were to spend all of his time doctoring, he would earn an additional $60 from treating patients and pay Tully a total of $30 ($10 times 3 hours), to do the equivalent gardening. In other words, both Dr. Caligari and Tully would be $30 ahead.

In this example, even though Dr. Caligari has an absolute advantage in both tasks and Tully, necessarily, has an absolute disadvantage in both tasks, it is still beneficial for both to specialize producing the service at which each has a *comparative advantage*. Dr. Caligari's and Tully's comparative advantage lie in producing that good at which each is the low opportunity cost producer. For every hour Dr. Caligari gardens, it costs him $60 in forfeited income. For Tully, because of his having zero medical skills, an hour spent gardening cost him no foregone income from doctoring. Tully is obviously the low opportunity cost producer of gardening, and Dr. Caligari is the low opportunity cost producer of doctoring.

Do not miss the importance of this discovery. The implications are profound. Comparative advantage allows us to benefit from the division of labor and exchange even if we are superior at producing every relevant good. Perhaps more importantly, comparative advantage allows those of us who are inferior at producing everything to also benefit from exchange. One of the beauties of the free market is that it is not an *all or nothing* proposition, a winner takes all game. The market is a social institution that allows people to find their respective niches based on each person's divinely created uniqueness. Even if I am the worst at everything, there will always be something at which I am the least worse. Therein will be my comparative advantage. I do not have to be absolutely the best at anything to benefit from the market economy. As long as people have a comparative advantage in something, which everyone does, they can benefit from specializing in producing that good and trading it for other things that increase their standard of living. And all of the benefits of the division of labor are due to our ability to exchange.

The history of our nation is filled with examples that bear out the truth of the social benefits flowing from voluntary exchange and the division of labor. Nineteenth century historian John Fiske admirably recounts the benefits to both the Native Americans and the early Plymouth colonists resulting from their trading with each other, especially emphasizing the blessings of trade for the Indians.

> There can, moreover, be little doubt that the material comfort of the Indians was for a time considerably improved by their dealings with the white men. Hitherto their want of foresight and thrift had been wont to involve them during the long winters in a dreadful struggle with famine. Now the settlers were ready to pay liberally for the skin of every fur-covered animal the red men could catch; and where the trade thus arising did not suffice to keep off famine, instances of generous charity were frequent. The Algonquin tribes of New England lived chiefly by hunting, but partly by agriculture. They raised beans and corn, and succotash was a dish which they

contributed to the white man's table. They could now raise or buy English vegetables, while from dogs and horses, pigs and poultry, oxen and sheep, little as they could avail themselves of such useful animals, they nevertheless derived some benefit. Better blankets and better knives were brought within their reach; and in spite of all the colonial governments could do to prevent it, they were to some extent enabled to supply themselves with muskets and rum.[4]

This interpretation of events is affirmed more recently by an anthropologist and historian who writes

> what were trifles to . . . Europeans were assuredly not to Indians, who also lept at the chance to exchange what was mundane and common (furs and skins) for what was not (goods of new technological and symbolic value). Some North American Indians greeted Europeans by waving pelts in their hands, signaling unambiguously the familiarity of exchange as well as their keenness to obtain the rare and useful objects Europeans possessed. Many were familiar with exchange from trading with other native people long before Europeans arrived; some had had previous (but unrecorded) encounters with Europeans. Regardless, many knew what they wanted. In Carolina in 1670, Sewee Indians greeted an English boat by '[running] up to ye middle in mire and watter to carry us a shoare where when we came they gaue us ye stroaking Complimt of ye country and brought deare skins some raw some drest to trade with us for which we gaue them knivues beads and tobacco and glad they were of ye Market.'[5]

Voluntary exchange has the further social benefit of promoting peaceful relations between trading partners. While no human institution can guarantee that man will never take up arms against man in acts of aggressive confiscation, if people understand how they benefit from voluntary trade, they have an incentive to promote the peace that is necessary for such exchange to take place.

This salutary effect of voluntary exchange in a fallen world is again illustrated by the history of the settling of our nation. Charles Larpenteur worked in the North American fur trade from 1833–72, conducting business for the American Fur Company in what is now Montana, North and South Dakota, and Nebraska. In his memoir, he leaves us an account of how his value as an agent of exchange spared his life, even after he made

4. Fiske, *The Beginnings of New England*, 200–1.

5. Krech, *The Ecological Indian*, 151–52.

a significant entrepreneurial error by acting unethically. Larpenteur and four helpers were sent with a large stock of goods for trading to a large Indian camp. He then tells the following story:

> After I had stored everything properly, I was invited into the lodges of the chiefs and leading men, to partake of a dish of pounded buffalo meat and marrow grease, as is their custom . . . That night, when the liquor trade commenced, the very devil was raised in camp . . . There was I, with only four other men, among about 300 drunken Indians with no other alternative but to trust to luck . . . Then came two . . . drunken Indians; one of them named Cougher and the other an individual who had killed his own father; both had pledged with others to murder me in the lodge and plunder my outfit. But it happened that I had a good old friend in camp, whose name was The Haranger, and who made such a fine speech that they abandoned the idea. This is about as near as I can interpret it: "What is it that I hear? Brothers and kindred, do you think you will need your trader no longer, now that spring is come and trade is over? You have your fill of everything, and now talk of killing your trader. Where will you go? Go north and starve? Give away your hunts for nothing? Why kill this poor white man? What has he done to you? No, brothers! Have pity upon him, upon me; spare his life." On his saying this which they understood to be the conclusion of his speech, a young man got up and handed him his knife, as a sign of approval and so the idea was given up.[6]

This passage illustrates a number of important facts we do well to remember. It shows us how the ethics of market participants are played out in market outcomes. Voluntary exchange does not guarantee ethical behavior. It does not ensure that everyone who trades will make wise or good decisions. On the contrary, in a free market people are free to act out their ethical framework as long as they do not aggress against another's property. Sellers sometimes try to reap profits by unethical or unwise means. However such actions often prove very costly in the end.

Nevertheless, this piece of history also teaches us that when entrepreneurs act unwisely, they often do so at their own peril. In this case, the fur trader Larpenteur placed his life needlessly at risk by making the entrepreneurial decision to engage in the liquor trade, which he should have known could have dire consequences for himself, not to mention to the Indians. The danger to his life was not due to his trading with the Indians *per se*, but rather was due to his choice to provide them with al-

6. Larpenteur, *Forty Years A Fur Trader on the Upper Missouri*, 200–2.

cohol. Larpenteur made a poor entrepreneurial decision that nearly cost him much more than merely material profits.

This bit of history further teaches us that a moral framework is important for a well-functioning market. Trade is easier to facilitate if buyers and sellers behave with integrity. Trading with the Indians would have gone much more smoothly if Larpenteur had not sold them alcohol. Things would have been much less uncertain for him and his trading partners would have acted less threateningly if he had foregone peddling liquor and stuck to trading goods for which there were less probable negative consequences.

This account further illustrates one of the themes that runs through the book of Proverbs. While by no means a manual to get rich, the book of Proverbs includes much instruction on true success and its relation to material blessing. One of the lessons that can be found in its pages is that those who generally act ethically have their needs more than provided. In describing the benefits of wisdom, Proverbs states, "Long life is in her right hand; in her left hand are riches and honor" (Prov. 3:16). Elsewhere, we read "Riches and honor are with me [wisdom], enduring wealth and righteousness" (Prov. 8:18). The negative principle also applies. Greed for unjust gain can cost people their life (Prov. 1:19). While good morals do not guarantee riches, Proverbs teaches that often he who fears the Lord and acts with wisdom is blessed with prosperity, while he who seeks to gain via wickedness will come to naught.

Additionally, Larpenteur's story teaches us of the importance of ethical constraint for a well-functioning market. Notwithstanding the foolishness of the liquor trade, mercifully, the Indians recognition of the benefit of exchange with Larpenteur constrained them from making a rash, murderous decision. The perceived benefits of exchange forestalled violent conflict. However, exchange would have occurred even more smoothly had not Larpenteur sought to enrich himself and his company by offering to sell liquor that he knew would be likely abused.

American history confirms economic theory. Even after placing his life in danger by providing the Indians the means to become drunk, the life of Charles Larpenteur was spared because his would be assassins understood the benefits of trade with him. Notwithstanding the fact that some people will buy and sell goods that may be abused, voluntary exchange is mutually beneficial and results in social progress by making possible the division of labor and encouraging peaceful relations.

THE IMPORTANCE OF PRIVATE PROPERTY

This chapter, then, has brought us into the realm of economics proper and identified a number of important principles relating to voluntary exchange. Free trade is not a zero-sum game where someone wins at another's expense. Voluntary exchange is a mutually beneficial endeavor that leaves both parties better off than they would be without the trade.

We also learned that the possibility of exchange opens the door to the great productivity gains available through division of labor. Exchange allows people to specialize in producing those goods at which they are relatively more efficient, in order to trade away their surplus for other goods they want, but are relatively less efficient at producing. This allows for every market participant's standard of living to increase, as people are able to have more goods at their disposal allowing them to achieve more of their ends. This process, furthermore, will continue as more and more exchanges take place and as people's wealth increases, subsequently extending the market for the goods that they demand.

However, all of this increase in standard of living made possible due to the division of labor brought to us by our friends at voluntary exchange is predicated by a big *if*. The above documented growth in wealth and standard of living is possible if, and only if, people are indeed able to trade what they have for things they want more. In other words, all of the benefits of voluntary exchange can be had only if people have the right to private property. If a society refuses to defend private property rights, they should not expect to be economically prosperous. To the extent that property rights are restricted, so will exchange be limited and the division of labor constrained.

This link between private property rights and economic prosperity is borne out by comparing the level of economic freedom in various counties with the respective per capita income of those countries. The word *freedom* here is used to refer to an environment in which people have extensive private property rights. A free society is one in which people are able to use their property as they see fit, so long as they do not infringe on the same right of other people. For example a free society would allow people to own guns to be used for hunting, self-defense, or however else they want, but would not allow people to use them to threaten or shoot others just because they want to. In a free society, not only are people protected from private criminals, but their property is also kept off limits from the state. A community featuring a thriving voluntary exchange network, then, is also one in which the state does not intervene in the economy by imposing price controls, confiscatory taxation, coercive wealth transfers, or manipulating the money supply.

Researchers who have investigated the effect of freedom and economic growth have discovered a clear positive relationship between economic prosperity and private property. Countries with the highest level of private property rights, described here as economic freedom, tend to be the wealthiest. Countries in which people have very restricted private property rights are also the poorest.

More Freedom Yields More Prosperity

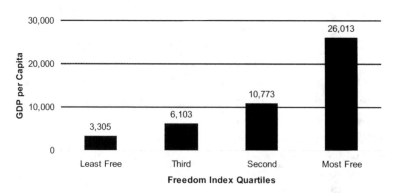

Figure 4.3. Those countries that scored in the highest quartile for economic freedom also had the highest per capita GDP. The Economic Freedom of the World Index is a statistic developed by the Fraser Institute and is designed to measure a country's level of economic freedom. Source: Gwartney, et. al, *Economic Freedom of the World*, 23.

As illustrated in figure 4.3, countries that are classified as most free tend to have larger per capita national incomes than those countries with less freedom. In 2005, those countries that featured a high level of private property protection had an average yearly income of $26,013 per person. As the level of protection for property declined, so did average income so that those countries considered least free reaped an average yearly income of only $3,305 per person.

The link between private property, exchange, the division of labor, and economic prosperity, is something we know is true because good economic theory tells us so. Economics tells us that the freer people are to use and trade their property as they see fit, the more economic prosperity there will be, because the more voluntary exchange is possible, and, hence, the more developed the division of labor will be.

It is important to note that the correlation between economic freedom and prosperity does not, by itself, imply that economic freedom is ethically good and morally desirable. The freedom-prosperity correlation is an example of what is sometimes called *positive* economics. It is a statement of what is. Such a statement does not, ultimately, tell us whether we should, in fact, protect private property rights.

Economics, by itself, cannot tell us what is good. It can only tell us that if we do *a*, then *b* will result. Patient readers who persevere to the end of this text will discover that sound economic theory tells us that at higher prices, the quantity of a good that people will want to purchase will be smaller; a higher minimum wage will make unemployment larger than otherwise; a price ceiling on meat will result in a shortage; if we embrace socialism, economic chaos results. However, none of the above findings, by themselves, tells us what should be our policy. The fact that raising the minimum wage will increase unemployment will not dissuade its increase if we decide that unemployment is actually a good thing. Or perhaps, even though politicians know that increases in the minimum wage will lead to some people losing their jobs, they decide that there are other considerations that warrant its increase anyway.

A CHRISTIAN VIEW OF PROPERTY

Once we move from examining the economic consequences of a certain policy to recommending that we should or should not follow that policy, we have moved out of the strictly economic realm and into the ethical. Questions of what we should or should not do are moral questions, not only economic ones. This distinction is important, because economic policy is never enacted in a vacuum. Economic activity always takes place within a certain institutional framework. We have already seen that the institution of private property has a great effect on the level of economic prosperity a society enjoys. However, legal institutions such as private property rights do not merely float in ether detached from the rest of reality. The particular social institutions in place at any given moment rest upon people's worldview: their view of man and ultimately their creator. Thoughtful readers will be able to see that the kind of economic institutions people have is ultimately determined by their theology.

For Christians especially, it is not enough to know that a certain means, such as private property, will achieve a certain end like economic prosperity. We know that the chief end of man is to glorify God and enjoy him forever. Notwithstanding the preferability of life and prosperity over death and starvation, the fundamental question regarding social institutions is not are they useful, but are they right.

Historically, Christians have recognized that private property is part of the Christian ethic and, hence, a Christian social institution. There are many warnings against greed, trusting riches, and the love of money, and exhortations to share with those less fortunate. Nowhere, however, is the possession of property condemned. In fact, the commandments against theft indicate that God's moral law requires private property.

The morality of private property was recognized by many of the patriarchs in the early church. For example, men such as Clement of Alexandria and Augustine endorsed private property and allowed for the possession of riches, while cautioning Christians not to become ensnared by their belongings. They urged Christians to use their property for the good of the community. Even Tertullian, no fan of merchants or money-making, allowed that owning property was morally permissible.

During the Christian Middle Ages, the towering theologian was Thomas Aquinas. He was a firm believer in the superiority of private property over communal property, basing his convictions on both natural and divine law. He argued that private property was both a necessary feature of man's earthly estate and a guarantor of societal order and peace. Pope John XXII in his papal bull, *Quia vir reprobus*, written in 1329, correctly explained that God's dominion over the earth is reflected by Man's dominion over his material possessions. As such, property rights are rooted in man's nature as created by divine law. The Thomistic tradition culminated with the late scholastics such as Miguel Salon, Juan de Medina, Pedro de Aragon, and Domingo de Soto who all defended the morality of private property.

Key leaders in the Protestant Reformation also argued for the legitimacy of private property. John Calvin, for instance, while commenting on the eighth commandment, wrote, "To sum up: we are forbidden to pant after the possessions of others, and consequently are commanded to strive faithfully to help every man to keep his own possessions." He further explained that this command specifically prohibits theft, fraud, and cheating others out of their property.[7]

7. Calvin, *Institutes of the Christian Religion*, 408–11.

The Reformed Christian confessions likewise indicate that Christian justice demands the protection of property from the aggression of others. For example, the Heidelberg and Westminster Longer Catechisms instruct us that God forbids theft, robbery, kidnapping, fraud, using false weights and measures, removing property markers, and extortion. God likewise requires the fulfilling of contracts, the payment of debt owed, and restitution of goods stolen.

Princeton theologian Charles Hodge, in his *Systematic Theology*, explains that in the eighth commandment, God instituted a divine right to property. "This commandment," writes Hodge, "forbids all violations of the rights of property. The right of property in an object is the right to its exclusive possession and use."[8] He goes on to elucidate that violations of the right of property include theft, fraud, and communism, regarding which he wisely notes, "the right of property, [is] ordained by God, and cannot be violated without incurring his displeasure and the certain inflictions of his divine punishment."[9]

Why does Christian wisdom so firmly advocate the right to property? The Christian view of property begins with the understanding that we hold property as stewards of God, who is the Creator and ultimately owns all that there is. He tells us that "Every beast of the forest is mine, the cattle upon a thousand hills," (Ps. 50:10) and that "The earth is the Lord's and the fullness thereof, the world and those who dwell within" (Ps. 24:1).

As owner of all there is, God has the right to confer the use of property on whomsoever and under whatsoever restrictions he pleases. In his sovereignty and providence, he can work to make some more prosperous and others less prosperous. Additionally, from a Christian perspective, only God has the right to determine how people can morally add to their property. As they accumulate property, God forbids people to aggress against the property of their neighbor. They must recognize and submit to the right of every man to be secure in his property. God communicates this to us through both general and special revelation.

Francis Wayland, a nineteenth-century Baptist minister who was the president of Brown University for 28 years, summed up the Christian case for private property in his treatise on ethics, *The Elements of Moral Science*. Wayland explained that the right to property is founded on God's will being made known to us in three ways.

8. Hodge, Charles. *Systematic Theology*, 421.
9. *Ibid.*, 421.

The first way God communicates the right to property to us is through our natural conscience. It seems that the ethic of private property is part of the law the Apostle Paul tells the Christians in Rome, is on our hearts (Rom. 2:15), and as such, is communicated to us by our conscience. All people, as soon as they begin to think—even at a very young age—perceive the difference between mine and yours. Particularly among children is this evident; the concept of *mine* does not have to be taught. This relation of property is expressed by possessive pronouns in all languages. People naturally feel that whoever violates their property rights is wrong.

Additionally, the created order also bears witness to the rightness of private property. There are observed consequences to either upholding or violating the right to property. History shows that both the existence and progress of society, and the human race itself, depends on the right to private property. In those lands without the right to property, people tend to labor only enough to manage their own individual subsistence; they tend neither to accumulate capital goods nor to plan for the future. In such societies, there is little accumulation of capital, no tools, no provision for the future, no houses, and no agriculture; those people who survive into adulthood tend to exist in the basest of poverty. Without private property, Wayland notes, the human race must perish or exist in wretchedness. Civilization progresses, therefore, in proportion to the right to private property being held inviolate.

God's supernatural revelation in his Word also treats the right to property as something not to be violated. The Old Testament contains many precepts against acts of theft. The eighth commandment (found in Exod. 20:15 and Deut. 5:11), of course, prohibits theft. Elsewhere, the Old Testament Law also explicitly prohibits theft and fraud (Lev. 19:11, 13). In Proverbs 30:9 we read that stealing profanes the name of God. The prophet Jeremiah lists stealing with sins of murder, adultery, swearing falsely, worshiping Baal, and walking after false gods (Jer. 7:9).

This prohibition against theft is reiterated in the New Testament as well. Our Lord Jesus affirms that not stealing is a commandment to be kept by the rich young ruler (Matt. 19:18; Mark 10:29; Luke 18:20). When exposing Ananias and Sapphira for lying to the Holy Spirit, Peter expressly told them that their property was theirs to do with as they saw fit (Acts 5:4). The Apostle Paul implies that not stealing is part of loving your neighbor as yourself and commands Christians to repent of their stealing, and instead provide for themselves and others through honest labor (Rom. 13:9; Eph. 4:28). When exhorting churches to contribute to the common charity fund, Paul never calls for coercive extraction of property, but for voluntary giving (2 Cor. 9:7).

The biblical prohibition of theft includes abstaining from fraud. In Leviticus, we read that God told Moses to tell the people, "You shall do no wrong in judgment, in measures of length or weight or quantity. You shall have just balances, just weights, a just ephah, and a just hin: I am the Lord your God, who brought you out of the land of Egypt. And you shall observe all my statutes and all my rules, and do them: I am the Lord" (Lev. 19:35–37). The condemnation of fraudulent weights and measures is repeated in Proverbs where we read, "A false balance is an abomination to the Lord, but a just weight is his delight" (Prov. 11:1); and "Unequal weights and unequal measures are both alike an abomination to the Lord" (Prov. 20:10). God considers it an abomination to cheat a trading partner. It is considered an act of stealing.

The encroachment of someone else's property by moving recognized property boundaries is likewise prohibited. The Judaic law states, "You shall not move your neighbor's landmark, which the men of old have set, in the inheritance that you will hold in the land that the Lord your God is giving you to possess" (Deut. 19:14). In Proverbs we find confirmation of this prohibition, where the writer straightforwardly mandates "Do not move the ancient landmark that your fathers have set" (Prov. 22:28). One chapter later he warns, "Do not move an ancient landmark or enter the fields of the fatherless for their Redeemer is strong; he will plead their cause against you" (Prov. 23:10–11). The context of both reminds us that God intends private property rights to be a legal protection for the poor, not only the wealthy and politically powerful.

Besides outright acts of theft and fraud, we read in the Bible that God prohibits the thoughts from which violation of the right of property proceeds. Jesus tells us that the motivation for theft and other sins come from the heart (Matt. 15:18–20). Not surprisingly God prohibits our coveting the property of others (Exod. 20:17, Deut. 5:21).

Scripture makes it abundantly clear that we are not to take from others what is theirs. Everyone is to be secure in the property that they own. The question remains, however: How do we rightfully come into ownership? Is private property a creature of the state? Is it primarily a legal institution?

Again God provides us an answer. Our property is God's gift. God tells us that "The earth he has given to the children of man" (Ps. 115:16). In his book *Practical Philosophy*, southern Presbyterian R. L. Dabney identified and answered the key question: Is this gift to man individual or general?

To answer this we need to keep in mind the chief end of man. Our whole life is to be directed toward glorifying God and enjoying him forever. We have the cultural mandate to be fruitful and multiply, to subdue the earth, fill it, and have dominion over it. This requires much use and transformation of the earth God has given us.

Dabney wisely noted that an existential fact of the created order is that use is individual. All action, and hence use of means, is undertaken by individuals. Therefore, if God has given man the earth to use for his glory, it must be that he has distributed particular units of the earth to individual persons so they can use them to fulfill God's mandate.

All human beings bear God's image. We are all given reason, free-agency, and responsibility. As such, we are all responsible to fulfill God's dominion mandate. Each of us, therefore, has an initial right to share in God's gift of nature. In obtaining property, which is God's gift to us, however, we must act justly, meaning that we may not transgress anyone else's right to their share in God's gift.

These facts guide us into understanding how people justly obtain private property. In the first place, God gives each person the gift of their own life and body. They, therefore, have ownership over their own labor.

Additionally, people come to own property by homesteading unappropriated land or resources. God gives the desire to use land, and if someone finds a plot of unused land, and there is no one to contest his claim, the person can take ownership of it. If someone first appropriates a natural object and transforms it with his labor, it becomes the property of the first user. If someone else subsequently interferes with this person's use of his possession, he is guilty of trespassing against the other's right of property.

A person also can come into ownership of a good by production. By using our own labor and other means that God gives us, we can produce goods that are ours. Just as the earth is the Lord's because he made it, when you take natural things that God has given you and transform them into a new good, that good you made is yours.

Finally, we can come into legitimate ownership through exchange. We can trade a good that we own to someone else in return for a good that he has. After exchange, the good we receive is our property. The same is true for a gift we receive, except that, because it is a gift, we do not have to give anything in exchange to the giver.

The New Testament instructs that the Christian ethic of property mandates the right to *voluntary exchange*. Christian ethics prohibits forced exchanges between people. This is implied by passages teaching that the moral right to property includes the owner's ability to use that property as he wants, provided he does not aggress against someone else.

In Acts 5, Luke documents the demise of Ananias and Sapphira for lying to the Holy Spirit. Ananias and his wife sold a piece of property and asserted they were giving all of the proceeds to the Church in Jerusalem, all the while holding back some of the revenue for themselves. The Apostle Peter confronted Ananias saying, "While it remained unsold, did it not remain your own? And after it was sold, was it not at your disposal? Why is it that you have contrived this deed in your heart? You have not lied to men but to God" (Acts 5:4). Peter expressly says that their property was theirs to do with as they saw fit.

Peter was affirming the same principle Jesus used in his parable of the laborers in the vineyard recorded in Matthew 20. In this parable about the Kingdom of God, Jesus told the story of a master who hired different laborers to work different amounts of time, but paid them the same amount of money. In response to complaints from the laborers who worked the longest, but received the same pay as those who worked the shortest, the master, who represents God, says, "Friend, I am doing you no wrong. Did you not agree with me for a denarius? Take what belongs to you and go. I choose to give to this last worker as I give to you. Am I not allowed to do what I choose with what belongs to me? Or do you begrudge my generosity?' (Matt. 20:13–15). While not the main point of the parable, the moral force of Jesus' teaching relies on the acceptance of the principle that the owner of property can do with it what he wants, as long as he does not violate someone else's property right.

If one comes into possession of property legitimately as described by one of the above methods, then with regard to human relationships, he has an exclusive right to the possession and use of that good. We also have a duty not to infringe on rights of others to be secure in their property. We do not have the right, for example, to use a gun we own to rob someone of their belongings. As already indicated, we are expressly forbidden to take from someone what is rightfully theirs. This would be a violation of their rights. The right to property does not imply a moral right to violate God's law. Ultimately everyone will answer to God for how they use the property he gives them. The right to private property we are explaining constrains aggression against property through human interaction.

The individual right to private property is further affirmed if we consider the biblical record. The recorded history of the first family reveals that God did not give property to man in general all at once. In order to begin fulfilling God's cultural mandate, Adam and Eve first appropriated *some* of the earth. They worked the land and it became theirs. Their sons, Cain and Abel, did likewise. Cain tilled the earth and raised crops that were his. Abel used land to raise animals that were his.

Personal reflection upon our own consciences, the created order, and careful study of God's Word conclusively demonstrates that not only is the social institution of private property generally beneficial, but also it is morally right. This conclusion has great consequences for Christian social thought. It is the basis for a free society that manifests voluntary exchange and eschews violent coercion.

It should also be noted that, not only does the divine right of private property forbid individuals from murdering and stealing from each other, but God also forbids rulers from doing so. He does not command us to not murder or steal, unless we work for the government. No indeed. God commands rulers to serve the Lord with fear (Ps. 2:10–12). In doing so, they should rule with justice (2 Sam. 23:3) and judge justly with everyone (Ps. 82:1–4). God tells the ruler to defend the poor and fatherless, but at the same time this does not imply that the ruler should be partial to the poor (Exod. 23:3; Lev. 19:15). The thirteenth chapter of Romans limits the state to the punishment of evildoers and the praise of those who do good.

By the term *the good* the Apostle Paul meant essentially keeping the peace. As biblical commentator John Murray notes, "Paul provides us with a virtual definition of the good we derive from the service of the civil authority when he requires that we pray for kings and all who are in authority 'that we may lead a tranquil and quiet life in all godliness and gravity' (1 Tim. 2:2)."[10] The English Standard Version renders "tranquil" as "peaceful." The charge God gives to the civil authority is to keep the peace and thereby allow believers to enjoy peaceful and quiet lives. This charge implies that the main function for any magistrate is to protect the innocent from aggression. The state is not authorized to become an agent of aggression who takes from the productive and give to the politically connected.

10. Murray, *The Epistle to the Romans*, 152.

If the civil authority is to fulfill its function of keeping the peace while acting justly and with the fear of the Lord, it should be clear that the magistrate must also use divinely approved means in so doing. In other words, just as all people are to respect the property of others by not stealing from or defrauding their neighbor, neither should the ruler. Just as private citizens should not steal, the state should not steal either. Scripture is clear that none of us should individually take what is not ours. Similarly, we should not collectively take what is not ours through the state. Good government is one that protects citizens from violence and does not intervene in the economy to take wealth from some people and distribute it to others. Only a social system with private property allows for mutually beneficial exchange and sets the stage for economic growth resulting from the division of labor. Only such a system allows us to obey God's mandate to have dominion over creation without starving ourselves or stealing from or killing each other.

GOD'S MANDATE AND GOD'S BLESSING

In the beginning God created the heavens and the earth and then created man and woman and placed them in the midst of the garden. He gave them dominion as stewards over all that he created. God gave us the command to be fruitful and multiply, to fill the earth and subdue it. This necessitates combining land, labor, and capital goods to produce those things that assist us in this life. However, after Adam's sin, the ground was cursed and the earth became much less forth-coming. The problem of scarcity became more constraining and the prospect of filling the earth could easily be seen as a threat to the survival of the human race rather than a command given by a benevolent God.

Fortunately, God made the universe in such a way that allowed for the discovery of methods to fulfill God's mandate in a peaceful way. He created us with rational minds capable of understanding the mutually beneficial results of exchange. He created us in such a way that human society helps us be more productive, not less. He also gave us his moral law that, among other things, laid the ethical foundations for the very social institution necessary for taking advantage of exchange and the division of labor: the institution of private property. Thanks be to God for his kindness to us.

SUGGESTED READINGS

Beisner, *Prosperity and Poverty*, 43–55, 149–59. Taken together, these two chapters provide a powerful argument that the Christian ethic of property implies that a free market is a necessary component of economic justice. Beisner also shows, consequently, that the Christian ethic of property requires firm limits on the state.

Dabney, *The Practical Philosophy*, 448–95. Dabney's excellent chapter on the duties and rights concerning private property from the perspective of a nineteenth century southern Presbyterian.

Heyne, *The Economic Way of Thinking*, 157–72. An outstanding defense of the productive service provided by the middle man and the speculator.

Hodge, *Systematic Theology*, 421–37. In this chapter, Hodge argues that God's commandment against theft mandates a divine right to private property.

Mises, *Human Action*, 157–64. In this important section Mises extends David Ricardo's law of comparative advantage to a full-fledged law of association.

Rothbard, "Freedom, Inequality, Primitivism, and the Division of Labor," 3–35. An outstanding explanation of the importance of the division of labor for the development and flourishing of society.

———. *Man, Economy, and State*, 79–185. Rothbard provides an excellent exposition of the principles and implications of voluntary exchange.

Wayland, *Elements of Moral Science*, 229–36. Wayland, an important nineteenth century northern Baptist and president of Brown University, provides a very clear exposition of the Christian ethic of private property from his very popular text book on ethics.

5

Indirect Exchange and Market Prices

IN THE PREVIOUS CHAPTER, you became familiar with the tremendous value there is to exchange made possible in an environment of private property. Voluntary exchange is mutually beneficial, because it only occurs when each party is able to achieve more of his ends by doing so. Trade takes place when the parties value the goods being traded in reverse order or, in other words, when they have what we call a double coincidence of wants.

Exchange is a universal experience throughout the world and throughout history because of the great variety in people and environments that God created. Everyone has different skills and aptitudes. Every plot of land has different qualities. Everyone has capital stocks made up of different goods in different quantities.

Specialization and the division of labor allow each person to develop his best skills and the inhabitants of each region to develop its own unique resources. Without exchange, this could not happen, because everyone would then have to produce everything he consumes. Exchange is only possible if there is private property. We cannot trade what we do not own.

INDIRECT EXCHANGE

All of the principles of exchange discovered thus far have been gleaned from thinking about people who trade different consumer and producer goods for other consumer and producer goods. Movie tickets were traded directly for other movie tickets. Mangoes were traded directly for beef. Trading consumer and producer goods directly with each other is known as *barter*. Any voluntary exchange that takes place in a barter economy is mutually beneficial and helps develop the division of labor. However, barter is not without serious limitations.

One of the constraints faced by people in a barter economy is what economists call the *indivisibility problem*. For some goods, if they are divided into smaller parts, the sum of the parts is not nearly as valuable as the undivided whole. Suppose, for example, that you are just beginning one of those high-protein, low-carbohydrate diets and go to your local grocery store to buy a dozen eggs for breakfast this week. If you specialized in producing butter, you could make the trade relatively easily. You and the grocer might come to an agreement to trade a dozen eggs for a half pound of butter and it would be easy for you to carve a half pound of butter from the slab you produced in your butter factory.

On the other hand, if you were a breeder of pedigree Lowchen dogs, trading for a dozen eggs would not be such a breeze. You would hardly want to make a one-to-one trade for a dozen eggs, because Lowchens are quite rare and extremely valuable. Moreover trading for eggs with the dogs brings us face to face with the indivisibility problem. What is the likelihood that the grocer would respond favorably to your suggestion, "Well, giving you a whole dog for a dozen eggs is too much. I would not give one up for less than 500 dozen eggs, but I cannot eat nearly that many before they go bad and I get sick of them. I tell you what I will do. I'll trade you the back left leg of one dog for a dozen eggs." How will the grocer respond? Most likely, he will respond with horror. He has no use for the left hind quarter of a Lowchen. You most likely do not have much use for a three-legged dog. The value of individual dog parts is so low that their sum will be much less than the value of the whole Lowchen. Therefore, in a barter economy, those people who would like to specialize in the production of relatively indivisible goods such as pedigree canines will find it much harder to get what they want via trade.

By permission of Leigh Rubin and creators Syndicate, Inc.

The indivisibility problem is a special case of the more general lack of the double coincidence of wants problem. Recall from the previous chapter that if both parties value each good in question in reverse order, exchange will occur and will be mutually beneficial. If, however, the parties are lacking the double coincidence of wants, they will not exchange, because one of the parties will not perceive himself to be better off after the trade. Take, for example, the value scales of Groucho and Harpo in the following cases:

	Case One			Case Two	
Groucho	Harpo		Groucho	Harpo	
(Beef)	(Mangoes)		(Beef)	Beef	
Mangoes	Beef		Mangoes	(Mangoes)	

In case one, Groucho and Harpo value the beef and mangoes in reverse order, so both are able to achieve more highly valued ends if they trade with one another. However in the second case, the double coincidence of wants does not exist. While Groucho would be willing to trade his mangoes for Harpo's beef, Harpo values his beef more than mangoes, so a trade would leave him worse off. Harpo is unwilling to trade, so no exchange takes place. This sort of problem occurs quite regularly in a barter system and greatly limits the development of the division of labor in a barter society.

As people tried to avoid the double coincidence of wants problem, they set in motion a process by which they moved from a barter economy into one characterized by indirect exchange. Indirect exchange occurs when goods are not directly traded for other goods desired, but are traded indirectly through some medium of exchange.

What is meant by *medium*? We do not mean the relative size between large and small. We can get at the meaning by consulting the Old Testament law. In Deuteronomy 18:11 God prohibits any medium from being numbered among his people. In fact the work of a medium was so abominable that in Leviticus 20:27 God commanded that any man or woman who was a medium should be stoned to death. What was this medium to which God was so opposed? In this context, a medium is one who engages in the occult practice of trying to contact the spirits of the dead. Mediums supposedly allow the living to communicate with dead people and the dead to communicate with the living. The medium is the go-between. Similarly, a medium of exchange is the go-between. We trade goods through the medium of exchange we call money.

The process describing the origination of the medium of exchange is the story of the emergence of money. If you never have asked yourself the question before, think now about why we use money to make our exchanges and how money came to be. Today, virtually everyone in a modern economy engages in an exchange in which at least one of the goods being traded is money. How did primitive societies move from barter economies to using money?

Because of the problem of the lack of the coincidence of wants, people sought out goods that could be used in exchange. Go back to the example above in which no trade was taking place. Groucho is hankering for some beef. In fact, he wants it so badly he could almost taste it. Almost, however, is not good enough. He wants beef so badly he is willing to trade his mangoes for a sirloin steak. Harpo however is having none of it. Harpo does not want to trade his beef for mangoes.

Is there any way for Groucho to achieve his end? There is if he can obtain something that Harpo *does* want more than his beef. Suppose Groucho discovers that Harpo wants butter. Suddenly things change.

Groucho now demands butter when he didn't before, and he demands butter, not for its use value to him, but for its exchange value. Groucho does not want to eat butter. He does not want to sauté a few bratwursts in its melted state. He does not want to rub it on himself before he lies out on the beach. He demands butter because he thinks he can trade it to Harpo for some beef. If he finds someone like Chico, who has butter but wants mangoes more, Groucho is in business. Suppose that Groucho does this and the value scales for Groucho, Chico, and Harpo are as shown below:

Groucho	Chico	Harpo
(Beef)	(Mangoes)	(Butter)
(Butter)	Butter	Beef
Mangoes		(Mangoes)

Butter appears on Groucho's value scale where there was none before. He values butter more than mangoes, because mangoes cannot help him achieve what he wants more, which is the beef.

Groucho could trade his mangoes to Chico for Chico's butter. This trade gets Groucho closer to his goal of consuming what he really, really wants. Once Groucho owns some butter, he could then trade it to Harpo for some beef. Groucho ultimately trades mangoes for beef, but through the medium of exchange—namely butter. After all of the trades are completed the preference rankings of our characters are as follows:

Groucho	Chico	Harpo
Beef	Mangoes	Butter
(Butter)	(Butter)	(Beef)
(Mangoes)		(Mangoes)

Note that as the result of this indirect exchange, all three people are able to obtain whatever they value the most. Groucho has the beef he loves, Harpo has his butter, and Chico has mangoes.

On a societal level the changes are even more profound. Because butter is now demanded by more people, it becomes more marketable, meaning that it is easier to sell. Not only did Harpo want butter, but Groucho also demanded butter after he discovered its exchange value. This greater demand for butter, in turn increased butter's marketability. People demanded even more of it, because its increased marketability made it even more valuable in exchange. The only way for people to obtain a more marketable good is to trade away goods that are less marketable. This process continues as people exchange less marketable for more marketable goods in pursuit of attaining their highest valued ends.

It is in this way that money emerges. As a particular good becomes recognized as being more marketable, more people demand it for its exchange value. As more people demand it for its exchange value, it becomes even more marketable. As it becomes more marketable, more people demand it and its value spirals upward. This process finally culminates with the most marketable good being used as a general medium of exchange in all markets. The general medium of exchange is called *money*.

Different societies at different times in history used different commodities for their medium of exchange. In early Colonial Virginia, for example, tobacco was used as money for some time. The medium of exchange was sugar for a period of time in the West Indies. Cattle were used in Ancient Greece. In parts of South America centuries ago, cocoa beans were used as money and at one time the value of cocoa was so high that a person there could buy a human slave for only two beans. Historically two metals displaced all other commodities as the medium of exchange: gold and silver. Now, because of government intervention in our monetary system, we use paper dollars as our medium of exchange.

This account of the emergence of money teaches two things. The first is that money is a commodity. It is an economic good. Hence, all of the laws of economics apply to money just as they apply to consumer and producer goods. Secondly, the good that ultimately became money became the most marketable good through voluntary exchange. Money then, is a product of the free market, not a product of the state.

Notwithstanding the biblical warnings against the love of money and being a slave to mammon, the development of the institution of money conferred several benefits to human society. Because the use of money in indirect exchange overcame the common lack of double coincidence of wants problem, the development of money greatly expanded exchange and the division of labor. This, of course, led to further increases in productivity, incomes, wealth, and standards of living.

Another, very important, albeit often unrecognized, benefit of the development of money is that a common medium of exchange allows entrepreneurs to calculate profit and loss. In a free market based on voluntary exchange of private property, it is the entrepreneur that oversees production. He is the one who obtains the use of the factors of production (land, labor, and capital goods), and transforms them into items to be sold on the market. Because such production takes time and the future is uncertain, entrepreneurial activity always entails a certain amount of risk. The entrepreneur must always forecast what the public will want in the future and this is not mechanically determined by what they want in the present. Not only must the entrepreneur forecast future market demand, but he must also calculate the most efficient, least costly, method of producing those goods that he thinks will be in demand. Entrepreneurs make such decisions by comparing the expected price of the product they are making with the sum of the prices of the different factors they could use to produce it.

In the absence of money, such comparisons are very difficult. Suppose Michelle, an entrepreneur, decides that the housing market looks promising and decides to enter the construction business to build four-bedroom, two-bath, and two-car garage houses. How does she decide whether she should use a metal frame or wooden studs, whether she should roof with slate or asphalt shingles, whether she should side it with wood, brick, aluminum, or vinyl? You're probably tempted to think, "That's easy, just compare the prices of the factors and use the cheapest, assuming that the buyer does not care." Without money, however, this would be practically impossible. In a barter economy, the prices of metal studs, two-by-fours, slate and asphalt shingles, wood, aluminum, vinyl siding, and bricks would all be expressed as an array of exchange ratios for all of the other goods that could possibly be traded with them. For example, the price of a metal stud might be one and a half two-by-fours, three packs of shingles, ten hours of unskilled labor and three and one-fourth gallons of gasoline. Each exchange ratio would merely be a ratio of physical quantities of different goods that could not be easily compared with other physical quantities of goods.

In fact, however, it is unlikely that there would be direct exchange between all of the various items on a carpenter's materials list. Therefore, no exchange ratios among the non-traded items would exist, and no calculation of profit and loss could be made. Even if this could be done, profit-or-loss calculations would require the comparison of not only the input prices to each other, but also the comparison of the input prices to the price of the house, which could only be expressed in terms of physical quantities of other goods.

With the advent of money, however, such calculations become much, much easier. In an economy using a medium of exchange, all goods are exchanged for money. Because all goods are traded for money, all prices are expressed in terms of money. Consequently, people can now compare the market price of one good against that of every other good. Entrepreneurs can now calculate how well building a house using particular factors will satisfy the demands of the consumers. They can compare the market price of their products with the sum of the market prices of all the factors they could use to produce their products.

Such calculation tells entrepreneurs whether it makes sense to build another house and if so, exactly what sort of house to build. If Michelle reckons that the most she can sell a house for is $120,000 and the absolute least it will cost her to buy all of the factors needed is $150,000, she knows that she would be foolish to even start. By building the house she would be assuring herself of a $30,000 loss on the project. She would be better off not building one at all. The high price of the factors relative to that of the house, also tells us that the factors she would have used for the house would be better used in some other line of production. Even Christ alluded to such economic calculation when encouraging his followers to count the cost of discipleship. He rhetorically asked, "For which of you, desiring to build a tower, does not first sit down and count the cost, whether he has enough to complete it?" (Luke 14:8)

Suppose that Michelle knows a couple that agrees to pay her $120,000 to build a house. Because of the existence of money prices, she is able to calculate that a metal framed house with slate shingles and brick would cost her $150,000 to make, but a wood framed house with asphalt shingles and aluminum siding would cost her $100,000 to produce. She then is able to make the rational choice to produce the less costly version, thereby freeing up metal studs, slate shingles, and bricks for those lines of production where they are more valuable.

Hence, it is the development of money that allows entrepreneurs to calculate profit and loss. Moreover, because in a free market, prices are determined via voluntary exchange according to people's subjective value scales, not only do entrepreneurs have the incentive to produce those things that earn them the largest profit, but those items that are profitable are so precisely because they are what people want, what they really, really want. It is the existence of money prices that makes our complex modern and very productive economy possible. The ability of entrepreneurs to use money prices to calculate profit and loss is one of the blessings of the emergence of money.

MONEY PRICES

The importance of free market prices for the coordination of our economy is why so much of economics is devoted to our next topic of study: price theory. As explained above, the price of a good is simply an expression of that good in terms of another good. It is a rate of exchange. Hark back to the previous chapter when we were rooting out the principles of exchange and mentioned in one of our examples that Groucho and Harpo would both be willing to exchange two bushels of mangoes for two pounds of beef. In this case, the exchange ratio would be two for two. In other words, the price of a pound of beef would be one bushel of mangoes. The price of a bushel of mangoes would be a pound of beef. What if, instead Groucho exchanged 1000 strawberries to Zeppo in return for two cows? The price of the cow would be 500 strawberries. Alternately, the price of one strawberry would be 1/500th of a cow.

With money being used for all exchanges, money prices serve as the common denominator of all exchange ratios. We do not say that the price of a ticket to a Major League Baseball game is 4 bushels of mangoes, because mangoes are not traded for baseball tickets. The price of a baseball game would be more like $35 per ticket. Likewise, college tuition is not expressed in terms of pounds of beef, because people do not trade beef for coveted spots in high-octane economics lectures. They pay thousands of dollars to do so, so the price of college tuition may be something like $15,000 per semester. This is certainly easier to deal with than supplying your instructor with 250 cows or 250,000 strawberries.

Note also that these prices are those that are actually paid in actual exchanges. Consequently, in a free market, all money prices for all goods are the result of actions determined by human value scales. Prices are determined by the value scales and actions of buyers and sellers, what economists like to call demanders and suppliers. In order to keep things straight, it is best to take them one at a time.

DEMAND

The word *demand* is an unfortunate term to use to describe the role of the buyer in economics, because it rightfully conjures up the image of a belligerent crab used to getting his own way, like the gum-chewing girl in *Willie Wonka and the Chocolate Factory*. However, thanks to the wild success of Alfred Marshall's *Principles of Economics*, economists have been using the term since 1890 and there seems little hope of changing it now. The concept of demand relates the different quantities of a particular good a person is willing and able to purchase at every given price. Because such purchases are actions, you might think that how many units of a good a person would demand at each price is ultimately determined by his value scale. You would be right.

Let's examine the case of Greg, who wants to experience the joy of contemplating the best of God's created order. He knows that one of the best ways to do this is to listen to great music. If he could just get his hands on a good recording of a Beethoven symphony, he reasons, his ability to experience such joy would be ample. Greg likes Beethoven so much that for him each Beethoven symphony recording in the catalog is worth listening to and equally serviceable for getting at what Beethoven was trying to express. A modern economy such as ours uses money as the medium of exchange for all purchases. Consequently, Greg, when considering how many Beethoven compact discs to buy will be comparing the value of CDs to different quantities of money. Suppose that, at the time he is contemplating purchasing Beethoven records, his value scale comparing the two is as given below:

<u>Greg's Value Scale</u>

$15
(First CD)
$14
$13
(Second CD)
$12
$11
$10
(Third CD)
$9
$8
(Fourth CD)
$7
$6
$5
(Fifth CD)
$4

Greg compares different dollar amounts with marginal Beethoven recordings. You will note that, not surprisingly, the law of marginal utility applies even to Beethoven albums so that if the quantity of Beethoven

CDs Greg obtains increases, the value he would place on each successive CD is lower. Greg values the second Beethoven CD he could obtain less than the first, the third he could obtain less than the second, the fourth he could obtain less than the third and so on.

From the information on Greg's value scale, we can derive how many Beethoven CDs he is willing to buy at any given price—in other words, Greg's demand for Beethoven recordings. For example using Greg's value scale, ask yourself how many Beethoven records he is willing to buy if he walks into Acme Music and the price on the CDs is $15 each. We can see from Greg's preference ranking that he values $15 more than even the first Beethoven CD. Consequently, at a price of $15, Greg demands exactly zero Beethoven recordings. What is the most that Greg is willing to pay for the first CD? That would be $14, the largest amount of money that Greg values less than the first CD he could buy. Therefore, we call $14 the maximum buying price for the first CD.

The preceding paragraph provides us with enough information to begin what economists call a demand schedule. A *demand schedule* is a table listing the various combinations of prices and quantity demanded at each price. According to our information, we know that at a price of $15 Greg demands no recordings, but at a price of $14 he demands one. The beginning of his demand schedule, then, appears as in table 5.1.

Table 5.1

Greg's Demand Schedule	
Price	Quantity
15	0
14	1

We can finish Greg's demand schedule by looking again at his value scale. We noted that Greg's maximum buying price for the first Beethoven CD was $14. The maximum buying price for a second CD is $12. Greg values $13 less than the first CD he could buy, but more than a second one. Therefore if the price for Beethoven albums charged by Acme Music is $13 each, Greg will still only demand one. If, however, the price were instead $12 per recording, Greg would be willing to spend $12 on the first and, since he values $12 less than a second Beethoven record, he would be willing to spend another $12 on the second CD. This lower maximum buying price for the second CD is due to the law of marginal utility, which remains in force as Greg accumulates recordings. Greg's maximum buying price for a third CD is $9, so at a price of either $10 or $11, Greg will only demand two. If the price were $9, Greg would demand three recordings and would demand the same amount if the price were $8 each. In order for him to demand four CD's, the price would have to be at least as low as the maximum buying price for the fourth CD, or $7. At a price of $7, $6, or $5 per recording, Greg would be willing to purchase four recordings. Finally, if the price per CD charged by Acme Music was as low as $4 per CD, then Greg would be willing to purchase five recordings. From the information in Greg's value scale, Greg's demand schedule, illustrated in table 5.2, can be derived.

Table 5.2

Greg's Demand Schedule	
Price	Quantity
15	0
14	1
13	1
12	2
11	2
10	2
9	3
8	3
7	4
6	4
5	4
4	5

Greg's demand schedule demonstrates a general relationship between the price of a good and the quantity of that good that a person will demand. Notice that Greg is willing to purchase a larger quantity of Beethoven recordings at a price of $4 than he is at a price of $15. On his demand schedule we find that at different hypothetical prices, the price and quantity demanded move in opposite directions. This is known as a negative or inverse relationship. This inverse relationship between the price of CDs and the quantity of CDs demanded is not something that we know merely from observation. Remember that Greg's demand schedule was derived from his value scale, and all value scales manifest the law of marginal utility. The law of marginal utility applies to all human action. Therefore, this inverse relationship between price and quantity demanded is not only true for Greg and his Beethoven CDs. It is true for all people and all actions, which is why we can call this relationship an economic law.

You may have heard that a picture is worth a thousand words. Visual learners reading this text might be happy to know that demand schedules can be easily expressed in graphical form. Using a graph with the price of Beethoven CDs on the vertical axis and the quantity of those CDs on the horizontal axis, we can derive a picture of the different quantities of CDs Greg would be willing to buy at different hypothetical prices. Using Greg's demand schedule, we can plot a point on the graph for each combination of price and quantity demanded, resulting in the demand curve shown in Figure 5.1.

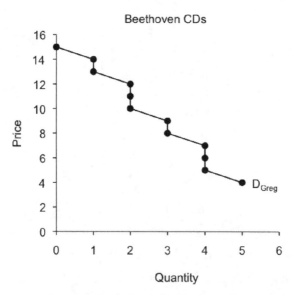

Figure 5.1. Greg's demand curve for Beethoven CDs.

The demand curve is a useful device because it shows us three things. First, it illustrates the generally inverse relationship between the hypothetical price of a good and the quantity of that good demanded. Additionally, the demand curve illustrates the maximum price a buyer is willing to pay for any given quantity and the maximum quantity the buyer is willing to buy at any hypothetical price. As such, Greg's demand curve is a graphical representation of the law of demand. It illustrates the generally inverse relationship between price and quantity demanded. Greg is willing to purchase more recordings at a price of $4 than he is at a price of $15. Note also that there are some parts of the curve that are vertical, indicating that at the hypothetical prices along those sections, Greg would still buy the same quantity. Greg, for example, would be willing to buy four recordings if the price per disk was $7, $6, or $5. At lower hypothetical prices, the quantity that Greg is willing to purchase either is the same or is larger. This is because if he accumulates more CDs, their marginal utility would be lower, so he would be willing to give away a smaller and smaller amount of cash for each recording.

Examining the value scale of only one buyer, Greg, has been enough for us to derive a demand schedule and curve and to gain an understanding of the relationship between price and quantity demanded. However, for most goods you and I can think of, there is more than one buyer. That is certainly true of Beethoven albums. Suppose that besides Greg, Marsha also wants to score some records by Beethoven and her value scale is as indicated below:

Marsha's Value Scale
$15
$14
(First CD)
$13
$12
(Second CD)
$11
$10
(Third CD)
$9
(Fourth CD)
$8
$7
$6
(Fifth CD)
$5
$4

Not surprisingly, for Marsha we find that, like Greg, Beethoven CDs exhibit diminishing marginal utility. Using the same method as we did for Greg's value scale, we can derive Marsha's demand schedule as shown in table 5.3.

Table 5.3

Marsha's Demand Schedule	
Price	Quantity
15	0
14	0
13	1
12	1
11	2
10	2
9	3
8	4
7	4
6	4
5	5
4	5

Again, as expected, Marsha is willing to buy more Beethoven records at lower hypothetical prices.

If we add another buyer, we can develop the demand for a simple three-person market. Suppose that Peter also appreciates the music of Beethoven. For brevity, only his demand schedule is reproduced in table 5.4.

Table 5.4

Peter's Demand Schedule	
Price	Quantity
15	0
14	0
13	0
12	1
11	1
10	2
9	2
8	2
7	2
6	2
5	3
4	3

If Greg, Marsha, and Peter are the only people looking to buy Beethoven recordings, we can use all three of our buyers' demand schedules to derive the demand schedule for the entire market. The market demand schedule is the total quantity of Beethoven CDs that will be demanded at each hypothetical price.

In order to accomplish this, all that must be done is summing the quantities of recordings that would be demanded by Greg, Marsha, and Peter at each hypothetical price. Reproducing their respective demand schedules in one table, we are able to derive the market schedule that is in table 5.5.

Table 5.5

Derivation of the Market Demand Schedule				
Price	Greg	Marsha	Peter	Market
15	0	0	0	0
14	1	0	0	1
13	1	1	0	2
12	2	1	1	4
11	2	2	1	5
10	2	2	2	6
9	3	3	2	8
8	3	4	2	9
7	4	4	2	10
6	4	4	2	10
5	4	5	3	12
4	5	5	3	13

At a price of $15 per CD, none of the three are willing to buy, so the market quantity demanded is zero. If the hypothetical price was $14 each, however, Greg would demand one while Marsha and Peter would still not be willing to buy. Consequently, the quantity demanded by the entire market at $14 per record is one. If the price for the CDs was $13 a piece, then Greg and Marsha would both demand one each, while Peter would still not be willing to buy. The quantity demanded by the market would then be two. You should be catching on. What is the quantity that would be demanded by the market at a price of $8? Well, Greg would be willing to buy three CDs; Marsha would be willing to buy four, while Peter would only be willing to buy two. The quantity demanded at a price of $8 would be three plus four plus two or nine CDs total.

Like the individual demand schedules, the different combinations of price and quantity demanded can be plotted on a graph, as in figure 5.2, resulting in a market demand curve.

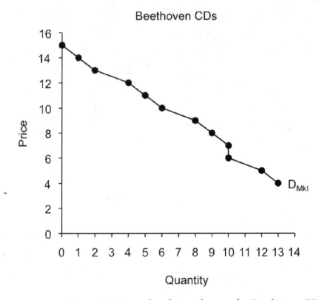

Figure 5.2. A market demand curve for Beethoven CDs.

Notice that the existence of more than one buyer reinforces the negatively sloping market demand curve. With more people entering the market to buy goods, such as Beethoven CDs, there are fewer vertical sections on the market demand curve than there are on any of the individual demand curves.

Now some of what has been learned about the relationship between the price that people must pay for a good and the quantity that they will be willing to buy can be summed up. Remember that the inverse relationship between the hypothetical price of CDs and the quantity of CDs buyers demand is not something that we know merely from observation. The market demand curve is derived merely by summing the different quantities that individual buyers—such as Greg, Marsha, and Peter—would buy at each different hypothetical price. Also remember that these respective demand schedules were derived from each person's value scale, all of which exhibit the law of marginal utility. Further recall from chapter three that the law of marginal utility is rooted in the very nature of man as a being created by God in his image, and therefore engages in purposeful behavior. Because all people bear this image, all people act according to their value scales that reflect the law of marginal utility. Therefore, we can say that the inverse relationship between the price of Beethoven albums and the quantity that people in the market would be willing to buy is not isolated to only Greg, Marsha, and Peter, but is also manifested by everyone in the market for Beethoven CDs. Because this relationship applies to the purchases of everyone, economists can call this relationship an economic law. This economic law is called the law of demand.

The *law of demand* tells us that, at the time of purchase, there is an inverse relationship between the hypothetical price of a good and the quantity of that good a person will buy. If the price were lower, people would demand either the same amount or a larger quantity of the good. They would never buy less. Although we do not establish this law from observation, it can be observed. The law of demand was evident in the automobile market the year following the terrorist attacks on September 11, 2001. In an effort to forestall falling sales General Motors came out with its "Keep America Rolling" campaign which featured several discounts like cash back or reduced interest financing (even 0% in some cases). These, of course, lowered the final price of buying a car and new car sales increased during the year following the terrorist attacks.

Even in the arts, an area to which many people do not like to think economics applies, we find evidence of the law of demand at work. In 2002 the British government decided to do away with admission fees to all of their national museums. What do you think happened to attendance? You guessed it. The number of visitors sky-rocketed! Britain's National Gallery, for example, saw a 79% increase in attendance for 2002, according to a London newspaper, *The Guardian*. An entertainment weekly, *Pollstar*, reported that the average ticket price of a rock concert increased 8.5%, from $47 to $51 from the year 2001 to 2002. At the same time, concert attendance at the top 50 concerts during the first half of 2002 dropped by 3%.

What else have we learned? For one thing, the market demand curve is simply a graphical representation of the law of demand. As such, it first communicates to us the maximum quantity that buyers are willing to buy at each hypothetical price. For instance, suppose that the price offered by sellers for Beethoven CDs is $11. If so, the market demand is such that five CDs would be demanded. If sellers wished to sell more, say ten recordings, what would they have to do? The answer, you can see from the market demand curve, is that they would have to charge a lower price of $7 each. You can see from the demand curve that five is the maximum number of Beethoven CDs people will demand at a price of $11. They simply will not buy any more. The price must be as low as at least $7 before they are willing to buy ten.

Secondly, the market demand curve tells us the maximum price buyers are willing to pay for any given quantity. For instance, if you wanted to find out the highest price people are willing to pay to buy a total of six CDs, the demand curve tells us that the answer is $10 per recording. If the price was higher, say $12, sellers should not expect to sell as many, because the demand curve again tells us that the most people are willing to buy at a price of $12 is four. So you can see that the demand curve shows us the maximum quantity people are willing to buy at any given price and the maximum price people are willing to pay for any given quantity.

You should be starting to see how economics helps us to understand how people's subjective values work to determine how many units of particular goods they will buy at different prices. However, buyers are not the only people who are party to the exchanges that yield market prices. Every exchange has a seller as well as a buyer, and no one forces a seller to sell his product at a particular price. In the free market, all exchanges are voluntary, so that the prices that are paid in exchange for the goods are mutually agreed to by both buyers and sellers. Consequently, sellers also have a role to play in determining the market price of goods in an economy. Therefore, we now turn our attention to supply.

SUPPLY

Before we get to an in-depth study of supply, however, some confusion should be cleared up. Since at least 1890 and the publishing of Alfred Marshall's extremely influential *Principles of Economics*, a number of people have been under the impression that, although demand for goods might be motivated by subjective valuations of acting people, the supply of goods is actually determined by the objective costs of production. This is an error born of confusion regarding the pricing process.

In the first place, suppliers are people too. The price they will accept is determined by their subjective valuations just as the price that buyers are willing to pay is determined by their subjective valuations. Likewise, the cost of any action, including production, is always subjective.

It can be argued, that, while it is true that costs of production are subjective in that they are motivated by the producer's value scale, we also know that what motivates producers is the quest for maximized profits and these profits are calculated monetarily. Therefore, one might argue that while cost is subjective in a broad sense, it is also objective in the sense that the cost of producing any good can be expressed in terms of objective money amounts, namely the sum of the prices of the factors of production necessary to produce the good.

Certainly the cost of production will have something to do with how much of a good will be produced. If the cost of producing a good is relatively high compared to the price sellers expect to receive for it, producing the good is less likely to turn a profit and less of it will be made. If the cost of producing a good is relatively low compared to its expected selling price, making it will be more likely to reap a profit, so more of it will be produced.

However, when explaining how people act to determine market prices, we need to remember that the prices being determined are for goods already in existence that sellers are ready to sell. In other words, these are goods that are already produced. Because goods available for sale are already produced, from the producers' point of view, the costs to produce them are sunk. They have already been incurred and cannot be changed no matter what price the seller can receive for the good. Therefore such costs do not enter into the decision of what price to accept for the good.

For example, suppose that it cost $15 to produce a copy of Bernard Hermann's soundtrack to the film *North-by-Northwest*, one of Hitchcock's best suspense films with a terrific ending atop Mt. Rushmore. The price that a seller will actually charge is not determined by the cost of production. Certainly the producer wants to receive a lot more than $15 for each one he sells. If he sells it for $25 per CD, the price would not be determined by the cost of production. On the other hand, it may turn out that people do not want this soundtrack nearly as much as the producer thought they would. He might find that he cannot sell unless he offers a selling price of $6. In this case, the price also is not determined by the selling price.

The costs of production are, in a limited sense, objective costs, because they are monetary costs which are objective amounts of money. These objective costs of production do form the basis for economic calculation. Producers use these costs to decide whether or not to produce the good in question. They compare the sum of the prices of the factors used to produce the good with the price they expect to receive when they sell the finished good. If the price of the finished good is more than the sum of the prices of the factors of production, they will go ahead and produce the good, expecting to reap a profit. If the price at which they expect they can sell the finished good is less than the sum of the prices of the factors of production, they calculate that they will earn a loss if they produce this good and therefore decide not to make it.

The important point regarding cost and supply, however, is one of time. Whether or not we consider the costs of production subjective in the broad sense or objective in the limited sense, these costs cannot enter in to the decision of supplying goods already produced, because they are sunk costs. Once the good is produced, the production costs are already incurred. They are sunk and, hence, irrelevant to the determination of the actual selling price for the good.

Having put the question of objective costs to bed, we can examine the principles that underlie what economists call supply, or the quantity of goods sellers are willing to sell at any given price. Selling a good for money is an action and, like any action, is dependent on the preference ranking of the individual doing the acting. It follows that the supply of goods is just as dependent on personal subjective value scales as is demand.

Take, for example, my grandfather Neak Pearman.[1] Granddad Neak made his living as a horse trader in the Sand Hills of Nebraska. His decision of how many quarter horses to sell at any given price was determined by how he ranked different horses in comparison to different amounts of money on his personal value scale. On any given day he would have a stable of a few horses for sale.

Suppose that on one particular day a potential buyer offered to buy horses from him. Further suppose that at the time of the offer he had four horses in his stock that were all equally serviceable, and his value scale was as follows:

---

1. His real name was Elmer James Pearman, but for reasons not entirely known to me, he went by the nickname Neak. People living in the Sand Hills of Western Nebraska were famous for their nicknames. Two of Granddad Neak's brothers were called Eck and Bink. Two of Neak's sons have carried nicknames throughout their lives. One has been tagged with "Spook" because of his rather nervous, high-strung nature and another goes by "Chub" for obvious reasons.

Neak's Value Scale

($2,000)
Fourth Horse
($1,900)
($1,800)
Third Horse
($1,700)
($1,600)
($1,500)
Second Horse
($1,400)
($1,300)
($1,200)
First Horse
($1,100)

The different amounts of money Neak could sell his horses for are in parentheses because he does not own those amounts of money. Each successive horse is shown without parentheses because he does own them. Remember also that he is in the business of selling quarter horses, so the first horse is the first horse that he would sell, the second horse would be the second horse he would sell, and so on. This explains why each marginal horse is ranked higher than the previous horse. The more horses Neak sells the smaller the total stock of horses that remain in his possession, and, according to the law of marginal utility, the more he would value each marginal horse.

Keeping these facts in mind, Neak's supply schedule for quarter horses, illustrated in table 5.6, can be derived from his value scale.

Table 5.6

Neak's Supply Schedule	
Price	Quantity
2,000	4
1,900	3
1,800	3
1,700	2
1,600	2
1,500	2
1,400	1
1,300	1
1,200	1
1,100	0

For instance, if someone would come to his house offering to buy a horse for $1,100, how many would he be willing to sell? The answer is zero. Why? Because he values each horse more highly than $1,100. He would not part with any of his horses at a price of only $1,100, so the quantity he would be willing to supply is zero. What price would have to be offered in order to induce him to sell his first horse? Well, if a buyer offered Neak $1,200, Neak's value scale is such that he would indeed be willing to sell his first horse at a price of at least $1,200. Consequently, Neak's minimum selling price for the first horse is $1,200. At a price of $1,100 per horse Neak would not be willing to sell any, but at a price of $1,200 each, he would supply one.

In order for Neak to part with more than two horses, however, what would have to happen? His value scale makes it clear that a higher price would have to be offered. If the buyer offered him a price of $1,300 per horse instead of $1,200 Neak would not be induced to sell his second horse, so the quantity supplied would still be one. How many horses would he be willing to sell at a price of $1,400 per horse? His value scale shows that he would be willing to sell his first horse for that amount, but he would not be willing to sell his second horse for another $1,400. In fact, Neak's minimum selling price for the second horse is $1,500. If the buyer offered to buy horses from him at $1,500 each, we can see from Neak's value scale that he values $1,500 more than each of the first and second horses, therefore, he would be willing to supply two horses at a price of $1,500 each.

What is the price per horse necessary to induce Granddad Neak to sell three horses? Checking his value scale shows that the minimum selling price for the third horse is $1,800, so at a price of $1,800 each Neak would sell three horses, while at prices of either $1,700 or $1,600 he would still be willing to supply only two. If the price offered increased to $1,900 per horse, it would not be high enough to induce him to sell the fourth horse. The minimum selling price for the fourth horse is $2,000, so at a price of $1,900 the quantity supplied is still three horses, while at a price of $2,000, Granddad Neak would be willing to sell four horses.

You will remember that Greg's demand for Beethoven CDs could be represented in a graphical form, as well as in table form. The same is true for Neak's supply of quarter horses. On the graph in figure 5.3, again with price on the vertical axis and quantity of horses on the horizontal axis, the supply curve is drawn illustrating all combinations of price and quantity of horses Neak is willing to supply.

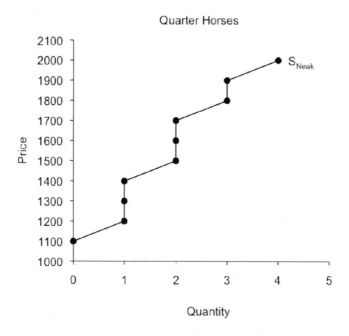

Figure 5.3. Neak's supply schedule.

Neak's supply schedule and supply curve both reveal a fundamental principle of economics. At the time of sale, if the hypothetical sale price would be higher, the seller would be willing to sell either the same or larger quantity of goods compared to that at a lower price. Hence, there is a generally positive trend in the supply curve. The higher the hypothetical selling price, the larger the quantity sellers are willing to supply. The lower the hypothetical selling price, the smaller the quantity sellers are willing to supply.

Similar to the demand curve, the more sellers that enter the market, the more smooth the market supply curve would be. As in the case of demand, a market supply curve can be derived from summing the individual quantities supplied at all prices. For instance, suppose that Neak's brothers, Eck and Bink also were in the horse trading business in the Sand Hills of Nebraska and had the individual supply schedules as given in table 5.7.

Table 5.7

Supply Schedule for Eck and Bink		
Price	Eck	Bink
2,000	5	5
1,900	4	5
1,800	4	4
1,700	3	4
1,600	3	3
1,500	2	3
1,400	2	2
1,300	1	2
1,200	1	1
1,100	0	0

The market supply curve is derived by summing the quantities of horses each horse trader is willing to sell at different hypothetical prices. Summing the quantity of horses willing to be sold by Eck and Bink with Granddad Neak at every hypothetical price yields the supply schedule in table 5.8.

Table 5.8

Market Supply Schedule	
Price	Quantity
2,000	14
1,900	12
1,800	11
1,700	9
1,600	8
1,500	7
1,400	5
1,300	4
1,200	3
1,100	0

Similar to the derivation of the demand curve, the price-quantity combinations can be plotted as points on a supply curve. Using the market supply schedule in table 5.8, we can derive the market supply curve for quarter horses shown in figure 5.4.

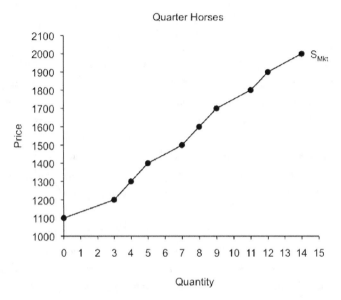

Figure 5.4. The market supply curve for quarter horses.

The market supply curve graphically represents the law of supply. The *law of supply* states that at the time of sale, there is a positive relationship between the price of a good and the quantity of that good supplied. The higher the price at which quarter horses could be sold the larger the quantity of horses that sellers would be willing to sell. Additionally, the lower the price at which sellers could sell their horses, the smaller the quantity of horses sellers would be willing to sell.

The supply curve communicates more than the law of supply however. It shows the maximum quantity suppliers are willing to sell at each hypothetical price. If, for instance, we want to know how many horses would be sold if the price was $1,600, the supply curve shows that, at this price, sellers would sell eight horses. For the sellers to be willing to sell a larger quantity, buyers would have to offer a higher price. The supply curve also reveals the minimum price necessary to induce sellers to sell any given quantity. If one is curious as to the minimum price that could be offered at which the sellers would sell eleven horses, the supply curve reveals that the lowest price at which sellers would sell a total of eleven horses is $1,800. At any price lower than that, they would be willing to sell a smaller quantity.

THE PRICING PROCESS

Having digested the principles underlying supply and demand schedules and curves, it is time to proceed to the romantic portion of economics. This is the part where we bring supply and demand together after all these years in order to determine prices. Before we examine just how prices are determined, however, it might be wise to remind ourselves exactly what a price is. Now I can hear you thinking, "Duh, a price is what we have to pay to buy something." This is true, but what is the significance of this fact? Prices are the result of voluntary exchange. As such, the price of a good is an exchange ratio. It is the amount of one good that you can trade for another good.

In a modern economy featuring indirect exchange, all goods are traded for money, so prices are denominated in the monetary unit. Therefore, in the United States, prices are always quoted in terms of dollars and cents. One does not find prices for compact discs at Wal-Mart quoted in terms of bushels of mangoes or pounds of beef. If a buyer pays $10.99 to Amazon.com in exchange for the *Deutsche Grammaphon* recording of Beethoven's *Symphonies No.5 and 7* featuring the Vienna Philharmonic conducted by Carlos Kleiber, the exchange ratio is $10.99 per CD.

And who decided that this amount of money would exchange hands for one compact disc? Some might be tempted to suggest that it is Amazon.com. After all, it is Amazon that posted $10.99 on its webpage next to the information about the CD, so it must be Amazon that decides the price of the good. Prices, however, are only prices if they are actual exchange ratios. A price is an actual exchange ratio resulting from an actual exchange. And how many parties are there to an exchange? Two, of course. This means that if someone actually pays $10.99 in exchange for Kleiber's rendition of Beethoven's *Symphonies No. 5 and 7*, this price must have been mutually agreed to. Both the buyer and seller must find the price agreeable if an exchange actually takes place. Therefore it is both the buyer and seller who determine the price of a good.

It is the purpose of this chapter to explain just how these actual, mutually-agreed-to prices are determined in the free market. As is often the case, we can learn quite a bit from looking at a simple example. Consequently, we are going to start with the simplest example possible, that of isolated exchange.

In an isolated exchange there is only one buyer and one seller. Suppose that we check in with Greg again; remember that Greg still wants to revel in the music of Beethoven, but has not yet been able to satisfy his end. He finally realizes that a Beethoven recording is not going to fall out of the sky into his CD player, so he decides to go to the local store to get one (assuming the music store has good taste, which is an heroic assumption these days). Suppose, however, that just as he gets to his car he is accosted rather abruptly by a gentlemen named Johnny Fever who is very animated and says to Greg, "Yo dude, check it out. I got a deal for you. Ludwig van Beethoven practically free!" Would Greg purchase the Beethoven CD from Johnny? As always, it depends on the value scales of the participants. Suppose that the value scales of Greg and Johnny are as illustrated below:

Greg's Value Scale	Johnny's Value Scale
$13	($13)
$12	($12)
$11	Beethoven
(Beethoven)	($11)
$10	($10)
$9	($9)
$8	($8)

What is the most Greg is willing to pay for the Beethoven recording? The answer is $10. We know this because according to Greg's value scale, $10 is the highest dollar amount to which Greg still prefers Beethoven. Consequently, $10 is Greg's maximum buying price. So, will Greg buy the CD from Johnny? Well, knowing Greg's maximum buying price is not enough. Remember it takes two to trade so we must also check out Johnny's value scale. We see from Johnny's value scale that the least amount of money he will accept in exchange for the Beethoven record is $12. We know this because this is the lowest amount of money that Johnny still prefers more than the Beethoven CD. Consequently, $12 is Johnny's minimum selling price. Now we know enough to answer our previous question with a resounding *no*. Greg will not make the purchase and Johnny will not make the sale. Why? Because Greg's maximum buying price is lower than Johnny's minimum selling price. The most that Greg is willing to pay is $10, while the least that Johnny is willing to accept is $12. They do not value the goods in reverse order, so no exchange will take place.

In order for people to trade, the minimum selling price of the seller must be lower than or equal to the maximum buying price of the buyer. Suppose, for instance, that things are different. Suppose that Greg's value scale is the same, but that Johnny's value scale is as illustrated as follows:

Greg's Value Scale	Johnny's Value Scale
$13	($10)
$12	($9)
$11	($8)
(Beethoven)	($7)
$10	Beethoven
$9	($6)
$8	($5)

Because Greg's value scale is the same, his maximum buying price is still $10. Johnny's ranking is different, however, so that his minimum selling price is now $7. Now will an exchange take place? The answer is *yes* because Greg is still willing to pay at least $10 while Johnny is willing to accept as little as $7. Greg's maximum buying price is greater than Johnny's minimum selling price. Ten is greater than seven, even in the new math. Therefore an exchange will take place and both Greg and Johnny will leave more satisfied, because they both walk away with a good they value more than the good with which they came.

We are not now, however, only interested in whether an exchange will take place. We are investigating how market prices are determined. So the question that presents itself is what price will Greg pay to Johnny for the CD? Again, the price will be determined by the value scales of the buyer and seller. Given Greg's and Johnny's value scales, we cannot say everything about what the price will be for the Beethoven recording, but we certainly can say something. Since Greg's maximum buying price is $10, we know that the price cannot be any higher than $10. At any higher price, Greg would walk away Beethoven-less. We also know that the lowest possible price is $7, because this is Johnny's minimum selling price. At a price lower than $7, Johnny would not sell. Consequently, based on the value scales of the buyer and seller, we know that an exchange of money for the Beethoven CD will take place and the price paid by Greg will be in a range from between $7 and $10 inclusive. It will be within a range at or above the minimum selling price of the seller and at or below the maximum buying price of the buyer. In this case, economic theory is unable tell us what the precise price will be. The exact price paid by Greg will be determined by negotiations between Greg and Johnny somewhere in that range.

Things become a bit more definite as we move to a different case where there is more than one buyer and one seller. Let's face it. Greg is not the only person alive who has discovered the musical genius of the man who wrote the *Waldstein Sonata, Eroica Symphony*, and his *Ninth Symphony* with its *Ode to Joy*. Suppose that Marsha also wants to be transported to musical Elysium and she is also walking to her car so she can go shopping in search of the Beethoven sound. Fortunately for us, Marsha's car is parked quite near Greg's so as she lilts up to her car she hears the negotiations taking place between Greg and Johnny. Marsha decides that she also wants in on this deal and is ready to pay what her value scale allows. "What does it allow?" I can hear you asking. Well suppose that Marsha's value scale is as indicated below:

Greg's Value Scale	Marsha's Value Scale	Johnny's Value Scale
$13	$11	($10)
$12	$10	($9)
$11	$9	($8)
(Beethoven)	(Beethoven)	($7)
$10	$8	Beethoven
$9	$7	($6)
$8	$6	($5)

The most Marsha is willing to pay for a Beethoven CD is $8. Suppose that she were to quickly offer to pay $8, her maximum buying price, to Johnny Fever. The presence of more than one buyer now raises two questions: Who will be the buyer of the CD and what price will be paid? Will Marsha be the one that walks away with a bit of Beethoven's best? Do not forget about Greg. His value scale reveals that he also is willing to pay at least $8 for a Beethoven CD. So, we have two people who are willing to pay $8 and only one recording for sale. Who will be the successful buyer in this situation? To answer that question we need to answer another:

what is Greg willing to do that Marsha is not? Greg is willing to pay more than $8. In fact, he is willing to pay at least $10 as we already saw. This makes Greg the most eager buyer. The *most eager buyer* is the person who is willing to pay the highest price for the good in question.

If Greg is in danger of losing the Beethoven recording to Marsha at a price of $8, he will immediately bid up the price. This means he will offer a higher price in an effort to ensure that he is the one who actually gets to buy the CD in the end. The answer to the first question (that of who will make the trade with Johnny Fever), is Greg, because he is the most eager buyer.

Now to the question of the price paid. We know that Greg has to pay a price higher than $8 to guarantee that he is the one who buys it. That means that the lowest price that Greg can pay and still be the one satisfied with the Beethoven is $8.01. Of course, we already know that he is actually willing to pay up to $10. Again economics cannot tell us in this case exactly which price will be paid. However, economics can tell us that it must be determined by bargaining in the range between $8.01 and $10.00.

Generally, in cases of more than one buyer and only one seller, economics teaches us the following principles. First, the person who buys the good will be the most eager buyer, and second, the price paid for the good will be less than or equal to the maximum buying price of the most eager buyer, and greater than the maximum buying price of the next most eager buyer.

Note also that the range of potential prices narrowed as more than one buyer competes for the good. The range of possible prices will tend to further narrow as more buyers are added. Suppose that additional buyers come on the scene, each with a different maximum buying price.

Suppose that three other buyers have come along and the respective value scales of all the potential traders are as illustrated below:

Greg	Marsha	Peter	Jan	Bobby	Johnny
$11	$9	$7	$10	$8	($7)
(Beethoven)	(Beethoven)	(Beethoven)	(Beethoven)	(Beethoven)	Beethoven
$10	$8	$6	$9	$7	($6)

Of our new potential buyers, Peter is less interested in Beethoven, having a maximum buying price of $6. The most that Jan is willing to pay is $9. Bobby has a maximum buying price of $7.

Given this new information, who will be the person who actually gets to dwell with Beethoven? Well, Greg is still the most eager buyer because he is still the person willing to pay the highest price, so Greg will still be the one that walks away with the Beethoven recording. How much will he pay, however? Remember that Marsha is willing to pay at least $8, just like Greg. So Greg would have to offer more than $8 so that he gets the CD instead of Marsha. However, is Marsha still the next most eager buyer? The answer is no because Jan is willing to pay up to $9, which is more than what Marsha is willing to pay. Now Jan is the next most eager buyer, so Greg will have to offer a price high enough so that Jan will drop out of the market for Beethoven. Consequently he will have to offer a price higher than $9 to ensure that he is the one who gets the record. He will have to offer at least $9.01. In this case then, as a result of the addition of four new buyers, one of which had a maximum buying price of $9, the price is even more precisely determined. Competition reduces the indeterminate range for the price actually charged for the good exchanged.

The case of more than one seller and only one buyer is exactly the converse of the above. Suppose that Greg is the only buyer wanting to buy a Beethoven album, but suppose now that in addition to Johnny Fever, a chap named Herb also has a Beethoven CD he wants to sell. We can ask similar questions to those previously asked. Will an exchange take place? Who will actually make the sale? What will be the price charged?

We know that an exchange will take place if the trade will be mutually beneficial. To know if this is the case, we must look at the value scales of the relevant people shown below:

Greg's Value Scale	Herb's Value Scale	Johnny's Value Scale
$11	($6)	($7)
(Beethoven)	Beethoven	Beethoven
$10	($5)	($6)

Greg's value scale shows that his maximum buying price is $10. Johnny Fever's minimum selling price is $7. That is enough to tell us that there will be an exchange because such a trade will be mutually beneficial. However, it is not clear that Johnny Fever will be the one making the sale.

In fact, it is even clearer that he will not. Herb is willing to sell his Beethoven CD at $7, but he is also willing to sell it at $6. The lowest Johnny Fever will go is $7, but what will Herb do to ensure that he is the person who is able to sell his CD? Herb will bid down the price. He will lower his selling price to $6 in order to entice Greg to buy the recording from him. In this case, Herb is the *most eager seller*, because he is the one who is willing to sell his good at the lowest price. Johnny Fever is the next most eager seller. We see then that Herb will be the person who is able to sell his CD and it will be sold at a price between $6.99 and $6.00. In general, in the case of more than one seller and only one buyer, the sale will be made by the most eager seller, and the price for which the good is sold will be greater than or equal to the minimum selling price of the most eager seller and less than the minimum selling price of the next most eager seller.

While we have already learned some valuable principles regarding price determination in relatively simple situations with only one person as either buyer or seller or both, the majority of exchanges do not take place in markets with either only one seller or one buyer. Most goods are traded in markets where there are numerous buyers and numerous sellers. We therefore turn attention to the case that is more common: a complex market economy based on an intricate network of exchanges between many buyers and many sellers.

As we begin examining the pricing process in such an economy we should remember that all people undertake action in order to increase their state of satisfaction. They want to obtain goods that are most highly valued to them because those goods serve their most highly valued ends.

In order to achieve their highest level of satisfaction, buyers want to pay the lowest possible price for goods they purchase, given their value scales. The less money they pay to buy a good, the more money they have left over to buy other goods. The more goods they buy, the more ends they can satisfy. Paying the lowest possible prices, therefore, allows buyers to attain the greatest satisfaction possible according to their subjective preferences.

Sellers, on the other hand, will achieve their highest level of satisfaction by selling their goods at the highest possible price. The higher the prices they receive, the greater their income. The more income they receive, the more money they have to purchase goods that they use to satisfy their ends. The more goods they obtain, the more ends they can achieve and, hence, the more satisfied they will be.

In a free market, buyers and sellers will want to agree on a price so that the most possible preferences are satisfied. Economic analysis shows that in any market, there is one price at which the most buyers and sellers satisfy their preferences: the price at which the quantity of a good demanded equals the quantity of the good supplied. At that price, everyone who wants to sell at that price can sell and everyone who is willing to pay that price can buy.

This price that will be paid and received in the market for a good is determined by the value scales of all buyers and sellers in that market. We can best get a handle on how prices are determined looking at a specific example of many buyers and sellers. Suppose that, as is very likely, there are many buyers interested in the Beethoven experience and many sellers interested in making a profit by satisfying those buyers. These different buyers have different value scales, which cause the different buyers to have different maximum buying prices. Likewise, the different sellers have different value scales that result in their having different minimum selling prices. We can begin by ranking the many buyers by their maximum buying prices (MBP) and the many sellers by their minimum selling prices (MSP). We do so in Table 5.9.

Table 5.9

Buyers	MBP	Sellers	MSP
Greg	$10	Jennifer	$10
Jan	$9	Venus	$9
Marsha	$8	Andy	$8
Bobby	$7	Johnny	$7
Peter	$6	Herb	$6
Cindy	$5	Bailey	$5
Alice	$4	Les	$4

Given the value scales of the different buyers and sellers, what price will result in the most preferences of both buyers and sellers being satisfied? Which price allows the most people to achieve what is highest on their value scales and the most mutually beneficial exchanges to take place? As indicated, it will be the price that clears the market, in the sense that at that price everyone who wants to buy can buy and everyone who wants to sell can sell.

In the example, the price that results in the most preferences being satisfied is $7. How do we know this? At that price there are four buyers who are willing to exchange their money for Beethoven CDs. Greg, Jan, Marsha, and Bobby all value a Beethoven record more highly than $7, while Peter, Cindy, and Alice do not, so they will not make any purchases.

At the same time, four sellers are willing to sell their Beethoven CD for $7. Les, Bailey, Herb, and Johnny all value $7 more highly than their Beethoven record while Jennifer, Venus, and Andy value their CD more than the $7 they could get in exchange.

Remember, for an exchange to take place, the parties must value the goods in reverse order compared to each other. The buyers must value the Beethoven CD and $7 in reverse order compared to the sellers. At a price of $7, there are four buyers who would want to buy a Beethoven CD and four sellers who would want to sell a Beethoven CD. Therefore at a price of $7 four CDs would be sold. At this price the quantity of CDs demanded equals the quantity of CDs supplied.

We know that this price will result in the most preferences being satisfied by looking again at the value scales of the market participants. At the price of $7 all buyers would be left owning the good that they value more highly. Greg, Jan, Marsha, and Bobby all would own a Beethoven CD, which they each value more highly than $7. At the same time, Bobby, Peter, and Cindy would still own their $7, which they value more highly than a Beethoven CD. None of the buyers would be left holding a good valued less highly than what was available in the market.

At the same time, at a price of $7, each seller would also end up owning the good that he or she values more highly. Les, Bailey, Herb and Johnny each would own $7, which they value more highly than their Beethoven CDs. Jennifer, Venus, and Andy would each still own a Beethoven CD which they value more highly than $7. None of the sellers would be left holding a good he or she valued less highly than what was available in the market.

There will be more mutually beneficial exchanges at a price of $7 than at any other price. Because the price at which quantity demanded equals quantity supplied is the price that results in more satisfaction, buyers and sellers have every incentive to agree to pay and receive this price. This is further understood if we consider how satisfied buyers and sellers would be at different prices.

Suppose that for whatever reason—your fault, my fault, nobody's fault[2]—someone offers to either buy or sell Beethoven CDs for $4. What would happen? To answer this question we must see how many buyers are willing to buy a Beethoven record at $4 per CD. The value scales of

2. For those readers for whom that phrase does not ring a bell, I recommend a good, albeit not great, Western movie *Big Jake*, starring John Wayne.

the many buyers are such that all seven of them will be willing to buy at that price. In other words, at a price of $4, there would be seven CDs demanded. How many would be supplied at a price of $4? It appears as if only one seller, Les, is willing to sell, so the quantity supplied at $4 is one. At that price, the quantity of records demanded would be greater than the quantity of records supplied. This is called *excess demand*.

If the price were $4 how many Beethoven records would actually be sold? If there is only one seller willing to sell, it does not matter how many more than that are willing to buy. Only one would be sold. One fortunate buyer would be happy, while the rest would be unsatisfied and frustrated. Six buyers would be left with their preferences unsatisfied because they value the CD more than $4, but are not able to make the exchange. They would be left possessing a good, $4, that they value less than what they hoped to trade for—the Beethoven CD. The other buyers are willing to pay the price, but are not able to buy because there are too few sellers to meet their demand.

What can Greg, Jan, Marsha, and the rest do to lessen the chances that they will be one of the frustrated buyers who go home with a fist full of dollars but no Beethoven? Because they are willing to pay a higher price, they can offer to do so. They are more eager buyers. Cindy says to the sellers, "Hey, what are you doing? Don't sell your CD to Alice for $4, I'll pay $5." Peter says, "No don't sell it to Cindy for $5, I'll pay $6!"

Suppose that the price buyers offer to pay is $6. What then? Note two things. At the higher price, there would be fewer eligible buyers. Neither Alice nor Cindy is willing to pay $6 for a Beethoven CD. We could call them the least eager buyers. In any event, the quantity of CDs demanded would be lower at the higher price, which is what you should expect if you understand the law of demand.

In addition to there being fewer buyers, the number of eligible sellers would be larger. At the higher price of $6, Herb and Bailey, who are unwilling to part with their Beethoven recordings at a price of $4, are willing to sell. At a higher price, the quantity supplied would be greater. At the higher price of $6, five CDs would be demanded and three would be supplied. There would still be frustrated buyers, but the number of frustrated buyers would not be as great as it would be at a lower price. Nevertheless, whenever excess demand occurs some buyers would be frustrated, so in order to minimize this possibility, what will the more eager buyers do? They will want to offer the price at which there are no frustrated preferences. They want to offer the price at which there is no risk of their not getting any Beethoven at a price they are willing to pay.

In the case of our example, the more eager buyers would want to offer to pay a price of $7 each. Why seven? At the higher price of $7, the number of willing buyers would be still lower as Peter would drop out of the market. The quantity demanded would be four; lower than it would be at a price of $6. At the same time, at the higher price the number of people willing to sell would be greater as Johnny would enter the market ready to sell his album. The quantity supplied would be four.

As explained above, at a price of $7, the number of CD buyers who would want to buy is equal to the number of sellers who would want to sell. In other words the quantity demanded would equal the quantity supplied. Every buyer who valued a Beethoven CD above $7 could buy and everyone who prefers $7 to the Beethoven CD could keep their money. There would be no frustrated buyers. At the same time, everyone who wanted to sell at a price of $7 could sell. Buyers would have no incentive to bid up the price above $7. The market would be at rest.

We have seen that in a free market of many buyers and sellers, the price will be bid up by the more eager buyers so that everyone who wants to buy at that price can buy. What would happen, however, if the price would happen to be above that market-clearing price? Would most buyer and seller preferences be satisfied?

Suppose, for example, that someone offers to sell their Beethoven CD for $9. The value scales of the buyers tell us only two buyers, Greg and Jan, would be willing to pay $9 for a CD. On the other hand, there would be six sellers willing to sell. In this case, the quantity supplied would be greater than the quantity demanded. This is called *excess supply*. The actual number of CDs sold would be two and there would be four sellers left unsatisfied holding the Beethoven. Although we cannot say for sure who they would be, we do know that they would be four of the six willing sellers. Four sellers would prefer to trade away their Beethoven CD to receive $9, but will not be able to because there are not enough willing buyers. Because these four would be willing to sell their CD at that price, but could not, the four sellers' preferences would be frustrated.

Like the potentially frustrated demanders facing excess demand, these suppliers have every incentive to avoid being frustrated by excess supply. One of them, Andy for instance, might say to the buyers, "I'll not be undersold. I'll sell this recording to you for $8." Not all of the sellers are willing to lower their sale price. Venus will not, because he is not willing to sell for less than $9. Jennifer is not willing to sell it at that price either. It is the more eager sellers who are willing to sell for less. At this lower price of $8, in addition to Greg and Jan, Marsha would also be willing to buy the CD, so the quantity demanded would be three, which is larger than it would be at the higher price. At the same time, Venus, who would be willing to sell the CD for $9 would not be willing to sell at the lower price of $8. The quantity supplied would be five, which is less than it would be at the higher price. There would still be excess supply, but it would not as great as when the price was higher. Because there would still be excess supply at a price of $8, however, there would still be frustrated sellers. The more eager sellers would have an incentive to offer a price even lower to ensure that they are the ones who are able to make the sale.

How low will the price be bid down? In this case the price will be bid down until it is $7. Why that? If the selling price is $7, notice that the quantity demanded would be four as Bobby would be willing to buy along with Greg, Jan, and Marsha. All four of them value the Beethoven CD more than $7. At the same time, the quantity supplied would be lower than at the higher price because Andy, who would be willing to sell at a price of $8, would be unwilling to sell for $7. Only Les, Herb, Bailey, and Johnny value $7 more than their Beethoven CD. At $7, then, the quantity demanded would be four and the quantity supplied would be four, which means that the quantity demanded and the quantity supplied would be equal This means that everyone who wanted to buy could buy, and everyone who wanted to sell can sell!

Now if everyone who wanted to buy a Beethoven album at a price of $7 could buy it, would they have any incentive to further bid up the price? Would there be any buyers who, while in the process of holding out the wad of seven ones, about to make the exchange, suddenly exclaim, "What am I doing?!?!? Here, I'd much rather pay $25 for that CD!" The answer, of course, is no. By the same token, if everyone who wanted to sell their Beethoven record at $7 could sell, the sellers would have no incentive to bid down the price. They have no incentive to say, "I know I can sell this for $7, but would you please only pay me $3?" Therefore, buyers and sellers will act to bid the price to $7 and once the price is bid there it will stay at that level until market conditions change.

The price at which quantity demanded equals quantity supplied has been called by many names by different economists. Because it is the price that is paid and received in the free market, this price is referred to as the *market price*. Because it is also the price at which there is no excess supply or demand, it is said to be the price that clears the market, or the *market-clearing price*. Finally, because it is the price at which quantity demanded equals quantity supplied, neither buyers nor sellers have an incentive to pay more or charge less. Therefore, at this price the market is at rest, so this price is called the *equilibrium price*.

The above determination of the market price can also be shown us-ing market demand and supply schedules on a graph of the market. Using maximum buying prices of the buyers and minimum selling prices of the sellers in table 5.9, market demand and supply schedules can be derived as shown in table 5.10.

Table 5.10

Market Demand and Supply Schedule		
Price	**Demand**	**Supply**
$10	1	7
$9	2	6
$8	3	5
$7	4	4
$6	5	3
$5	6	2
$4	7	1

Using these market demand and supply schedules, we can plot mar-ket demand and supply curves. Because we are trying to determine the market price, we can place both the supply and demand curves on one graph such as we see in Figure 5.5.

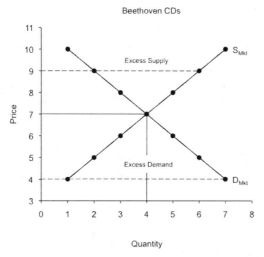

Figure 5.5. Graph of the Market for Beethoven CDs.

In so doing, we can immediately see what would happen in the market for Beethoven CD's if the price was set at $4. Remember that the demand curve tells us the maximum quantity buyers will buy at any given price, while the supply curve tells us the maximum quantity sellers will sell at any given price. The graph clearly shows that at a price of $4 the quantity demanded would be greater than the quantity supplied, which means that there would be excess demand. If the price is $9, however, the graph plainly shows that the quantity supplied would be greater than the quantity demanded, which means that there would be excess supply. The only price at which there would be neither excess demand nor excess supply is where the quantity demanded is equal to the quantity supplied. The graph rather neatly shows the equilibrium price—the one price that clears this market—to be where the demand curve and supply curve intersect. In this case they intersect at a price of $7. At that price, four Beethoven CDs will be sold.

How does this pricing process play itself out in the well-developed economy in which we live? In the vast majority of markets, the numerous buyers and sellers do not get together to haggle over prices. That is because the number of buyers and sellers are so large that no personal negotiation takes place.

When people go into the local Borders store and see a price on a Beethoven CD that they think is too high, they do not go find the manager and see if he will accept less. There is no direct negotiation of prices.

You might be tempted to think, therefore, that it is the suppliers who set prices. However, do not forget, that when economists speak of market prices, they are talking about the prices that are actually paid and received. For an exchange ratio to be a real market price, an exchange must have actually taken place. So, if people make their way back to Borders' music department looking to buy a recording of Beethoven's *Symphonies No. 5 and 7* but the price charged by Borders is above their maximum buying price, they will simply turn around and walk out.

As explained earlier, Borders has every incentive to offer to sell its Beethoven CDs at the equilibrium price. Their decision to offer their merchandise at a specific price is the result of an entrepreneurial judgment on their part. They must forecast what they think the equilibrium price will be on that day. If they predict that they can sell all of the copies of Beethoven's *Symphonies No. 5 and 7* they want to at a price of $15.99 that is the price at which they will offer to sell them, and that is what the sticker on the CD will say. If, however, there are large numbers of potential buyers, for whom the price on the CD is above their maximum buying price, these potential buyers will turn around and walk out. It may turn out that Borders does not sell anywhere close to what they thought they would at their offer price. Perhaps they had five copies in their inventory and they thought they could sell all five, but only sold one. They made an entrepreneurial error. They invested in buying five CDs from the manufacturer thinking they could sell them at $15.99 and found that they could only sell one. Meanwhile, the CDs will gather dust until Borders can sell them.

The question for Borders is what does it do now? In the free market, sellers are continually rethinking their pricing decisions. If Borders finds at the end of the day that their Beethoven CDs are not moving, it has another chance to sell them tomorrow and must make another pricing decision. If it expects that demand for these CDs will be higher the next day, it may keep the price at $15.99. If it expects that demand will remain lower than it initially thought, it may lower the price in order to stimulate sales. If it anticipates that demand for the CDs will be greater the following month because it will include the 200th anniversary of the composition of a particular symphony, and therefore people will hear about it in the news and become more interested in the music, it may keep the price at $15.99, thereby building up their inventory. Its rationale would be that at the higher price people will not be willing to buy them this month, so as it receives a couple copies each week, Borders' stock will build. It might be easier to build inventory now rather than try to do so when more retailers are demanding Beethoven records from the manufactures. Because CDs are small and very durable, they are easy and relatively cheap to store. It also might be cheaper to store them in the stockroom rather than ship them back to the warehouse and then ship them back to the store when the demand increases.

If Borders, on the other hand, offers to sell at a price lower than equilibrium, the Beethoven CDs will fly off the shelves as customers buy them in great haste. Borders will continue to run short of those CDs until they catch on and raise the price at which they offer to sell them.

Sometimes, selling goods at way below the equilibrium price can result in very extreme behavior. The August 17, 2005 *Cincinnati Post* ran an Associated Press story with the headline, "Thousands Show Up For Computer Sale." It documented a stampede of people that resulted when the Henrico County school system sold one thousand Apple iBook laptop computers for $50 instead of the normal price of between $999 and $1,299. People began arriving at the store before two in the morning. The gates were opened at seven a.m. and pandemonium ensued. People rushed forward and an elderly gentleman was pushed to the ground. Someone was beaten with a folding chair. Seventeen people were reported injured with four needing to go to the hospital. One woman even urinated herself, because she did not want to lose her place in line.

Certainly, the Henrico County School System just wanted to get rid of the laptop computers and thought they were providing a social benefit by allowing people to buy a used computer at a price much lower than the current market price. It is true that 1,000 buyers left satisfied. However, approximately 4,500 had their preferences frustrated, not to mention being pushed and thrown to the ground.

Our analysis of price determination has allowed us to discover one more economic law: the *law of one price*. This law states that for any good, there will be one and only one market price which is paid and received. If different sellers attempt to charge different prices for the same good, buyers would immediately withhold buying from the sellers asking the higher prices and make their purchases from sellers asking the lower prices. This would immediately alert sellers to adjust their prices accordingly. Those sellers who feared not making sales would quickly lower their prices and those who realized they could make larger revenues charging higher prices would quickly raise their prices until one price would rule in the market.

THE BEAUTY OF IT ALL

Understanding the process of price determination is part of the beauty of economics. Beginning with the true axiom that humans act, in an earlier chapter, we derived the law of marginal utility. In this chapter we have seen how that law is manifested in the actions of suppliers and demanders, culminating in a market price. The beauty of the free market is that each equilibrium price is mutually agreed upon by both the buyers and the sellers who participate in the exchanges. As such, it is a voluntary price. No buyers are forced to buy when they do not want to buy and no sellers are forced to sell when they do not want to sell. Nor is anyone prohibited from buying or selling for that matter either. No one is sent to the gulag; no one's church gets burned down; no one's spouse gets shot. Every act to buy or not buy, and every act to sell or not sell is voluntary and peaceful.

Another wonderful benefit of free market prices is that the market price is the result of the free actions of people motivated by their subjective value scales. Therefore, the price that is arrived at via voluntary exchange is also the price that results in the most satisfied preferences.

Additionally, remember that in a free market, entrepreneurs use market prices to calculate the expected profits or losses of potential investments. They do so by comparing the prices they expect they can receive for the products they are producing and selling with the sum of the prices of all the factors used to produce those products. If the price of the final product is high enough to more than cover all of the costs of production, the entrepreneur will make a profit. Notice further—and this is the beauty of it—that because the prices of goods are determined by supply and demand, which in turn are manifestations of people's subjective values, entrepreneurs who produce goods that they can sell for a relatively higher market price earn profits precisely by providing those goods that consumers actually want.

SUGGESTED READINGS

Böhm-Bawerk, *Positive Theory of Capital*, 215–35. An excellent early exposition of how prices are determined in voluntary exchange.

Mises, *Socialism*, 95–109. The outstanding exposition about the importance of money prices for economic calculation.

Rothbard, *Man, Economy, and State*, 187–98, 233–49. These two sections provide an excellent exposition of the emergence of indirect exchange and the determination of monetary prices.

———, *What Has Government Done to Our Money*. The best introduction on the origin of money as a result of voluntary exchange and also how government intervention in the realm of money has led to numerous problems.

6

Market Changes

You DISCOVERED IN THE previous chapter that the prices of all the goods in a free market are determined by supply and demand. Be careful to note, however, that when economists use the phrase, *supply and demand*, they are (or at least should be) merely talking about the actions of suppliers and demanders who are motivated by their subjective value scales. There are no such things as autonomous market forces that drive the economy. You will never hear the echoing, distant voice of Obi Won Kenobi telling you to, "Use the supply and demand, Luke." Supply and demand are merely manifestations personal value scales. Nevertheless, economists often use verbal shorthand in their conclusion that prices are determined by supply and demand.

The law of one price tells us that in every market for a particular good, there will be only one price for which the good is bought and sold. This is the market-clearing price that results in the most preferences being satisfied. This market price is determined by individual decisions to buy or refrain from buying and to sell or refrain from selling at various prices. We have seen that, to avoid excess demand, the most eager buyers bid up the price until the quantity demanded equals quantity supplied. To avoid excess supply, the most eager sellers will lower their price until the quantity supplied equals the quantity demanded. Once this equilibrium price is agreed upon, the market is at rest, because there is no longer any incentive for anyone to bid the price up or down.

Experience tells us, however, that market prices for economic goods do not indefinitely stay the same. In other words, once the market price for a particular good is determined, it does not remain constant over time. For instance, during the summer of 2002, regular gasoline was selling for a market price of $1.60 per gallon. During the summer of 2003, however, the same grade of gasoline could be bought for $1.39 per gallon. By the end of the summer of 2005, it had sharply risen again to $3.19 per gallon. During the summer of 2008, gas prices peaked at $3.99 per gallon. One of the beauties of economics is that it can not only help us to understand how prices are determined, but it can also help us understand why and how prices change.

Prices are determined by the demand of the buyers and supply of the sellers. Therefore, we can reasonably conclude that the market price for a good will change if there is a change in supply or demand. Demand will change if the maximum quantity buyers will buy at any given price changes. Supply will change if the maximum quantity sellers will sell at every given price changes.

Because both the demand schedules of buyers and supply schedules of sellers are determined by their subjective rankings of different units of goods compared to varying amounts of money, market prices are also the result of subjective value scales. Consequently, anything that alters the subjective value people place on either the goods or money will lead to changes in supply or demand and hence market prices. Before beginning our in-depth study of market changes, however, it will be helpful to drink more deeply of demand theory for just a moment.

DEMAND ELASTICITY

We'll briefly digress to discuss what economists call *demand elasticity.* Demand elasticity is a measurement of how responsive buyers are to a change in price. The law of demand tells us that as the hypothetical price of any good increases, the quantity of that good demanded will decrease. The elasticity of demand determines by *how much* quantity demanded would decrease if the hypothetical price increased, as well as how total expenditure changes when there is a given change in price.

Changes in total expenditure resulting from a change in price will be determined by how sensitive buyers are to price change. Suppose that a CD store sells its copies of Bob Ekelund's, *Reverie,* a recording of solo piano pieces, at a price of $10 each. At that price it can sell eight of them. This price/quantity combination is illustrated in figure 6.1 as point O.

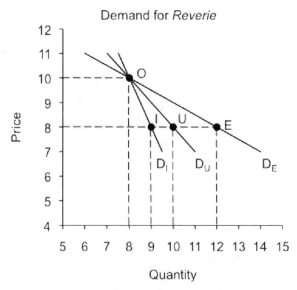

Figure 6.1. Different demand elasticities between prices of $8 and $5.

The store's total revenue from these CDs is $80. In an effort to increase his total revenue, the store owner may consider lowering the price of the CDs to $8. Why would the seller do this? He may know the law of demand that says at the lower price more would be purchased. However, whether such a drop in price actually would increase total revenue depends upon how sensitive buyers are to the lower price.

Suppose, for example, that the music store does lower the price to $8 and the preferences of buyers are such that the demand curve for Ekelund's *Reverie* is demand curve D_I. What would happen to the store's total revenue on *Reverie* if the store charged the lower price? While at a price of $10, eight CDs would be sold, at a price of $8, nine records would be sold. If the demand schedule is as illustrated by demand curve D_I, total revenue at a price of $8 would be $72 which is lower by $8. It would be unwise for the store to decrease the price to $8, because it results in less revenue, not more.

On the other hand, suppose that the demand schedules of the buyers are such that the demand for Ekelund's *Reverie* is illustrated by demand curve D_U. In this case, at the lower price of $8, buyers would demand 10 CDs. Total expenditure on these goods would be $80. Decreasing the selling price from $10 to $8 would have no effect on the store's total revenue.

Finally, suppose that the store decreases the price to $8, but the preferences of buyers generate demand curve D_E. With this demand schedule, if the price charged would be $8, 12 CDs would be sold and the stores total revenue would be $96. In this case it would be wise to drop the price to $8, because in doing so, the quantity of CDs demanded would increase enough so that total expenditure, and hence revenue, would increase.

We have here three different demand scenarios. Each of the three scenarios features identical price decreases, from $10 to $8, but the consequences regarding total revenue are all different. Why? The answer lies in how sensitive buyers are to the change in price.

At lower prices, more will be demanded. The crucial question is how much more will be demanded. In the above example, when demand is like we see in our first case illustrated by demand D_I buyers are relatively unresponsive. The quantity demanded increases, but only by two, so that the total amount spent on the CDs falls. If as a result of a price change, total revenue changes in the same direction as price, economists say that demand is *inelastic*. Notice that when demand is inelastic, the demand curve is relatively steep compared to the other cases.

In our second case, the decrease in price has no effect on the store's total revenue. This is because the negative change in the price is exactly offset by the positive change in quantity demanded. In fact the percent change in price is equal to the percent change in quantity demanded. If one was to express this in a ratio of percent change in quantity demanded to percent change in price, the resulting ratio would equal one, which is why demand in this case is called *unit elastic*.

In our third case, a decrease in price results in an increase in total revenue. This is because the percent increase in the quantity demanded is greater than the percent decrease in the price of the CDs. The negative price effect is more than made up for by the positive quantity effect, so total revenue increases when the price falls. In this case, there is an inverse relationship between price and total revenue. If, as a result of a price change, total expenditure and the price of a good change in different directions, economists say that demand is *elastic*. Notice that when demand is elastic, the demand curve is relatively flat compared to the curves exhibiting inelastic or unit elastic demand over the range of the price change.

If a seller attempts to increase his total revenue by raising its selling price, it must hope that demand is inelastic. This is precisely what the U. S. Postal Service was thinking in the spring of 2002. On March 23, 2002 the *New York Times* reported that the U. S. Postal Service decided to raise its rates for a first class stamp from 34 cents to 37 cents, part of an average 7.9 per cent increase for all first class mail. The reason given for the rate hike was so that U. S. Postal service could receive "an immediate influx" of increased revenues to help it recover from the recession that followed the terrorist attacks of September 11, 2001 and the anthrax scare that followed the attacks.

Why did postal officials expect that revenues would increase when they should know that higher prices lead to decreased quantity demanded? It is because they thought that their customers would be relatively unresponsive to the 3 cent price change, so that the demand for first class mail was inelastic. It turns out that they were right. The proposed increase in first class postage rates mentioned above took effect on June 30, 2002. From September through November, the total number of first-class pieces delivered was 2.7% lower compared to the same quarter the year before, while the total revenue received on first class mail increased by 4.3% compared to the same quarter the year before.

What have we learned then? When there is a change in price, if buyers are relatively responsive to the price change, there will be a relatively large change in quantity demanded. If this is the case, then economists say that demand is elastic. When demand is elastic, the slope of the demand curve will be relatively flat, and the total expenditure on the good will move in the opposite direction as the price. If buyers are relatively unresponsive to a change in price, so that there is a relatively small change in quantity demanded, economists say that demand is inelastic. If demand is inelastic, the demand curve is relatively steep, and total expenditures on a good will move in the same direction as the change in price. If demand is unit elastic, a price change will have no effect on total expenditure on a good.

We can see then, that those U. S. Postal Service officials who expected revenue to increase as a result of their price increase were hoping that the demand for first class postage would be inelastic. They were expecting that their customers would be relatively unresponsive to the price increase, so that, even though the quantity of first class stamps bought might be fewer, the decrease would not be great enough to lead to a decrease in total revenue.

Now that you have a grasp of what is elasticity of demand and its relationship to total expenditure, it may be interesting to consider the question: why is it that for different goods and at different times, buyers are more or less responsive to changes in price? What is it that determines demand elasticity? Remember that the demand curve is simply a graphical representation of a demand schedule which is derived from individual value scales. Ultimately, whether demand is elastic or inelastic is determined by the preferences of the buyers. The question becomes: what can influence how responsive buyers are to a change in price?

The primary answer is the availability of close substitutes. If buyers tend to think that all colas taste basically alike and the price of a twelve ounce can of Pepsi increases from fifty cents to seventy-five cents, buyers will be relatively responsive to that price change. A relatively large number of buyers would abstain from Pepsi at the higher price. Why? Because there are a number of other cola products that will also allow them to experience what they used to call the "pause that refreshes." They could drink Coca-Cola, RC Cola, or even Sam's Choice, as well as many other store brands. If the price of Pepsi increases to seventy-five cents, then a large numbers of buyers will seek out these substitutes, so the quantity of Pepsi demanded will fall a rather large amount.

If, on the other hand, buyers tend to be very loyal to a particular brand, because they do not think that all colas taste the same, then fans of Pepsi may not be very responsive to a twenty-five cent price jump. This could be because, in their eyes, Coke, RC, and all the rest are not viewed as close substitutes. To them, nothing gives them the joy of cola except Pepsi, so they will not be as willing to give up Pepsi in the event of a price increase. If this is the case, although there would be some drop in quantity demanded if the price of Pepsi increased, the drop would be relatively small.

This helps to explain why our friends at the U. S. Postal service might have expected that the demand for their services was inelastic. While it is true that there are an increased number of substitutes for parcel delivery, overnight or otherwise, and there are even substitutes for personal communication once carried out only via the mail, such as e-mail, instant messaging, and the telephone, the fact remains that the U. S. Postal Service has a government mandated monopoly in providing first class mail service. Because of this legal monopoly, other firms, such as Federal Express and UPS, are prohibited from entering the first class letter delivery industry. Consequently, customers do not have any of close substitutes to choose from when sending a first class letter.

CHANGES IN DEMAND

Now that we have completed our brief review of demand elasticity, we turn to the matter at hand, which is, as you may remember, to explain what will lead to a change in price. As noted at the beginning of this chapter, prices are determined by demanders and suppliers. Consequently, a good's market price will change if there is a change in either demand or supply.

We will examine such changes one at a time, beginning with changes in demand. To understand the effects that a change in demand has on price, it will be helpful to identify just what economists mean by the phrase *change in demand*. Remember, demand is the different quantities of a good that buyers will buy at each hypothetical price. As such, demand is not only one quantity that would be demanded at only one price; it includes every price and every quantity in a person's demand schedule. It includes every point on the demand curve. The demand curve demonstrates the maximum quantity buyers will buy at *every* given price. So a change in demand is illustrated by a change in the entire demand curve.

A common error is to mistake a change in quantity demanded with a change in demand. The two distinct changes are each shown in figure 6.2.

A change in quantity demanded refers to what would happen if the price at the time of purchase increases or decreases. Given the buyer's value scale, a different quantity is demanded at different hypothetical prices. A move between point *A* and point *B* shows the buyer buying a lower quantity due to a higher price. At the higher price, the buyer wishes to purchase a different *quantity demanded*.

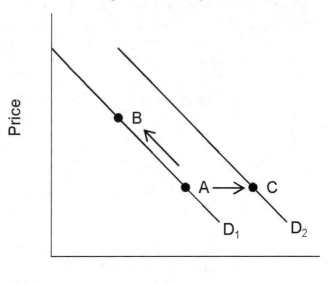

Changes with Respect to Demand

Figure 6.2. A change in quantity demanded versus a change in demand. A movement from point *A* to point *B* shows a decrease in quantity demanded. A movement from point *A* to point *C* shows an increase in demand, because point *C* is on an entirely different demand curve.

A move from point *A* to point *C*, however, illustrates a change in demand because it is the result of a change in the entire set of price-quantity combinations that makes up a buyer's demand schedule. A movement from point *A* to point *C* shows that the buyer is willing to buy a larger quantity at the *same* price. Therefore, if economists say that *demand* has changed, they are not talking about moving from one point to another point on the same demand curve, nor are they talking about a change from one price and quantity combination in the demand schedule to another. They are talking about a change in *every* maximum quantity buyers will buy at any given price. They are talking about an entirely different demand curve.

What happens in the market if, for whatever reason, buyers are willing to buy a larger quantity of a good at every price? In other words, what happens to the market price and quantity sold if the demand for a good increases? What we have already learned about price determination can help us answer this question.

Suppose we look again at the market for Beethoven recordings. The market price from the example in the last chapter was $7. At that price, everyone who wanted to buy could buy and everyone who wanted to sell could sell. Suppose now that, for whatever reason, the market demand for Beethoven CDs increases. Such an increase is shown by demand and supply schedules in table 6.1.

Table 6.1

Old Market Schedules for Beethoven Recordings			New Market Schedules for Beethoven Recordings		
Price	Demanded	Supplied	Price	Demanded	Supplied
10	1	7	10	5	7
9	2	6	9	6	6
8	3	5	8	7	5
7	4	4	7	8	4
6	5	3	6	9	3
5	6	2	5	10	2
4	7	1	4	11	1

As shown in figure 6.2, when demand increases, the entire demand curve shifts to the right. This is because buyers are willing to buy more at every hypothetical price.

What would happen in the market for Beethoven CDs if the demand for them would increase? It is relatively easy to see from the graph in figure 6.3 that the market price and quantity sold would both increase.

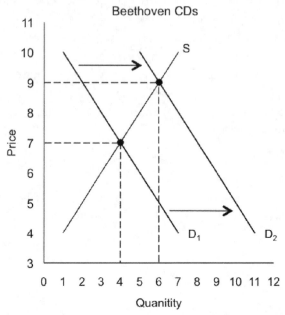

Figure 6.3. An increase in demand results in a higher market price.

You can tell this by looking at what the new equilibrium point would be. At the point where the new demand curve would intersect the supply curve, the price is higher than the original and the same is true with the equilibrium quantity.

Economics, however, is not about finding intersection points on graphs picturing hypothetical demand and supply curves. It is about understanding and explaining human action in exchange. Good economics tells us more than that the price and quantity sold would be higher than it would be without the increase in demand. Good economics explains *why* and *how* they would be higher.

The market price would be higher because, if there was an increase in demand, the quantity buyers are willing to buy at the prevailing price would no longer be equal to the quantity sellers are willing to sell at the prevailing price. In fact, if the price did not change when demand increases, the quantity demanded would be greater than the quantity supplied at the prevailing price, so there would be excess demand for Beethoven recordings. Remember that if there is excess demand for a good, there will be frustrated buyers. These are buyers who are quite willing to buy the good, but cannot because there is not enough supplied at the prevailing price. As explained in last chapter, in an effort to reduce the possibility of frustration, the most eager buyers would bid up the price in order to insure that they are the ones who get to buy. In our example, the most eager buyers would bid up the price to $9 per CD, the price at which everyone who wants to buy can buy. So we see why an increase in demand results in a price higher than if demand remained the same. It is not due to some mysterious force drawing the price upward. It is due to the actions of the most eager buyers bidding up the price.

What about the change in quantity, however. Why does it also increase? *Because the graph says so* is not a very satisfactory answer. Remember that the law of supply tells us that there is an inverse relationship between the hypothetical price of a good and the quantity of a good that sellers are willing to sell. If the price of Beethoven recordings is higher, sellers become more willing to part with their stock and so supply a larger quantity. Therefore, if the market price for Beethoven CDs would increase due to an increase in demand, so would the market quantity.

What would happen if demand for Beethoven recordings decreases? It turns out that the effects would be the exact opposite. A decrease in demand means that buyers are willing to buy fewer CDs at every given price. This is illustrated in the market schedules in table 6.2.

Table 6.2

Price	Old Market Schedules for Beethoven Recordings		Price	New Market Schedules for Beethoven Recordings	
	Demanded	Supplied		Demanded	Supplied
10	1	7	10	0	7
9	2	6	9	0	6
8	3	5	8	1	5
7	4	4	7	2	4
6	5	3	6	3	3
5	6	2	5	4	2
4	7	1	4	5	1

A decrease in demand is also shown on the market graph by a leftward shift of the entire demand curve as illustrated in Figure 6.4.

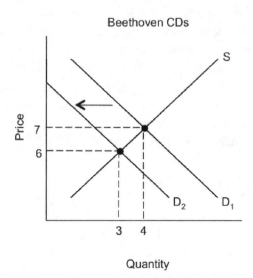

Figure 6.4. A decrease in demand results in a lower market price and smaller quantity sold.

The market graph clearly shows that if the demand for Beethoven recordings would decrease and the price remained the same, the quantity of CDs demanded would be less than that supplied. This, you will remember, is called excess supply. If there is excess supply, some sellers will be frustrated because they prefer to sell at the prevailing price, but there is not enough demand for them to sell all that they want to sell. When this occurs, the most eager sellers will bid down the price in order to insure that they are the ones that make the sale. They will continue to bid down the price until everyone who wants to sell can sell. This occurs at the new market price where the quantity demanded is again equal to the quantity supplied. So we see that, if demand would decrease, the market price would also decrease.

The graph also tells us that a decrease in demand would result in fewer CDs sold. This is due, again, to the law of supply. The excess supply is mitigated as the price is bid down because of the direct relationship between the price of a good and the quantity of the good that sellers are willing to sell. At lower prices of Beethoven records, sellers are less willing to sell and the quantity of CDs offered for sale in the market decreases. Consequently, a decrease in demand would result in both a lower market price and quantity.

Anything that changes the relative position of different amounts of money and different units of a good on a buyer's value scale will lead to a change in demand. Economists have identified a number of things that affect just how many units of a good that people will buy at each price. In the following sections we will examine those things that affect demand including income, tastes and preferences, expected prices, population, and the market for related goods.

Income

How do changes in income affect the demand for a good? Remember that the demand schedule is derived from buyers' value scales, which rank different units of a good against different quantities of money. Suppose that Nova Dawn's income increases. Maybe she gets a raise, sells her car, or takes a second job. In any case, the quantity of money she has on hand would increase. There are three things that people can do with an increase in their stock of money: they can spend the new money on consumption goods, save and invest it, or add to the cash balance that they hold. In most cases, people do all three. With part of the new money, they purchase additional consumption goods, some of it they add to their savings account, and some of it they add to the amount of money they have on hand for purchases.

The reason that they increase their spending can be found by considering the law of marginal utility. As explained in an earlier chapter, money is a particular commodity chosen as the common medium of exchange. Because money begins as a commodity, it is an economic good to which all laws of human action apply, including the law of marginal utility. The law of marginal utility tells us that with a larger stock of money, the marginal utility of that money is lower. Therefore, the value of different marginal units of goods will rise relative to different quantities of money. So if the income of buyers increases, this will tend to lead to an increase in the demand for goods. If Beethoven connoisseurs find their incomes growing, they will be willing to buy more compact discs at every hypothetical price than they would have been without the higher income. There tends to be a positive relationship between consumer income and demand.

Tastes and Preferences

Another factor that affects buyer demand is simply the subjective appeal that a good has. My own taste for the music of Ludwig van Beethoven, for example, has not always been what it is now. In sixth grade, I named as my favorite musical group Village People. As Ronald Reagan replied when charged with hypocrisy for supporting Franklin Delano Roosevelt in Roosevelt's first election, "I did many foolish things in my youth." In any event, I had no taste for Beethoven. All I knew of Beethoven was his *Pastoral Symphony* and did not really understand it. Only when I reached high school did my appreciation for Beethoven grow considerably.

No doubt you can guess that as people's subjective appreciation for a good waxes, they will demand more. As their appreciation wanes (as mine drastically did for Village People) so will their demand. In other words, there is a direct relationship between a person's taste for a good and his demand for a good. As a person becomes more enamored with a good, he is willing to buy more of it at each price. As someone finds a good more distasteful, he is less willing to buy at each price.

Expected Prices

Action is always forward looking. We act in the present in order to reap the end in the future. Our expectations regarding the future consequences of our own present actions greatly influence which actions we ultimately decide to undertake. Additionally, our expectations regarding the actions of others also affect what we decide to do. If an automobile manufacturer, for example, forecasts that next year people will completely stop buying sports utility vehicles, he would be foolish to make any more. In fact, the manufacturer would not build any more, because he would understand that building further SUVs would be a losing investment.

Not only are businessmen guided in part by expectations of the future, but buyers are to an extent as well. Once, shortly after my wife and I were married, she received a Land's End catalogue providing what appeared to be rather strange information on the cover. Land's End, an apparel company, was announcing that this catalogue was the last opportunity to order their Hyde Park Oxford dress shirts at its present price. Beginning with the next catalogue, the price for the shirts were going to be $4 higher.

At first blush this seems like an odd way to advertise. It is very rare that one sees a television commercial for Crazy Joe's Used Cars where he's exhorting people, "Come on down, we're *raising prices*!" Higher prices are not normally something one thinks is wise to publicize. However, here was Land's End doing just that. The company's strategy was based on their knowledge that a wife's demand for goods in the present is not only affected by present circumstances, but also by her expectation of future prices. Land's End was trying to stimulate demand in the present by correcting expectations of potential buyers.

The people at Land's End knew that, in general, expected future prices are positively related to present demand. Those considering buying a dress shirt, or a Beethoven CD for that matter, are more likely to buy that good in the present if they expect that the price in the future will be higher. Expectations of higher prices in the future tend to increase demand in the present. On the other hand, if buyers expect that future prices will be lower than present prices, they will tend to wait and put off buying until the prices drop. In other words, expectations of lower future prices will tend to reduce present demand.

Population

One thing that nearly every local chamber of commerce loves to see is a new factory opening up on the outskirts of town. This is because a new factory means more jobs that need to be filled by people moving into town. More people moving into town means more people who need to buy housing, clothing, food, and other goods.

Population represents the number of potential buyers and is directly related to market demand. If, for example, a new factory employing 200 people opens up in Glenwood, Iowa, a large part of those new jobs will be filled by people moving into town. These people will increase the demand for goods sold in Glenwood. As population increases, demand for the goods sold in Glenwood also increases.

On the other hand, suppose the Black Plague makes a comeback and comes through Glenwood and kills off one-third of the population. This will not increase the demand for much of anything, except perhaps mortuary services. As the population decreases, there are fewer people who will be able to buy economic goods, and the quantity of goods demanded at every price will decline.

Market for Related Goods

The market price for one particular good is not isolated from the rest of the economy. It is an exchange ratio that is affected by many other interconnected relationships on the part of both buyers and sellers. As such, changes that occur in the market for related goods can also impact the demand for the particular good with which we are concerned.

Identifying how changes in the market for related goods affect the market for specific goods requires very complex reasoning compared to the other factors that affect demand. This is because there are different sorts of changes that can occur in the market for related goods and each of these changes can affect prices for other goods differently. We can separate related goods into two categories: complementary goods and substitutes. For the sake of clarity we will examine them one at a time.

From the economic point of view, a complement is not the nice thing your friend tells you about your taste in textbooks. That is a *compliment*. When economists use the word *complement*, they are referring to complementary goods. Goods that are complementary in consumption are those goods that are used together to achieve a particular end. Because of the popularity of peanut butter and jelly sandwiches, peanut butter and jelly are good examples of complementary goods. You must use both to make the sandwich. Other good examples of goods that go together are golf balls and golf clubs, tennis balls and tennis racquets, and flashlights and batteries.

The one characteristic that all complementary goods have in common is that they must be used together in achieving the desired end. If Greg, for instance, wants the ultimate music experience and goes out to buy a CD player, the player, by itself will not be enough to fulfill his desire. He does not want merely a piece of electronics equipment he can show his friends. He wants a device that can be used to give him hours of musical bliss. The player, however, cannot do this unless he also has on hand a CD that he can play with the player. He needs both the CD player *and* the CD to achieve his end.

How will changes in the market for the complementary good affect the market of the good we are studying? It depends on why the market for the complementary good is changing. One change that could occur in the market for the complementary good is a change in the stock of the complement. Suppose, for example, we are interested in how changes in the CD player market will affect the market for Beethoven recordings. Suppose that the stock of CD players available for purchase becomes noticeably greater. This will lead to a decrease in the price of CD players. The law of demand tells us that more CD players will be bought at a lower price. With the increased number of CD players being purchased, this increases the demand for CDs. As mentioned above, people do not want

CD players only to have something to look at. In order for CD players to serve their purpose, their buyers must also demand CDs. Some of these buyers will demand Beethoven CDs. Consequently, as the stock of CD players (the complementary good) increases, prices for CD players fall, and the demand for Beethoven CDs will increase.

If the stock of CD players available decreases, the opposite will occur. The price of CD players will increase, resulting in a decreased quantity demanded. Fewer people will buy players, so fewer people will demand CDs. Therefore if the stock of CD players decreases, causing the price of them to increase, the demand for Beethoven recordings will decrease as well. There is, therefore, a direct relationship between the stock of a complementary good and the demand for the good we are considering.

As stated, however, changes can occur in the market for a related good for different reasons. A change in the demand for CD players would also change the market price and quantity in the CD player market. Suppose for example, that due to an increase in income or tastes and preferences, people increase their demand for CD players. If people are demanding more CD players, their end is most likely to enjoy high quality musical sound. A CD player, by itself, will not accomplish this end. They will also need CDs to play in the player. Therefore, if the demand for a complementary good increases, this will tend to increase the demand for the particular good being studied. On the other hand, if there is a decrease in demand for CD players, there would likely also be a decrease in demand for CDs. In other words, we can identify a direct relationship between the demand of a complementary good and the demand for the other good.

The other type of related good is called a substitute. As the name implies, a *substitute* is a good that can be used in place of another to achieve one's end, even if not perfectly. Examples would include chicken and pork. Years ago the pork industry used a catchphrase, "Pork, the other white meat" to try to convince people that pork was a close substitute for chicken. Different brands of tennis balls also provide an example of substituted goods. When playing a match with your friend, you could use either Wilson or Penn tennis balls. Either one would serve the end of playing tennis.

The determining factor regarding how close two goods are substitutes for each other is, of course, the subjective preference of the person using them. As with all economic decisions, it is the preferences of individuals that determine whether Coke, Pepsi, and RC Cola are substitutes for each other, or if only one can provide them that unique refreshment made possible by a cola soft drink.

Because substitutes can be used to satisfy the same end, a change in the market for one will affect the market for the other. How a change in the market for a substitute will affect the market of another good will depend on what is changing in the market for the substitute. One such change is a change in the stock of the substitute. Suppose, for example, that there is a change in the stock of a substitute for Beethoven recordings. Suppose that people who like Beethoven also like the music of Sibelius and view both as substitutes for each other. If the stock of Sibelius CDs would increase, the market price for Sibelius records would be lower.

How would this affect the market for Beethoven? We do not yet have enough information, because of the nature of the relationship between goods and their substitutes. Because both Beethoven CDs and Sibelius recordings can be used to satisfy the end of musical refreshment, music buyers do not have to buy both CDs to achieve their end. Either one can satisfy the buyer quite nicely. Consequently, when examining how a change in the stock of a substitute affects the market for another good, the elasticity of demand for the substitute becomes very important.

If the stock of the substitute good, the Sibelius CDs, were to increase, the price of Sibelius recordings would be lower. If the demand for Sibelius is elastic, the total outlay spent on Sibelius CDs would increase. This would leave less money for buyers to spend on other goods, including Beethoven albums. Consequently, if the stock of a substitute increases and the demand for that substitute is elastic, the demand for the good in question would be lower than if the stock of the substitute does not change. On the other hand, if the demand for Sibelius is inelastic, a decreased price would result in a drop in total expenditures on Sibelius records. This would leave more money left in the hand of buyers which they could spend on other goods, including Beethoven CDs. Therefore, if the stock of a substitute would be higher and the demand for that substitute is elastic, buyers would increase their demand for the good under observation. If demand for the substitute is unit inelastic, a change in price would have no effect on total expenditure on the substitute, so there would also be no effect on the demand for the good in question.

If the stock of the substitute good decreases, the exact opposite would occur. If the stock of Sibelius recordings declines, the price of those recordings would increase. If the demand for Sibelius CDs is elastic, total outlay on Sibelius would decrease and demand for Beethoven would increase. If demand for Sibelius records is inelastic, total expenditure on

Sibelius music would increase and demand for Beethoven CDs would decrease. Also, if the demand for Sibelius was unit elastic, total outlay on Sibelius music would not change, so there would be no change in the demand for Beethoven recordings.

In general then, the effect of a change in the stock of a substitute on the demand for another good depends on the elasticity of demand for the substitute. If the demand for the substitute is elastic, there is an inverse relationship between the stock of a substitute and the demand for another good. If the demand for the substitute is inelastic, there is a direct or positive relationship between the stock of the substitute and the demand for another good. If the demand for the substitute is unit elastic, a change in the stock of the substitute has no impact on the demand for the other good.

The market for the substitute good could also change, of course, if there is a change in demand for the substitute. If, for example, buyers demand more Sibelius recordings, this will tend to result in a higher market price for a larger quantity sold. Total expenditures on Sibelius CDs would necessarily increase, no matter what their elasticity of demand was. Consequently, an increased demand for the substitute would necessarily result in less money being spent on other goods including Beethoven records, so the demand for Beethoven would fall.

If demand for Sibelius CDs decrease, total outlay on Sibelius will fall, leaving more money for buyers to spend on other goods including Beethoven records. Demand for Beethoven music will increase. In general we see that there an inverse relationship between the demand for a substitute and the demand for another good.

The cause and effect economic relationships that make up the core of demand theory are summarized in table 6.3.

Table 6.3

Demand: Cause and Effect	
Change in...	Effect
Price of good	Change in quantity demanded
Consumer income	Change in demand
Consumer tastes	Change in demand
Expected prices	Change in demand
Population	Change in demand
Market for a complementary good	Change in demand
Market for a substitute good	Change in demand

A change in the price of any consumer good will result in a change in the quantity of that good demanded by consumers, given their subjective values at the time of the price change. A change in any of the other items listed will affect the relative place of marginal units of the good compared to different amounts of money and, hence, will result in a change in demand.

CHANGES IN SUPPLY

As indicated earlier, because market prices for goods are ultimately de-termined by the subjective preferences of buyers and sellers, market price changes are the result of both changes in demand and changes in supply. Just as there is a distinction between a change in quantity demanded and a change in demand, the same holds true for supply. This distinction is illustrated by the graph in figure 6.5.

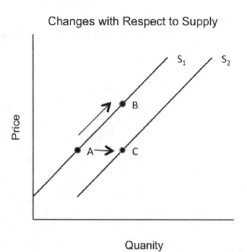

Figure 6.5. A change in quantity supplied is illustrated by a movement along a supply curve. A change in supply is illus-trated by a shift to a different supply curve.

Because of the law of supply, a change in quantity supplied would occur if there were a change in price. Such a change in quantity supplied is illustrated on a graph as a movement along the supply curve from point A to point B. Remember that a supply curve shows the maximum quantity that sellers will sell at every hypothetical price. At a higher price, sellers would be willing to sell more, resulting in the positively sloped supply curve.

When economists use the phrase, *change in supply*, however, they are not talking merely about selling a different quantity of a good at a different price. A change in supply means that sellers are willing to sell a different quantity at every hypothetical price. Graphically, this is illustrated not by a movement along the supply curve, but by a shift in the entire curve itself. A movement from point A to point C indicates that sellers are willing to sell a larger quantity at the same price. Such a movement, consequently, illustrates an increase in supply.

How will a change in the supply of a good affect that good's market price and quantity sold? Let us return to the case in which the preferences of the potential buyers and sellers of Beethoven CDs were such that the market clearing price for the recordings was $7 each. Then suppose that, for whatever reason, the quantity of CDs sellers are willing to sell at every price increases. In other words, there is an increase in supply of Beethoven recordings. Such an increase in supply is shown in the supply schedules in table 6.4.

Table 6.4

Old Market Schedules for Beethoven Recordings			New Market Schedules for Beethoven Recordings		
Price	Demanded	Supplied	Price	Demanded	Supplied
10	1	7	10	1	11
9	2	6	9	2	10
8	3	5	8	3	9
7	4	4	7	4	8
6	5	3	6	5	7
5	6	2	5	6	6
4	7	1	4	7	5

How will this affect the market for the Beethoven CDs? The market graph in figure 6.6 shows that the market price will decrease and the quantity sold will increase.

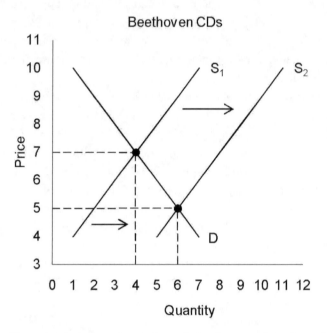

Figure 6.6. An increase in supply results in a lower market price and larger quantity sold.

As with changes in demand, however, the question is why? Economics helps us to understand why exactly the market price would be lower if there was an increase in supply. Going back to our example, if the initial market-clearing price is $7 and there was an increase in supply, a price of $7 will no longer clear the market. The market would no longer be at rest. If sellers persisted in selling for a price of $7, there would be excess supply of Beethoven CDs because the quantity of records supplied would be greater than the quantity supplied. Excess supply, of course, leaves some sellers frustrated, because they are willing to sell at the prevailing price but cannot because the quantity demanded is not great enough. Therefore, in order to ensure that they are the ones able to make the sale, the more eager sellers would bid down the price. As this occurs, the quantity supplied decreases and the quantity sold increases, shrinking the excess supply. The most eager sellers have an incentive to do this as long as they see the potential of having their preferences frustrated. They will cease to bid down the price as soon as their frustration is no longer an issue. Frustration will cease to be an issue as soon as the threat of excess supply is gone, which will occur as soon as the price drops to the point at which the quantity supplied equals the quantity demanded. Once it reaches this point, everyone who wants to sell can sell. There is no incentive to further bid down the price. At this new lower price, the market will once more be at rest and this new price will be the equilibrium price. We see then that an increase in the supply of a good will result in a lower market price.

What about quantity sold? The quantity sold in the market would be greater because sellers are willing to sell more at every hypothetical price and at the lower price buyers would be willing to buy more.

The effects of a decrease in supply are just the opposite. A decrease in supply is illustrated by a leftward shift of the supply curve. We can easily see from the market graph in figure 6.7 that a decrease in supply will result in a higher market-clearing price and fewer units of a good sold.

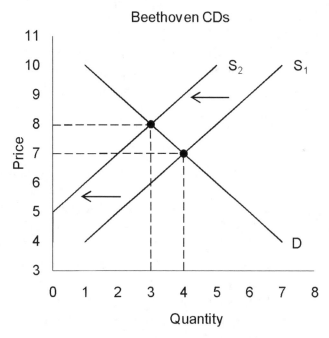

Figure 6.7. A decrease in supply results in an increase in market price and a smaller quantity sold.

A decrease in supply would move the market out of its restful condition. Because of sellers' willingness to sell fewer Beethoven records at every price, there would be excess demand for CDs if the price remained at $7. The quantity of recordings demanded would be greater than that supplied. In this case, there would be frustrated buyers who are willing to buy, but cannot because of the lower supply. The more eager buyers, in an effort not to return home empty handed, would bid up the price until everyone who wants to buy can buy. This would occur when the price reached the point at which the quantity supplied equals the quantity demanded. At this price buyers have no incentive to further bid up the price. The sellers are likewise happy, because everyone who wants to sell can sell. This new price is the new market price, and no one would have frustrated preferences. At this new price, fewer CDs are sold, because, as the price increases, fewer buyers remain in the market. Consequently, a decrease in supply will result in fewer goods sold at higher prices.

We see, then, how changes in supply will affect the market price and quantity sold for any economic good. What factors contribute to such changes in supply? Economists have identified several, and most of them are related to the stock of the good in existence. Anything that affects the number of Beethoven CDs that exist will have some effect on the quantity that sellers are willing to sell at each hypothetical price. These factors include investment in production, the price of factors of production, technology, the number of sellers, and the market for goods related in production. Sellers' expectations of future prices will also affect present supply.

Investment in Production

Investment is the act of buying factors of production for the purpose of producing a good that can be sold for income. Investment is never spending in general, but always spending on acquiring the use of particular factors of production. So for increased investment to affect the market for Beethoven CDs, the investment must be in the purchasing of factors used to produce the CDs. If there is an increased investment in the production of Beethoven recordings, then naturally more of them will be produced. As the stock of the CDs increase, sellers will be willing to sell more of them at any given price, so the supply of records will increase.

If there is, however, a decrease in investment in the production of Beethoven recordings, fewer will be produced. The stock of recordings will be smaller and sellers will be less willing to sell their records at any given price. The supply of Beethoven CDs will decrease. We can see, then, that there is a direct relationship between the quantity of investment in the production of a good and the supply of that good.

Price of Factors of Production

The prices of the factors used to produce a good will affect that good's supply because those prices affect how many factors can be bought or rented with a given sum of investment. Suppose for example, that a recording company has $2 million that it can devote to the production of Beethoven CDs. That firm spends its $2 million on buying land for the factory, hiring workers, and buying various capital goods. If the price of land, wages paid to labor, and the prices of those capital goods increase, the recording company can buy less land, hire fewer workers, and buy fewer capital goods with $2 million. Because the business has access to fewer factors of production, fewer CDs will be made and the supply of them will decrease. If the price of factors used to produce a good increase, the supply of the good will decrease.

On the other hand, if the prices paid for factors of production decrease, the supply of the good being produced will tend to increase. If prices for the factors of production drop, CD manufacturers can afford to buy or rent more factors and hence can produce more Beethoven recordings, which will increase their supply. Consequently, there is an inverse relationship between the price of the factors used to produce a good and the supply of the good.

Technology

The United States Department of Agriculture estimates that in 1987 it took the average American farmer three hours to produce one hundred bushels of wheat, compared to 275 hours it took the average American farmer to do the same thing in 1830.[1] That is a 3,233% increase in farming productivity over 157 years. What accounts for such growth? It is not that farmers today work any harder than they did in the early nineteenth century. In fact, one might argue that in terms of physical labor, they have it easier, yet they are still more productive. The answer lies in modern farmers taking advantage of modern technology.

Since the 1830s there has been regular advance in the number and quality of tools and equipment available for farmers. Even as recently as the 1700s, the most advanced farming tools were the wooden plow and handheld hoe. The first half of the nineteenth century saw the introduc-

---

1. The productivity statistics along with information regarding farm technology were found at United States Department of Agriculture, "Farm Machinery and Technology," 6.

tion of the iron plow and mixed chemical fertilizers. From 1862–75 the wide transformation from hand power to the use of horses spawned the first agricultural revolution in the United States. In the first half of the twentieth century gasoline powered tractors came into extensive use, with the first light tractor suitable for wide use marketed in 1926. As an old Ford tractor advertisement promised, using tractors, "means less work . . . more income per acre." These developments, in addition to more extensive use of higher quality fertilizers, enabled the American farmer continually to increase the number of bushels produced per unit of time.

Not surprisingly, increases in technology have led to greater total output. Better tools and equipment have increased dramatically the stock of agricultural products available for consumption since the early 1800s. Farm output increased by more than 45% in only the twenty years from 1980 to 2000. This increased agricultural output naturally led to increases in supply.

We find that there is a direct relationship between the level of technology and the supply of a good. This is because technology impacts how much output we can get with the same investment. Likewise, Technology affects how much can be produced with the same quantity of factors.

Technology can best be thought of as *ways of doing things*, or *technical know-how*. People often associate technology with machines. Recently there has been quite a push on the part of certain school administrators to increase the amount of technology in the classroom. Almost everyone who hears this phrase, including the administrators in question, immediately think of computers. However, Microsoft PowerPoint is not the only form of technology. Chalk and a chalkboard is also technology.

Technology, as its name implies, is technical know-how. But production technology is not merely knowledge. It is knowledge bound up in physical capital goods. For a tractor to help the farmer more efficiently do his job, it is not enough for the farmer merely to know what a tractor is and what it can do. The farmer must actually possess the physical tractor and use it with his land, labor, and other capital goods. For technology to be useful in production, it must be bound up in capital goods. As technology increases, so does the amount of output that can be produced with the same amount of labor.

When contemplating technology's effect on supply, keep in mind that the observed changes in technology must be related to the *production* of the particular good being analyzed. For instance, suppose that someone invents a better machine to press Beethoven CDs. This is a technological advance that will make it possible to produce more records with the same amount of land, labor, and complementary capital goods. Consequently, the supply of Beethoven CDs will increase. On the other hand, it is doubtful that such an increase in recording production technology will have any noticeable effect on the ability of farmers in Iowa to produce soybeans.

Additionally, a change in technology that affects the quality of a good will not increase the supply of that good. It may affect its desirability, and hence the demand for it however. The availability of better, higher-speed computers will not necessarily affect the supply of computers at all. Such technological changes only affect the preferences of buyers. They do not directly affect those of the sellers. Keeping this in mind, we see that when there is an advance in technology that affects the production of a good, this will result in an increased supply of that good. If a better CD producing machine is invented, this will help producers make more Beethoven records, increasing their supply.

It is true, in a free market especially, that there is a tendency for the level of technology to increase over time. This trend, however, is in no way irreversible. A loss in knowledge is also possible. Although perhaps not as likely, it could happen. Stanley Jaki has documented in his book, *The Savior of Science*, that science, and the technology that follows, came to a halt after initial spurts of growth in the ancient Greek, Chinese, and Indian cultures. A loss of knowledge is even possible. One imagines that the burning of the famed library of Alexandria set knowledge back in Cleopatra's Egypt to a significant degree.

Suppose that there is indeed technological retrogression related to production of compact discs. This would make it more difficult to produce Beethoven CDs. Manufacturers could make fewer with the same quantity of factors of production. The stock of the recordings would fall, resulting in sellers supplying fewer. Consequently, we can identify a direct relationship between the level of technology related to the production of a good and the supply of that good.

Number of Sellers

Just as changes in population and, hence, the number of potential buyers affects demand, the number of sellers affects supply. This helps explain why different people sometimes react differently to a new store coming to town. Suppose in a small town that already has three grocery stores, another opens up. Grocery buyers tend to like this occurrence because more groceries will be supplied to them. At the same time, often the established grocers in town do not care for the addition, for the same reason. There will be an increase in the supply of groceries available, which means more competition for them.

In any event, the number of sellers is directly related to the supply of a good. If there is an increase in recording manufacturers who decide to produce recordings of the music of Beethoven, this will increase their supply. On the other hand, if some of them go out of business, or if they merely stop producing such records, there will be a decrease in supply.

Expected Future Prices

Just as the buyers' expectations of future prices effect their present demand, so do the expectation that sellers have regarding future prices affect their current supply. If, for example, producers of recordings expect that next year there will be large increase in the price of Beethoven CDs, they will be less eager to sell their stock now and more willing to wait to sell until next year when they will get the higher price. Therefore, current supply will decrease. It should be noted that for the present supply to increase, the expected future price would have to be high enough to more than offset both any storage costs that may accrue and the producers' time preference. However, on the margin, higher expected future prices will tend to increase current supply. On the other hand, if sellers expect that, in the future, the price will decline, they will be more eager to sell as many records as they can before the price drops. Consequently, supply in the present will increase.

Goods Related in Production

Just as economists have identified certain relationships between goods in consumption, they have also recognized that the production process of some goods is such that changes in the market for good A could result in direct changes in the supply of good B.

Suppose, for example, that one production process yields two goods, both of which can be sold by the producer. These two goods are called complements in production. Examples of complements in production are chicken wings and leg quarters. Back in the 1980s the taste of Buffalo wings took hold with the American public and continues to this day. For those of you who just tuned in, Buffalo wings are fried or baked chicken wings often served in a spicy sauce. As the name implies, when making Buffalo wings only the wings are needed. It would be silly to make something called Buffalo wings with a thigh or a drumstick. Buffalo thighs or Buffalo drumsticks are completely different goods. The increased demand for wings, however, changed the way chickens were marketed. On the one hand, it would seem cruel for poultry farmers to merely chop off the wings of their chickens and sell them, leaving them with a coup full of amputee birds. On the other hand, it seemed a great waste to kill an entire chicken to harvest only the wings. Meat departments in grocery stores began selling family packs of only wings, but also large packages of only leg quarters and packages of only breasts. When the price sellers could get for chicken wings increased, that encouraged them to produce more wings. It also, however, encouraged them to increase their supply of thighs, legs, and breasts.

Suppose, on the other hand, that someday the interest in chicken wings wanes and the price of wings plummets. Poultry producers will be less likely to produce wings by themselves, which will also lead to a decrease in the supply of legs, thighs, and breasts by themselves. Consequently, there is a direct relationship between the price of a complement in production and the supply of the good in question.

Another type of good related in production is the substitute. Substitutes in production are goods whose production uses the same factors. If producers use their factors to make one of the goods, they must forego the production of the other good. One can think of corn and soybeans as substitutes in production. If a farmer uses his 400 acres to plant corn, he cannot at the same time plant soybeans on that land. He must choose to plant one or the other.

Suppose the farmer thinks that the price of corn will increase. He will be more likely to plant corn and less likely to plant soybeans. Therefore, an increase in the price of corn, the substitute in production, will tend to result in a decrease in the supply of soybeans. On the other hand, if farmers think that prices for corn will decrease, they will be less likely to plant corn and more likely to plant its substitute, soybeans. Consequently, a decrease in the price of corn will tend to increase the supply of soybeans, because they are relatively more attractive to the farmer.

We can see, then, an inverse relationship between the price of one good and the supply of its substitute in production. When the price of a substitute in production increases, the supply of the other good decreases. When the price of a substitute in production decreases, the supply of the other good increases.

The cause and effect economic relationships that make up the core of supply theory are summarized in table 6.5.

Table 6.5	
Supply: Cause and Effect	
Change in...	Effect
Price of good	Change in quantity supplied
Investment in production	Change in supply
Price of factors of production	Change in supply
Technology	Change in supply
Number of sellers	Change in supply
Expected prices	Change in supply
Market for a complementary good	Change in supply
Market for a substitute good	Change in supply

A change in the price of any consumer good will result in a change in the quantity of that good supplied by sellers, given their subjective values at the time of the price change. A change in any of the other items listed will affect the relative place of marginal units of the good compared to different amounts of money and, hence, will result in a change in supply.

THE ANATOMY OF A PRICE CHANGE

Now that we have surveyed several factors that can lead to a change in both demand and supply, we can put what we have learned to work in determining how a change in the market for a particular economic good will affect its market clearing price and quantity. The process we use to determine how such changes result in different prices is a methodical six-step process. By following the six steps you should always be able to figure out how a given change in the market will affect the market price and quantity.

Suppose, for example, that you are hired by the Recording Industry Association of America (RIAA) as a market analyst to keep tabs on the market for CDs. Your superior gives you a report indicating that the incomes of consumers in the age demographic most likely to buy compact discs are increasing. You are assigned to prepare a report explaining the effect of such a change on the CD market. Carefully proceeding through the following six steps will allow you to provide the RIAA with a smashing report and perhaps win you that raise that you think you have so long deserved.

Step One: Identify What is Changing

The very first thing that you must make clear is what, exactly, is the catalyst for the market change. In our example the change is an increase in consumer income.

Step Two: Determine Whether the Change Affects Demand or Supply

After you have correctly identified what has changed, you must determine whether this change is something that affects demand or something that affects supply. From what we have seen earlier in this chapter, an increase in consumer income is a factor that affects demand.

Step Three: Determine Whether the Supply or Demand Will Increase or Decrease

Once it is determined that either supply or demand will change, it is necessary to figure out just how it will change. It has already been determined that an increase in consumer income will result in a change in demand. To determine its ultimate effect on the market price and quantity of CDs, however, one must know whether an increase in consumer income will increase or decrease demand. As was noted earlier in the chapter, an increase in consumer income is generally directly related to demand, so an increase in consumer income will increase demand.

Step Four: Shift the Appropriate Curve in the Appropriate Way

Supply and demand graphs can be very helpful in that they can help you to determine how a shift in demand or supply will affect the market price or quantity relatively quickly. One thing you will want to remember, then, is that the market graph of supply and demand is your friend. The graph will never lie to you. Spend time practicing with it. You might even want to share your new economic tool with your friends and family. If you correctly use a graph, it will always respond with the correct answer to your problem. How do you correctly use the graph? By making sure you shift the correct curve in the correct way; used correctly, the graph will reveal the answer.

We have already determined that an increase in consumer income will increase demand. To help us determine what affect an increase in demand has on the market price and quantity of CDs, you can use a supply and demand graph, but you must make sure that you use your graph to illustrate what will really happen if there is indeed an increase in consumer income. In other words, if you want to illustrate an increase in demand, you must shift the demand curve. This would be the appropriate curve to shift. It is not enough to just shift the correct curve, however. It must be shifted in the correct way. Notice that the quantity of goods increases from left to right on the horizontal axis. Therefore, when illustrating an increase in demand, you should shift the demand curve to the right as illustrated in Figure 6.8.

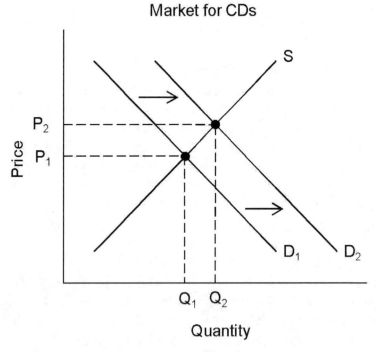

Figure 6.8. An increase in income for consumers of CDs will result in an increase in demand for CDs. The increased demand will cause the market price and quantity of CDs sold both to increase.

A decrease in demand would be illustrated by a shift of the entire demand curve to the left.

Step Five: Determine What Happens to the Market Price and Quantity

If you shift the appropriate graph in the appropriate fashion, the graph immediately will tell you what effect a particular change in the market for a good will have on that good's market price and quantity. If an increase in consumer income leads to an increase in demand as shown as a rightward shift in the demand curve, the market price for CDs will increase. Additionally, because sellers are willing to sell more at a higher price, the quantity sold will increase as well.

Step Six: Fully Think Through the Process by which the Price and Quantity Change

As promised, the graph does not lie. An increase in demand will result in an increased price and quantity sold. Remember also, however, that economics is not about what happens to an intersection point if a line moves. It is about how humans act to engage in the exchange of scarce goods in order to satisfy their ends. In other words, to be a good economist, you will need to do more than master a supply and demand graph. You must understand and have the ability to explain what actually happens—to explain what the graph represents. Economic theory gives you the principles necessary to fully explain how an increase in consumer income affects the final market price for CDs and the quantity of CDs sold.

If there is an increase in consumer income, this will lead to an increase in buyer demand. Consumers will be willing to buy more at any price. The demand for CDs will increase, causing the prevailing price to clear the market no longer. Because of the increase in demand, the quantity of CDs demanded would exceed the quantity supplied at the prevailing price. There would be excess demand. As with any case of excess demand, some buyers would be frustrated, because they would be willing to buy CDs at the prevailing price, but there would not be enough supplied. To ensure that they are the ones getting to buy, the more capable buyers will bid up the price until everyone who wants to buy can buy. This is illustrated by the intersection of the new demand curve and the supply curve. At P_2, everyone who wants to buy can buy and everyone who wants to sell can sell. At this new intersection point, the market is at rest again. The new equilibrium price is higher than the previous one.

The quantity sold will also increase, because of the positive relationship between price of a good and the quantity of that good that sellers are willing to sell. As the price is bid up, more sellers are willing to sell. Because of this increase in quantity supplied as a result in the higher price, the quantity sold in the market will increase along with the price. We are able to use economic theory to trace out the step-by-step changes that transmit higher consumer incomes into a higher price for CDs and larger volume sold.

MAKING SENSE OF PRICE CHANGES

This chapter has focused on the fact that the market, like the world, is in a state of flux. People change. Their values, circumstances, and environments all change. The recognition of such changes should keep us from being surprised when the market prices for goods and services do not remain constant over time.

The beauty of economics is that it allows us to make good sense of these changes. We have documented many factors that will affect either supply or demand. We have also identified just how particular changes in these factors will affect the market for a given good. Finally, you have become acquainted with a six-step process that will allow you to fully understand just how particular changes in relevant factors result in particular changes in the market price and the market quantity for a particular good.

This is powerful analysis. It not only allows you to make better sense of the world around you, it will also help you to understand the consequences of actions that hinder this process. Prices are not always allowed to change by the powers that be—the state. For various reasons government authorities sometimes attempt to place a legal maximum or minimum price on certain markets. The supply and demand analysis that has been explained thus will allow for determining the consequences of such price controls, an issue taken up in chapter fifteen.

SUGGESTED READING

Rothbard, *Man, Economy, and State*, 142–53. An outstanding analysis of changes in market prices.

Shapiro, *Foundations of the Market-Price System*, 116–37. In this chapter Shapiro provides a very good introduction to the nature and implications of price elasticity of demand.

7

Production, Capital, and Income Maximization

As is explained in the Introduction, one of the justifications for studying economics is that economic truth helps us to fulfill the cultural mandate in our world of aggravated scarcity, without either starving to death or killing one another. In Genesis 1:28, man is told to be fruitful and multiply, fill the earth, subdue it, and to have dominion over the other living creatures. To even begin to fulfill this mandate requires at the very least survival. One cannot exercise dominion if one is dead or dying of starvation. Obeying God in our cultural task also requires personal development and the development of the potentialities of nature.

Fulfilling the cultural mandate requires both survival and economic development, and that requires the consumption of economic goods. Survival requires that we consume food, clothing, and shelter. Human nature is such that it desires beauty, decoration, and a certain amount of comfort. We therefore desire additional goods that will accommodate such desires.

An existential fact of human existence is that we can only increase our stock of consumption goods by producing them. This in turn implies a fundamental economic principle: we produce in order to consume. We do not consume in order to produce. People engage in production, because, in so doing, they can obtain more consumer goods with which to achieve more ends.

These acts of production are never acts of new creation as if producers are God. Human production never results in new matter that previously did not exist. Production transforms matter in such a way that it is capable of satisfying a want. When a chef produces a plate of delicious buttermilk pancakes, she does not create them *ex nihilo* by wiggling her nose. She instead transforms matter that is not capable of directly satisfying a want into a good that is directly satisfying a desire to relieve hunger in a particularly satisfying way. When a customer walks into a diner wanting a tasty breakfast, he does not ask for one-fourth cup of flour, one-fourth cup of buttermilk, one-sixteenth cup of sweet milk, one-fourth raw egg, two teaspoons of butter, a half-teaspoon of sugar, one-eighth teaspoon of baking powder, and one-sixteenth teaspoon of baking soda. These do not satisfy the desire of the customer. He does not want a number of wet and dry ingredients. He wants buttermilk pancakes. The chef's act of production transforms all of the above factors into a plate of her finest. It is the final product that satisfies the desire of the breakfaster eager to consume.

Consumer goods are the result of industrious people applying various producer goods toward the end of producing the resultant products. The total quantity of consumer goods is, consequently, limited by the stock of factors of production in existence. One way people have increased the number of goods that can be produced with a given stock of factors is by taking advantage of the division of labor. A well-developed division of labor has resulted in a complex production process in which every member of society contributes a fraction of the total output. As indicated in chapter 2, in our modern economy, every consumer good has a relatively complex structure of production that incorporates many stages of production.

CAPITAL GOODS AND THE STRUCTURE OF PRODUCTION

Our complex structure of production is only made possible by the use of capital goods. As indicated earlier, capital goods are produced means of production. They are producer goods that must, themselves, be produced. Like consumer goods, capital goods also do not spontaneously generate. They cannot be wished into existence. They must be manufactured. Therefore, to have capital goods available to use requires lengthening the production process. Before a carpenter can use a hammer to shingle a house, the hammer must first be produced. Before a baker can use a Kitchen Aid mixer to whip up eggs to be used in a flourless chocolate cake, the mixer must first be produced. This is what sets capital goods apart from land and labor—the original factors of production. Without capital goods, the structure of production would be very short, incorporating only one stage: the mixing of land and labor. Such a short production structure would necessarily produce only very simple goods, such as fig leaves for clothing, food foraged and produced with bare hands, or huts made with branches found lying on the ground.

The use of capital goods is what makes possible our higher standard of living. Capital goods contribute to a higher standard of living because they increase the productivity of land and labor. Remember from the previous chapter how the development of plows, tractors, and fertilizers has increased the agricultural output of both farmland and farmers. In order to increase the output of human effort and land, it is necessary to lengthen the production process in order to first produce capital goods so that they can be used to produce more consumer goods.

As noted in chapter 2, all action takes time. The act of production is no different. Producing goods that people can then use to satisfy their ends also takes time. Because of time preference, people prefer to satisfy their ends sooner rather than later. Therefore, people prefer to produce goods using the shortest production processes possible. When making a batch of delicious buttermilk pancakes, mothers would rather make them in a half hour rather than taking half-a-day.

The shortest production processes, however, do not satisfy all our ends. In order to satisfy additional ends, we must resort to production processes that are longer. People embark on these longer production processes when the additional satisfaction that results from the additional output outweighs the cost of waiting longer to reap the results of their production. These production processes must be more physically productive in order for people to be willing to choose them voluntarily.

Longer production processes are more productive because they allow for the use of capital goods. People choose to lengthen the production process because the use of capital goods increases the productivity of land and labor, and does so in two ways. First, using capital goods allows us to produce more of those goods we could produce without the use of any capital goods. Second, using capital goods enables us to produce goods that we otherwise could not obtain at all. We will look at each way in turn.

In order to see that the availability of capital goods allows us to produce more goods than we could without them, a very simple example will again be used to discover some profound economic principles. Suppose our old friend Groucho finds himself on a desert island that is inhabited by a number of mango trees. He quickly realizes that he can successfully harvest mangoes and that they will be a staple of his diet. Fortunately for him he likes mangoes and sets out picking them. Further suppose that by applying his own effort—his labor—to the mango trees given by nature, he can pick 20 mangoes per hour. During a ten hour work day he can harvest two hundred mangoes.

Suppose further that he is really taken by the taste of the mango and would like nothing better than to increase the quantity he could pick in a day. Unfortunately, he realizes he is working absolutely as hard as he can, and the bounty of the mango trees he has picked so far are representative of all mango trees on the island. He wracks his brain trying to think of a way to increase his output. Is there anything he can do or try to more quickly bring in the mangoes? Then it occurs to him. If only he could get a stick. Then he could use it to whack the mango trees, shaking them so that the ripe fruit will quickly fall to the ground. He could then simply pick the mangoes off the ground, thereby increasing the quantity of fruit he could produce for consumption every day. After thinking a bit, Groucho estimates that, with the stick, he could increase his mango production to fifty mangoes per hour, boosting his output to five hundred every day. Groucho's productivity with and without the use of a stick is summarized in Table 7.1.

Table 7.1

Groucho's Productivity Harvesting Mangoes		
	Without Capital Good	With Capital Good
Mangoes per Hour:	20	50
Mangoes per Day:	200	500

Note, however, that in order to increase his mango harvesting, neither branches nor sticks in general are helpful to Groucho. What he must possess in order to increase his productivity is not just any stick, but a stick suitable for mango production. Some sticks are too large. Some too small. Some are too crooked. Some are too rough for his hands. Branches that still contain a plethora of smaller branches and leaves are simply too unwieldy and have too much wind resistance for much whacking capability.

In order to have a capital good suitable for mango production, Groucho must find a branch given freely by nature and use his labor to transform it into a capital good, a stick suitable for mango harvesting. He must find a branch of usable size and prune off all of the leaves and excess smaller branches. He may also want to smooth it so as to prevent him getting slivers as he beats the trees for the mangoes. This effort of producing a stick requires the lengthening of the production structure, because, in order to use a stick to pick mangoes he can consume, he must first produce the stick.

This act of producing the stick, like all actions, takes time. Suppose that in order to obtain a stick suitable for harvesting mangoes, it takes him ten hours to find and shape an appropriate stick. In other words, in order to produce a capital good, Groucho must forego ten hours of picking mangoes by hand, ten hours of leisure, or some combination of both, and devote those ten hours toward producing the stick. He can use a stick to increase future production and consumption of mangoes only by sacrificing consumption in the present. Sacrificing present consumption is what economists call *saving*.

All acts of saving can be divided into two categories. On the one hand are actions we can call plain saving. Plain saving is the act of storing up consumer goods in the present so that they can be consumed in the future. People engage in this sort of saving for any number of reasons. They may withhold goods from consumption for a rainy day in order to provide for consumption in case of emergencies. They might do it because, for whatever reason, they simply prefer to consume a particular set of goods on a particular day in the future rather than now.

A college student who receives a case of Martinelli's Sparkling Apple Cider from his parents on Monday may decide to wait until the weekend to drink any so that he can enjoy it with the girl he is courting. As he refrains from drinking the Martinelli's, he is engaging in plain saving. For him to be willing to save in this manner, he must value drinking Martinelli's with his girlfriend in five days more highly than drinking the cider by himself the day he received it. In essence the case of Martinelli's that he refrains from drinking in the present is changed from the category of consumer good and transferred into that of capital good. This capital good requires only waiting time to be changed back into a more highly valued consumer good, cider drunk with one's intended object of affection.

The other type of saving that more directly affects production is capitalist saving. Capitalist saving affects the production process by restricting consumption in order to lengthen the production structure to include capital goods. This is what we find Groucho doing. Suppose that, in order to produce a stick, he forgoes ten hours of picking mangoes. He restricts present picking and consumption of mangoes, so that he can lengthen the production process enough to obtain a stick useful for mango picking. Groucho transfers the labor time that he could use to pick mangoes by hand toward the accumulation of another producers' good—the stick. Economists refer to the transfer of saved resources to the accumulation of factors of production as *investment*.

In order to have use of capital goods, consequently, it is necessary to engage in investment in the production of capital goods. Such investment requires saving. If resources are consumed, those same resources cannot be used for the formation of capital goods. If Groucho spends all of his time and labor picking mangoes and then consuming them, he cannot obtain a stick. He can only do so if he saves. What is true for Groucho and his stick is true for humans in general. Saving is necessary for investment.

This is not to say that the saver and the accumulator of capital goods are always the same people. You can most likely think of examples of successful producers who are able to purchase capital goods, rent land, and hire laborers without saving up all of the money that they spend. How do they come up with the money? They borrow. They either borrow from banks or sell bonds. A bond is similar to a home mortgage in that it is a financial instrument entitling the holder to regular interest payments throughout the term of the bond plus the entire principle when the bond matures. Some might think that these examples run counter to our principle that investment requires saving. However, before jumping to that conclusion, pause for a moment and think about where the lenders get their money to lend. This must be money that the lenders do not spend on consumer goods. In other words, in order to lend, the creditors must first save. So even if those investing in obtaining factors of production do not spend their own money, but borrow it instead, *someone* must save in order to have money to lend to them. Therefore, from the perspective of the economy, *someone* must save if investment is to occur.

The use of capital goods not only allows for increased production of those goods that we can produce without capital goods. More importantly, using capital goods enables us to produce goods that we otherwise could not obtain at all. Many goods simply cannot be produced without the use of capital goods. If Groucho, for example, gets tired of a diet of only mangoes and decides he wants to add a bit of protein to his meals, he may want to eat some fish along with his fruit. In order to catch the fish, he must have a pole, hook, and a fishnet, or perhaps even a bow and arrow. None of these tools can be found in nature. All of these instruments must first be produced. In order to produce a fish dinner, then, Groucho must have capital goods.

Most of the goods that we have today would not exist without capital goods. Computer manufacturers need circuit boards, plastic, screens, and production facilities. Automobile makers need windshields, tires, metal parts, welders, electronic units and factories, along with land and labor. Compact disc producers need polycarbonate plastic, recording devices, aluminum, and a physical plant. None of the producer goods listed are available in nature. They must be produced. In order to make computers, automobiles, or compact discs at all, we must have use of capital goods.

Using capital goods advances people in time toward their objective of producing consumer goods. However, in order to use capital goods, they must first be produced, and production of capital goods requires a lengthening of the production process. Think about Groucho and his mangoes. If Groucho immediately sets out to pick mangoes by hand, he can pick and consume 200 mangoes the very first day and another on the second day. If, however, he had a stick, he could pick and consume 500 mangoes in one day. In order to produce a stick, however, he must refrain from picking and consuming a day's worth of mangoes. In deciding whether to produce the stick, he must compare the value of 200 mangoes consumed today with the value of 500 mangoes consumed one day in the future.

TIME PREFERENCE AND CAPITAL FORMATION

Capital goods are great things, because by increasing the productivity of our land and labor, they allow us to achieve our ends more quickly than we could without them. However, you may have noticed that people do not spend all of their time and resources producing only capital goods. They do not even spend all of their time producing period. If it is true that using capital goods allows people to reach their objectives in shorter periods of time, what keeps people from investing more and more resources in the production of capital goods? Why do they ever stop investing in the production of capital goods? The answer is related to their time preference for present goods. The production of capital goods takes time. The use of capital goods advances one in time because they increase our productivity. Having capital goods increases the output of consumer goods in the future, which is weighed against the loss of consumer goods in the present.

All production is the result of human action. Consequently, all general principles that apply to human action necessarily apply to production. Like all action, production takes place in the flux of time. Therefore, when deciding whether to save and invest in the production of capital goods, people not only consider the degree that they desire the good, but also think in terms of sooner and later. A person deciding between consuming mangoes and producing a stick to help in the future production of mangoes is not only choosing between two different goods at the same time. That person is choosing between a certain quantity of mangoes in the present and a different quantity of mangoes consumable in the future. The issue of time again enters the decision making process.

As already mentioned in chapter 2, people prefer to have their ends met sooner rather than later. Why is that? Because of the universal existential fact of life called time preference. Time preference is defined as everyone's preference for goods in the present above those same goods in the future. Positive time preference is a fundamental preference implied in all action. For example, even if Groucho recognizes that by restricting consumption in the present he can eat more mangoes in the future, he will not forego consumption forever. He does have to eat mangoes in the present to live. He does have to consume clothing and shelter. In order to even have a future, people must have their biological survival needs met today.

The category of positive time preference, however, applies not only to the realm of physical survival. It applies to all action. Anyone who acts to achieve any end could have waited to achieve that end at some point in the future. The fact that people act at all implies that they want the end served by that action satisfied sooner rather than later. This explains why a person who has the opportunity to earn $1,020 a year from now by investing $1,000 now might instead choose to hold onto his $1,000. He values $1,000 in the present more than $1,020 a year from now. If people did not have positive time preference, no one would forego a positive future return no matter how low.

Another factor people take into account when deciding to save and invest in the production of certain capital goods is the capital good's duration of serviceableness. In the early part of this text we considered the wisdom of Mick Jagger as it applied to scarcity. "You can't always get what you want." Alas, Mr. Jagger's wisdom regarding pithy statements related to economics seems to end there, for in terms of the durability of capital goods, time is definitely not on our side. Time preference is not only an issue regarding investment in the accumulation of capital goods because it takes savings to produce such goods initially. Positive time preference concerns us also because of the fact that capital goods do not last forever. All capital goods are perishable. They wear out and at some point they become unusable, which means that if we want to continue to use capital goods indefinitely into the future, we need more saving and investment.

The duration of serviceableness is not the same for all goods. Every particular capital good has its own unique period of usefulness. Wheat used in making bread flour, for instance, is a capital good that has a very short duration of serviceableness. It can be used once and only once. Its usefulness is over as soon as it enters the grinder. It cannot be preserved once it is ground into flour. On the other hand, the delivery truck that transports bread to grocery stores has a relatively long duration of serviceableness. It may last for perhaps 4,500 routes over fifteen years. Bread trucks are more durable than the wheat used to make the flour used to make the bread.

To see the importance of the limit on a capital good's duration of serviceableness, let us take another look at our friend Groucho, the mango man. Remember that by fashioning his capital good—the stick—he was able to increase his daily mango production to five hundred mangoes a day. Unfortunately for Groucho, his stick will not last forever. Every time he smacks a mango tree, a small, perhaps so small it is unnoticeable, bit of the stick is worn away. Suppose that Groucho can use his stick for ten days until it completely disintegrates into toothpicks. After it wears out, he is back where he started. As indicated in table 7.2, Groucho's rate of mango production would fall from five hundred a day back to two hundred a day.

Table 7.2

	Capital Stock and Productivity							
	Without Capital Goods		Consuming Capital		Maintaining Capital		Accumulating Capital	
Day	Sticks	Mangos per Day	Sticks	Mangos per Day	Sticks	Mangos per Day	Sticks	Mangos per Day
1	0	200	1	500	1	450	1	400
2	0	200	1	500	1	450	1	400
3	0	200	1	500	1	450	1	400
4	0	200	1	500	1	450	1	400
5	0	200	1	500	1	450	1	400
6	0	200	1	500	1	450	1	400
7	0	200	1	500	1	450	1	400
8	0	200	1	500	1	450	1	400
9	0	200	1	500	1	450	1	400
10	0	200	1	500	1	450	1	400
11	0	200	0	200	1	450	2	700
12	0	200	0	200	1	450	2	700
13	0	200	0	200	1	450	2	700
14	0	200	0	200	1	450	2	700
15	0	200	0	200	1	450	2	700

If he wants to continue to harvest more mangoes every day, he must get another stick.

If he wants to maintain his standard of living, he must, during the first ten days he has use of the first stick, work at producing another stick. What does this require, however? Well, what did Groucho have to do in order to obtain his first stick? He had to restrict his consumption and direct his time and labor toward the production of the stick. He must do the same in order to obtain a second stick. In other words, maintaining his standard of living requires further saving and investment. If he chooses not to make another stick during the life span of the first, Groucho is consuming capital. A person consumes his capital when he is not saving and investing enough to replace his capital goods after they wear out. If Groucho does not save and invest at all from Day One through Day Ten, his stock of capital goods decreases from one stick to none. His first stick has been used up. It has been consumed and it has not been replaced.

If Groucho does recognize the need for further savings and investment if he wants to maintain his standard of living past Day Ten, he will set out to produce another stick. Remember that making a stick requires ten hours of work. Suppose he spreads out his work on making a second stick throughout the life of the first stick. He could forego an hour of his leisure time every day a devote it to stick making. This would be an act of saving in that he is restricting his consumption of leisure. Suppose, however, that every day he decides to devote one hour he could use picking mangoes toward the production of another stick. As seen in table 7.2, Groucho's daily output of mangoes falls to 450 per day because he foregoes the 50 he could pick during the hour he is now spending making another stick. At first we might be tempted to think, what a bum deal. He produced his first stick so that he could eat more mangoes and now we want him to do with less.

Remember, however the purpose of the additional saving and investment. Even devoting an hour a day to stick production, he can, with the first stick harvest 450 mangoes a day, which is a lot more than the two hundred he could pick by himself. More importantly, look at the row for Day Eleven in table 7.2. By foregoing fifty mangoes every day for the first ten days, he is able to maintain his rate of daily mangoes on Day Eleven. At the end of Day Ten he lays down his first stick as it has worn out, but without missing a beat, he picks up his newly finished stick and continues right on as if nothing has changed. From Day One through Day Ten his stock of capital goods was equal to one stick. On Day Eleven it equals . . . one stick. It has remained the same. When a person's stock of capital goods remains the same over time this is referred to as *maintaining capital*. If a person saves and invests just enough so that his capital stock does not decline from one period to the next, he is maintaining his capital. Maintaining capital results in the maintaining of standards of living.

If Groucho is not willing to sacrifice the consumption of fifty mangoes a day during the first ten days, he will, on Day Eleven be left picking mangoes by hand to the tune of two hundred a day. In this case, we see that, because Groucho has not saved during days one through ten, his capital stock declines over time. This decrease in his capital stock over time is called *consuming capital*. It is referred to as capital consumption for two reasons. First, because of depreciation the capital is actually being used up. Secondly, the stock of capital goods decreases because of too much consumption and not enough saving on the part of Groucho. Over time the consumption of capital leads to decidedly lower standards of living.

What if Groucho decides he really likes mangoes and thinks that, while 450 a day is good, even more would be better? If there was only a way he could increase his production even more. His synapses fire on a champion idea. His mango production increased with the use of one stick. If he could use two sticks it should increase even more. Instead of whacking the trees with only one stick, he could take one in each hand and really have at it, beating them about the leaf and branch. Of course, these sticks do not grow on trees. He must take branches that do grow on trees and make them into sticks suitable for mango picking.

If he wants the use of two sticks, he must increase the labor time he takes away from mango picking and consuming and direct it toward stick production. Instead of spending only an hour a day making mango harvesting sticks, he must now spend two. As table 7.2 indicates, in order to take advantage of the use of two sticks on Day Eleven, Groucho will have to work on a new stick for two hours per day during the first ten days, and will only be able to pick and eat four hundred mangoes per day during the life of the first stick. However, on the eleventh day, Groucho can pick up, not one, but two sticks. Increasing one's stock of capital goods is called *accumulating capital.*

Accumulating capital goods allows people to increase their productivity from one period to the next. On Day Eleven Groucho is able not to pick only 450 mangoes, but is instead able to pick seven hundred a day. The availability of two sticks allows him to produce even more than he could with only one stick. As the result of increasing his stock of capital goods from one stick during the first ten days to two sticks from Day Eleven and beyond, Groucho finds his standard of living increasing. The number of mangoes he can eat a day increases. Of course, the quantity of mangoes he could trade away for other goods he also wants increases as well.

What can we learn from Groucho's example? Note that both accumulating and maintaining capital requires saving and investment. Consuming capital requires that good men do nothing. If people do not save, they will consume their capital. In fact, even if they save, they can still consume their capital if they do not save and invest at a high enough rate to replace their capital stock as it wears out. If, for example, Groucho devotes an hour every other day to the production of a stick, he would be saving and investing some from Day One through Day Ten, but not enough to have a stick on Day Eleven. Even though he saved some, he did not do enough to maintain his capital.

The decision whether to save and invest for the accumulation or maintenance of capital or to consume capital depends on how much the actor values present goods over future goods. Groucho, for instance, must decide whether to restrict mango consumption in the present in order to increase mango consumption in the future. His decision depends on two things.

The first is the value of the good being produced. A stick is useful to Groucho for the production of mangoes. He will be more likely to save and invest in the production of a stick if he values mangoes. If, at some point, he gets tired of eating mangoes for breakfast, lunch, and supper, his demand for mangoes will decidedly wane. If he is sick of mangoes, he will not be overly concerned with how to produce more of them. He will be less likely to save and invest in a production process that will allow him to increase future production of a fruit that he finds distasteful.

Additionally, the decision whether to save and invest is affected, not surprisingly, by his time preference. We can see how Groucho's time preference affects his action by looking at the hypothetical value scales below:

Case One: Higher Time Preference	Case Two: Lower Time Preference
500 Mangoes in Present	500 Mangoes in Present
200 Mangoes in Present	(500 Mangoes in 2 Days)
(500 Mangoes in 2 Days)	200 Mangoes in Present

In both of the hypothetical cases, Groucho values more mangoes in the present to fewer mangoes in the present, because with more mangoes he can satisfy more ends.

Whether Groucho will save and invest in future mango production is not determined by the fact that he values present goods more highly than future goods. Everyone does. Instead, his choice is determined by *how much* he values present goods over future goods. Everyone has positive time preference, but the intensity of time preference varies among people and can vary for the same person over time. Comparing Groucho's value scales in Case One and Case Two, we see that in Case One Groucho's value scale illustrates a higher rate of time preference compared to that in Case Two. A higher rate of time preference means that the actor values present goods over future goods more intensely. Groucho's value scale in Case One indicates that he values five hundred mangoes in the present more than two hundred mangoes in the present. No surprise there. Additionally, Groucho values five hundred mangoes in the present more highly than five hundred mangoes two days from now. Again this is what positive time preference leads us to expect.

When contemplating whether to save and invest in stick production, however, Groucho's trade off is not merely between more mangoes and fewer mangoes. It is not even between five hundred mangoes now and the same quantity in two days. His trade off is between two hundred mangoes now and five hundred in two days. Remember that people act in the present to reach an end in the future. In order to be able to harvest five hundred mangoes in two days, Groucho must decide to save and invest in the present. In other words, Groucho must in the present form an opinion of how he values the five hundred mangoes two days in the future. Because he must make this judgment in the present, his rank of the five hundred mangoes two days from now is called his present value of a future good. The good in question, five hundred mangoes two days from now is a future good, because he will not get it until the future. However, Groucho evaluates this future good in the present as he compares it with the quantity of mangoes he could harvest without further saving and investment.

Groucho's value scale in Case One is such that he values two hundred mangoes in the present more highly than five hundred in two days. This means that if he has to choose between consuming two hundred mangoes in the present versus consuming five hundred in two days, he will choose to consume the two hundred now. Groucho's value scale reveals that in Case One Groucho would not save and invest in the production of a stick. He would continue to pick mangoes by hand, reaping two hundred a day.

If Groucho's value scale is that depicted in Case Two, we see that Groucho has a lower rate of time preference compared to Case One. In Case Two, Groucho values five hundred mangoes two days from now more highly than two hundred mangoes in the present. This means that Groucho would be willing to forgo two hundred mangoes in the present so that he could consume five hundred mangoes in two days. In both cases, we see that Groucho's saving and investment decision is decided by whether the present value of the future good or the present value of the present good foregone is greater.

Comparing the investment decisions made by Groucho in Case One and Case Two, we arrive at another general economic principle. There tends to be an inverse relationship between a person's time preference and his saving and investment. If a person has a relatively high rate of time preference, he is more present oriented. He is, consequently less likely to save and invest, and less likely to maintain and accumulate capital. Conversely, if that person has a relatively low rate of time preference, he is more willing to put off the gratification of present consumption. Therefore, the lower the rate of time preference, the more likely a person will save and invest in the accumulation of capital.

It is important to recognize that even though the principles regarding time preference, saving, and investment were derived from the simplest of examples, the same principles apply to the complex modern economy. For example, suppose you read in the *Wall Street Journal* that the Spanish energy company De Falla offered to buy a rival company for $17 billion. Or perhaps you read in the *Financial Times* that Acme Microprocessors invested $200 million in five companies producing wireless internet technology. Maybe you read in *The Economist* magazine a story reporting that Come Fly With Me Business Jets expanded its line of airplanes to include more price ranges. Where did these firms get the money to buy natural gas operations, invest in telecommunications technology, and buy more jets? From the same place Groucho obtained the resources used to produce a stick: saving. Either they had to save money out of their profits for increased investment or they had access to the savings of others by selling shares of ownership in their company, called stocks, or by borrowing. Either way, for such investment to take place, saving was necessary.

This general principle also provides another clue regarding economic development. It has already been shown in chapter 4 that, as the division of labor develops, people tend to become more productive, resulting in economic expansion. However, the division of labor is not the only thing that determines actual economic development. The division of labor will only take us so far. The use of capital goods also increases the productivity of labor and land. As indicated above, in order to use capital goods, they must be produced as the result of saving and investment. Societies made up of people that have generally higher time preferences will tend to consume more and save less. Over time their stock of capital goods will shrink. Labor will become less productive over time, resulting in lower incomes, lower standards of living, and economic decline. On the other hand, societies made up of people that have generally lower time preferences will tend to save and invest more. Over time their stock of capital goods will increase resulting in increased productivity, higher incomes, and greater economic prosperity.

A NOTE ON ECONOMIC FUNCTIONS

The person who saves income and invests in the formation of capital is, reasonably enough, called a capitalist. As a capitalist, the saver can invest in his own production process in hopes of a positive future return, or he can loan money to other producers in return for an interest payment. Interest is a monetary payment to capitalists for giving up the use of their money in the present while it is either loaned out or paid to owners of factors until the capitalists receive it back in the future. In order to willingly part with their money in the present, in addition to the return of their initial investment, capitalists require an interest payment due to their positive time preference. We will discuss the rate of interest in more depth later in the text.

The person who actually drives the production process is called the entrepreneur. Entrepreneur is a French word that literally means *one who undertakes*. In other words, the entrepreneur is an undertaker. Because, perhaps, that word has certain negative connotations related to mortality, economics has retained the use of the French. The entrepreneur is the one who undertakes production by obtaining the use of the services of factors of production and combines them to produce a final product that they can sell for income. As stated earlier, all production takes time, so, compared to the act of production, the final product will be finished and can be sold only at some point of time in the future. One thing we know for sure about the future is that it is uncertain. There is no guarantee that once the final product is completed, the entrepreneur can find a buyer. To be successful, then, the entrepreneur must correctly anticipate future market conditions. The primary function of the entrepreneur, then, is forecasting the future in the face of uncertainty.

It should be noted that when economists use the terms capitalist and entrepreneur, they are speaking more about economic functions and less about particular people. This is because we live life as whole persons. We necessarily perform several economic functions as we seek to fulfill the dominion mandate God has given us. No person, with the possible exception of a child or invalid, is only a consumer. In order to consume, people must produce. They must either make their own consumer goods, or they must produce goods that can be traded for money that can be spent on consumer goods.

Likewise, the functions of the capitalist and the entrepreneur are often performed by the same person. In the first place, all action takes time, with the end being reaped at some point of time in the future. Every action—not only production—is speculative. Diners who go to their favorite Italian restaurant in order to eat their favorite fettuccini alfredo, for example, may find themselves unexpectedly disappointed with the restaurant if the chef has a bad night, was forced to work with inferior fettuccini, cream, butter, or parmesan cheese, or if the owners have changed chefs completely.

Additionally, all capitalist saving and investment necessarily has an entrepreneurial aspect. If a capitalist saves income and lends it to a producer, there is no guarantee that he will get his money back. The borrower might use it to make a good that no one wants to buy, in which case the producer will end up bankrupt, unable to pay his debts, and the capitalist will lose his investment. A capitalist must, consequently, make a speculative judgment regarding which borrowers will be successful in the use of their loans, so that they will be able to pay back the capitalist in the future.

Likewise, laborers, those who contribute their human effort to the production process, also have entrepreneurial considerations. Not every type of labor is equally valuable in the market. Different stocks of capital goods and natural resources and different consumer preferences will lead to different types of labor being more valuable in different locations and times. When a college student chooses a major to specialize in, with the aim of obtaining a better career after graduation, that student must make an entrepreneurial decision. Those who enter nursing programs, for instance, do so, in part, because they expect to find gainful employment in the nursing profession after they successfully complete their degree. It is always possible that, because of a decrease in demand for nursing (however unlikely that may seem), or a large increase in nursing graduates, registered nurses will not be able to find jobs in nursing that pay a wage they find acceptable.

Just as every action includes an entrepreneurial aspect, the task of the entrepreneur, successfully forecasting the final products that will be in demand in the future and undertaking to produce them, is often mixed with capitalist considerations. Many entrepreneurs rent land, hire workers, and buy capital goods with money that they have saved from their own income. In this case the entrepreneur and the capitalist is the same person. Even if the entrepreneur borrows all of the money he spends on factors of production, however, his success requires that he at least maintain the capital placed at his disposal. He must invest it in the line of production that will result in financial solvency. Again, he must play, in part, the role of the capitalist. Additionally, many entrepreneurs apply their own labor to their production activity, so it is possible that a capitalist-entrepreneur is also a laborer.

We see then, that although we are able to economically separate the different functions performed in the production process, they are often integrated in the same person. Regarding the terms, however, *entrepreneur* refers to the person who acts based on his forecast of an uncertain future. The *capitalist* is the person who saves and invests in order to receive interest in exchange for doing without a portion of his money for a specific period of time. The *laborer* is one who exchanges his labor for compensation.

PRODUCTION AND MAXIMIZING INCOME

Notice from the above taxonomy of economic functions that in a free market there is only one ultimate source of income: productive activity. In a market economy income is not distributed like differently sized slices of a giant income pie. Income is earned by supplying productive services. *Productive services* are services for which people are willing to pay. I may be very skilled in standing on my head while whistling "Dixie," but if people will not voluntarily pay me money to do so, I cannot earn any income from that talent.

© United Feature Syndicate, Inc.

The different economic functions described above were different productive tasks in which people engage in an effort to reap income. Most people earn income by supplying productive labor. Landowners supply the productive services of their land. Capitalists supply the productive services of present money in exchange for future money. Entrepreneurs supply the productive service of undertaking production to satisfy an uncertain future demand. To the extent that these actions are truly productive, they will generate a stream of income for the suppliers of the services.

People try to achieve as many of their ends as possible. The more ends they can fulfill, the more satisfied they will be. The goal of the producer is to make a product that can be sold in the future for monetary income. Given their value scales, producers try to obtain the highest monetary income possible.

Increasing their monetary income simply enables people to increase their ability to achieve their ends. Higher income enables people to expand their acquisition of exchangeable goods. The more income people receive, the more housing, automobiles, clothing, and food they can buy for their families. They can build a larger family library and purchase more music recordings. They can afford to take more and longer vacation trips. This does not mean that all producers care only about money. They merely recognize that, as the Preacher put it, "money answers all things" (Eccl. 10:19).

Additionally, higher monetary incomes allow people to engage in more charitable giving. Increased incomes will result, other things equal, in higher church tithes. It will make it more possible to donate to educational institutes, classical music radio stations as well as those that specialize in Christian programming. It enables people to give more to arts organizations and associations devoted to curing particular diseases.

INCOME ALLOCATION

There are three categories into which people can allocate their monetary incomes. They can spend it on consumption, meaning that they use it to buy various consumer goods, such as housing, transportation, food, entertainment, clothing, and other household items.

Alternately, they can save their income and spend it on investment. Saving is the restriction of consumption for a purpose to be attained in the future. Therefore, saved income is always directed toward investment. This investment could take the form of purchasing the use of factors of production, or it could be in the form of a loan directly to an entrepreneur. Saved income could also be loaned indirectly through the banking system. If you have saved some of your income and deposited it in a certificate of deposit, you are an investor. You are also a capitalist.

In addition to spending income on consumption and saving and investment, people can also add to or subtract from their cash balances. A person's cash balance is the amount of money he has on hand to make purchases as they come up. People hold money because among goods, money is uniquely suitable to adjusting action to the uncertainties of the future. If everyone knew exactly when and where their purchases were going to occur, they would not need to hold cash except at the precise moment when the purchase was made, and then not for very long. Experience teaches us, however, that people often carry varying amounts of money in their purse or wallet for extended periods of time. This is money that they could deposit in a savings account where it would earn interest. They hold the money because they know that they might want to make an unexpected purchase. Perhaps while walking through the campus snack bar a student is hit with a sudden urge to buy an ice cream sundae. If he did not plan to make this purchase and did not have any cash on hand, he would be left with a belly empty of ice cream.

During my years in college, my father once drove three friends and me back to campus after Thanksgiving break. As we were pulling into a truck stop to get gasoline, our car's water pump broke. Dad did not have a credit card or even enough cash to pay for the repair. The five of us had to pool our money together to afford the repair that allowed us to continue our trip. After he returned home, my father sent us all the money we had contributed. However, he did more than that. He began carrying a bill in a large denomination in the back of his wallet so that if a similar unexpected situation arose, he would be able to make the required purchase. Dad chose to hold more money because of the uncertainty of the future.

How people allocate income affects the production process. We can gain a better understanding of income allocation and its relationship to the production process by looking at an example. Take the case of industrious student Frank N. Furter. Toward the end of his junior year at State College, he had a vision. He pictured himself making the world a better place through hot dogs. He would serve hungry people in need of restoring themselves during the midday meal. In return, they would fund most of his senior year tuition by purchasing the best hot dogs in town during the summer. To accomplish his lofty goal, Frank would have to obtain the services of a number of factors of production and put them to work throughout the summer. This, of course, required funds that he could invest. In turn, this meant that he had to save. Fortunately, as indicated by Furter's income and expenditure statement, in table 7.3, he did not spend all of his income on consumption.

His available income came from two sources. He spent the summer before his junior year painting houses. At the end of the summer he had the handsome sum of $4,900. Additionally, he made $1,100 as a writing tutor at his university during his junior year. His total income for the year added up to $6,000.

While Furter did not spend all of his income on consumption, he did spend quite a bit. He had to pay for housing, food, clothes, books, a bit of Bach, Beethoven, and Sibelius, plus occasional social outings with his sweetheart. Over the course of the school year, these expenditures added up to $5,000. He further decided he wanted to keep $335 on hand in the form of cash.

This left him with $665 to invest in his hot dog business. After researching what it takes to make it in the rough and tumble world of wieners, he decided he would need a mobile hot dog cart, a place to set it up, a person to help him during the busy lunch time, propane gas to heat the cookers and warmers, and hot dogs themselves along with buns, and condiments. An inventory of weekly expenses on these items can be seen in table 7.3.

Table 7.3

Frank N. Furter's Income and Expenditure Statement				
Income		*Expenses*		
Painting Houses	$4,900	Consumption		$5,000
Writing Tutor	$1,100	Addition to Cash		$ 335
Total	$6,000	Cart	$250	
		Labor	$125	
		Rent	$ 25	
		Energy	$ 25	
		HD, B, & C	$240	
		Investment		$ 665
		Total		$6,000

After checking with several dealers, he was able to rent a used hot dog cart for $250 a week. He hired a high school student to help him for $125 per week ($25 per work day). He knew the owner of a local grocery store in a high traffic area of town, who let Furter set up his cart in the parking lot of the store for rent of $25 per week. It took $25 worth of propane to run the hot dog cookers and warmers every week. This left him $240 to spend on the hot dogs themselves, hot dog buns, and condiments, such as ketchup, mustard, cheese, onions, pickle relish, and chili. Furter calculated that for every hot dog he sold, it would cost him 60¢ for the hot dogs, bun, and toppings. His $240 could buy him inputs allowing him to make and sell 400 hot dogs in the first week. Frank N. Furter had an income of $6,000. He saved $665 and invested it in factors necessary to make and sell hot dogs to consumers for one week.

What makes saving and investing in production somewhat interesting is that all of the above described expenditures must be committed before Frank sells a single hot dog. Furter must invest in the present in order to receive income in the future. Once his investment is made, it cannot be undone. His investment is sunk. Even if nobody comes, for instance, he cannot take his hot dog cart back to the dealer and tell him that, even though he did have it for a week, no one bought any hot dogs so he should not have to pay rent. He cannot take four hundred cooked hot dogs back to his meat supplier just because they did not sell. Making the investment in advance of his sales is the risk that Furter must take.

Ultimately, then, Frank must forecast the future demand for his hot dogs in order to decide whether he should really enter the hot dog business. It is this demand estimation that led Furter to buy enough factors to make 400 hot dogs to begin with. Suppose that Frank figures that the demand for his hot dogs is as indicated in Table 7.4.

Table 7.4

Demand for Hot Dogs		
Price	Quantity	Total Revenue
$3.50	0	$ 000.00
$3.25	100	$ 325.00
$3.00	200	$ 600.00
$2.75	325	$ 893.75
$2.50	400	$1,000.00
$2.25	440	$ 990.00
$2.00	480	$ 960.00
$1.75	510	$ 892.50
$1.50	530	$ 795.00
$1.25	550	$ 687.50
$1.00	560	$ 560.00

From the demand schedule, Furter can also estimate the total revenue he could expect to make at each hypothetical price he could charge for each hot dog. For example, if consumer demand is as estimated, charging $3.50 for each hot dog will result in no sales, so his total revenue for the week would of course, be nothing. He would not take in a dime. Not a very good way to go about making money for college. If, however, he would charge a lower price, more people would purchase his product and increase his revenue. At a price of $3.00, for instance, two hundred hot dogs would be purchased and Furter's total revenue would then be $600. By taking the price and multiplying it by the quantity demanded at that price, Frank can calculate the total revenue he would bring in at each price.

As discussed earlier, Furter's goal is to make money to help him cover his college expenses. The higher his net income is at the end of the summer, the more satisfied he will be, because more of his college costs will be paid. Therefore, Frank will try to charge the price for his hot dogs that allow him to sell the quantity that will result in the highest net income. To do this, it is not enough to estimate his total revenue. He must also consider his expenditures. It is possible for a producer to reap high revenues and still go broke because his expenditures are even higher. Therefore, Frank must estimate both his total expenditures and total revenue he would reap at the different hypothetical prices he could charge for each hot dog. Furter's expenditures have been tallied along with his revenues in table 7.5.

Table 7.5

| | | | | | Total | |
| | | Total | Fixed | | Expend- | Net |
Price	Qty.	Revenue	Costs	Hot Dogs	iture	Revenue
$3.50	0	$ 000.00	$425.00	$000.00	$425.00	$(425.00)
$3.25	100	$ 325.00	$425.00	$ 60.00	$485.00	$(160.00)
$3.00	200	$ 600.00	$425.00	$ 120.00	$545.00	$ 55.00
$2.75	325	$ 893.75	$425.00	$195.00	$620.00	$ 273.75
$2.50	400	$1,000.00	$425.00	$240.00	$665.00	$ 335.00
$2.25	440	$ 990.00	$425.00	$264.00	$689.00	$ 301.00
$2.00	480	$ 960.00	$425.00	$288.00	$713.00	$ 247.00
$1.75	510	$ 892.50	$425.00	$306.00	$731.00	$ 161.50
$1.50	530	$ 795.00	$425.00	$318.00	$743.00	$ 52.00
$1.25	550	$ 687.50	$425.00	$330.00	$755.00	$ (67.50)
$1.00	560	$ 560.00	$425.00	$336.00	$761.00	$ (201.00)

Hot Dog Demand, Revenue, and Expenditures

Frank can divide his expenditures into two sources. On the one hand is the money he will spend that is unrelated to the precise number of hot dogs he makes and sells. To sell any at all he must commit money to pay for rent of his cart and space, labor, and his energy. The sum of these investments must be made just to sell one hot dog, and it will be the same if he sells one thousand. However, the sum estimated to cover the hot dogs, buns, and condiments will, of course, depend on how many he makes and sells. It costs him 60¢ for each dog. Combining the two sources of expenditures, Frank can calculate a column listing his total expenditure for each quantity he sells. To sell 440 hot dogs, for example, it would cost him $425 for the cart, site, labor, and energy, plus $264 for the hot dogs themselves, for a total expenditure of $689.00.

Furter's net revenue is equal to his total revenue minus his total expenditure. Once he has calculated his total revenue from his estimated demand schedule and he has calculated his total expenditure, he can calculate the net revenue he expects to reap for the week at every hypothetical price he could charge for his hot dogs. Given the market prices of the factors of production that he must pay and his expectations regarding the demand for his hot dogs, the best price that Frank could charge is $2.50 each. At this price, he would sell four hundred hot dogs, bringing in $1,000 in total revenue. He would have to spend a total of $665 to produce the four hundred dogs. Investing in the production and sale of four hundred hot dogs would generate net revenue of $335.00.

Note that this is not only a positive return, but it is the highest possible return he can expect to receive given his demand estimation. He could plan to produce and sell only 325 hot dogs in his first week. He could charge a price as high as $2.75 and still sell 325. Because he would sell fewer hot dogs, his total expenditures would decrease to $620. Although his expenditures would decrease, total revenue would as well, because he is selling fewer hot dogs. His total revenue would decrease to $893.75. Subtracting his expenditures from his revenue results in net revenue of $273.75 for the week. This is indeed a positive return on his investment. However, it is not the best he could do. He could bring in even more net revenue by selling more at a price of $2.50.

Similarly, if he wanted to sell more than four hundred at a lower price, Frank could do so. If he lowered his price to $2.25, he could sell 440 for total revenue of $990. His total expenditure, however, would also increase to $689, resulting in net revenue totaling $301. Again this is a positive return, which is certainly better than not being able to cover his expenses. However $301 is not the best he could do. By charging a price of $2.50 and selling four hundred hot dogs, he generates the highest net revenue possible, allowing him to achieve the most ends he can with his income. Any other production decision would leave him less satisfied.

The decision to buy factors of production allowing him to make and sell four hundred hot dogs a week and the decision to sell those hot dogs for a price of $2.50 a piece was determined by Frank's calculation of his expenditures and his estimation of the future demand for his product. Once the investment decision is made, he cannot take it back. Once he has committed money to the owners of the land, labor, and capital goods, the prices of those goods become irrelevant for his action that week. In other words, once costs are sunk, they become irrelevant for action. Costs are always the value of the alternative that must be foregone as the result of action. In order to act to obtain the land, labor, and capital goods necessary to produce four hundred hot dogs, Furter must sacrifice $665. This is money he could spend elsewhere on consumption, other investment projects, or use it to add to his cash balance. Committing to hot dog production requires that he gives up all other alternative uses.

Once the investment is made, the alternative courses of action change and hence costs are different. The act of selling the hot dog does not entail the sacrificed alternative uses toward which the $665 could be put, because these are not alternatives anymore. Once the money has been committed to hot dog production, it is impossible for him to spend the same $665 on clothes, for example, so we cannot say that the cost of selling four hundred hot dogs is the clothes he chooses to forego. This change in costs will not affect his selling decision if demand turns out as he predicted, but if the seller's forecast of the future demand for his product is less than precisely accurate, he must reconsider his selling decision or he will fail to earn the greatest net income.

If the demand for Furter's product turns out to be not as expected, then what should he do? Once an investment is sunk, the wisest thing for a seller to do is to charge the price that allows him to bring in the greatest total revenue. For instance, suppose that instead of the demand for Frank N. Furter's hot dogs being as forecast for his first week, it turns out to be as illustrated in table 7.6:

Table 7.6

Hot Dog Demand, Expenditure and Revenue				
Price	Quantity	Total Revenue	Total Expenditure	Net Revenue
$3.50	0	$000.00	$665.00	$(665.00)
$3.25	0	$000.00	$665.00	$(665.00)
$3.00	0	$000.00	$665.00	$(665.00)
$2.75	100	$275.00	$665.00	$(390.00)
$2.50	200	$500.00	$665.00	$(165.00)
$2.25	325	$731.25	$665.00	$ 66.25
$2.00	400	$800.00	$665.00	$ 135.00
$1.75	440	$770.00	$665.00	$ 105.00
$1.50	480	$720.00	$665.00	$ (55.00)
$1.25	510	$637.50	$665.00	$ (27.50)
$1.00	530	$530.00	$665.00	$(135.00)

Because his initial investment has already been made, his total expenditure no longer varies with the quantity of hot dogs he sells. It is the same $665 for every quantity.

Frank's best option is to find out which price results in his greatest total revenue. Given the lower than expected demand and his sunk investment, it is clear that the best thing to do is not to charge $2.50 for each hot dog. He could do this. He could say to himself, "I decided to get into this business thinking I could charge $2.50 for a hot dog and that's what I'm going to do." If he proceeds in this manner, however, he will have no net income at the end of his summer to help him with his college expenses. If he indeed would charge $2.50 he would sell not four hundred, but only two hundred hot dogs, generating total revenue of only $500. With total expenditures of $665, such a turn of events results in a negative net revenue of $-165. This means that, not only will he not have net income, but he would not even take in enough to get back the money he spent on the factors of production. If he tried to make another go at selling hot dogs the following week, he would have fewer funds to use. He would most likely have to draw down his cash balance to invest for another week's production. If he continues down this lonesome road, he will end up in bankruptcy. In any event, charging $2.50 does not seem to be Frank's wisest decision.

What should he do? Which price results in Frank selling the quantity of hot dogs that generates the highest total revenue? You can see in table 7.5 that his highest total revenue would be had if he lowered his selling price to $2.00 and sold four hundred. If he sold four hundred hot dogs at $2.00 each, this would generate total revenue of $800. With total revenue of $800, Furter would clear positive net revenue of $135. This is a greater amount than he could net by selling at any other price. If he tried to charge just a little more, say, $2.25 per hot dog, his total revenue would decrease to $731.25. Because he subtracts from this the same sunk $665, his net revenue would be only $66.25. On the other hand if Frank charges $1.75, just a bit lower then $2.00, his total revenue would be only $770. Again, because he subtracts the same $665 to calculate net revenue, his net revenue is only $105. This is more than $66.25 to be sure, but less than the $135 he could net by charging a price of $2.00. The moral of the story is that, after the investment has been made, an entrepreneur will maximize net revenue by charging the price that allows him to sell the quantity that will maximize total revenue.

What happens if things are even worse? What if there is no way for Furter to earn a positive net return at all? Suppose that the actual demand for his hot dogs is a lot worse than he expected, making it impossible for him to even cover his expenses. Suppose the demand for his hot dogs turns out to be as illustrated in table 7.7. What should he do then?

Table 7.7

Hot Dog Demand, Expenditure and Revenue				
Price	Quantity	Total Revenue	Total Expenditure	Net Revenue
$3.50	0	$000.00	$665.00	$(665.00)
$3.25	0	$000.00	$665.00	$(665.00)
$3.00	0	$000.00	$665.00	$(665.00)
$2.75	0	$000.00	$665.00	$(665.00)
$2.50	0	$000.00	$665.00	$(665.00)
$2.25	100	$225.00	$665.00	$(440.00)
$2.00	200	$400.00	$665.00	$(265.00)
$1.75	325	$568.75	$665.00	$ (96.25)
$1.50	400	$600.00	$665.00	$ (65.00)
$1.25	440	$550.00	$665.00	$(115.00)
$1.00	480	$480.00	$665.00	$(185.00)

It turns out that, once Frank makes his investment of $665, the best thing he can do is charge the price that maximizes his total revenue, even if he cannot cover his expenses. In doing so, even though he earns a negative net revenue, in charging the price that will maximize his total revenue, he minimizes his loss. Given his demand curve, Frank's best option is charging $1.50 for each hot dog. If he charges $1.75, he will take in $568.75 in total revenue, netting a loss of $96.25. If he charges $1.25, he will reap total revenue of $550, netting him $-115. By charging $1.50 his total revenue is maximized at $600. Subtracting the same $665 from his total revenue, he would reap net revenue of $-65. This is still a negative net return, but it is surely a smaller loss than either $96.85 or $115. Charging the price that maximizes total revenue allows Furter to cover as much of his expenses as possible, given his demand. This is the best Frank can possibly do. Consequently, other things equal, after an investment is made, the entrepreneur will do best to charge the price that maximizes total revenue. In doing so, he will either maximize his positive return or minimize his negative return.

If, however, the price at which he is able to sell his hot dogs is not even enough to cover his hot dog expenses, it would be better for him to shut down altogether. He would still be out the weekly rental for his cart, rent, and energy, but he would not incur any more expenses for condiments and hot dogs he could not sell.

The decisions that any producer must make in order to forecast and react to future demand are many. It is only those who are able to successfully make them on a regular basis who are able to remain in business, serving the public. This kind of makes you want to treat them with more respect, doesn't it?

PRODUCTION, CAPITAL, AND INCOME MAXIMIZATION

In our modern economy, producers engage in production to earn income they can allocate to achieve as many of their ends as possible. To do so capitalists must save and invest in the formation of a capital structure. Entrepreneurs use capital funding to obtain factors of production. They buy or rent land and capital goods. They hire labor. Workers sell their own human effort in return for wages. In a free market, all of these do their part and are rewarded according to how they serve society as expressed by people in their role as consumers.

To achieve their goals, entrepreneurs and capitalists must reap positive net returns on their investments. People often refer to all positive return as profit. You will note, however, that the word profit has been, conspicuously absent from this chapter. That is because not all positive net revenue is profit. When a bank pays savers a positive return in exchange for your deposits over a period of time, they are paying interest. In the next chapter, we will examine the relationship and distinction between net revenue, interest and profit.

SUGGESTED READING

Heyne, *The Economic Way of Thinking*, 199–213. Heyne provides a realistic explanation of pricing decisions and income maximization for the firm.

Mises, *Human Action*, 476–520. Mises' chapter on "Action in the Passing of Time" explains the importance of time preference and its implication for the formation and maintenance of capital.

Rothbard, *Man, Economy, and State*, 47–70, 206–31. The outstanding introduction to the principles of capital formation plus Rothbard's discussion of producers' expenditures and how producers seek to maximize income.

8

Interest and Profit

I N OUR MODERN MONETARY economy, the objective of producers is to make a product that sellers can sell so that they earn positive net income. The greater their net income the more ends producers can satisfy. They can expand their business, hire more workers, extend the market, and develop the division of labor, leading to increased standards of living. In this chapter we will examine the nature of this positive return that producers are seeking and distinguish between the economic categories of interest and profit. We will see that interest is the positive return sought by people in their role as capitalist and profit is the positive return sought by people in their role as entrepreneur.

THE ORIGIN OF INTEREST

We begin our investigations by examining the origin of interest. One might say that this part of the book will be very *interesting*. When you think of interest, you most likely think of the amount of money people have to pay back for loans used to pay for things like houses, cars, college tuition, and credit card bills. Or you might remember that the money the bank pays you for depositing your money in a savings account or certificate of deposit is also called interest. In both cases, what is referred to as interest is an amount of money paid to someone for the use of their money for a period of time. However, we will see that the key to explaining interest is not in the phrase *the use of their money*, but in the word, *time*. The origin of interest is in time preference.

As noted earlier, all production takes time. To engage in production, people must sacrifice present goods in order to obtain the use of land, labor, and capital goods which are used to produce a good which will be sold in the future. All production, then, involves exchanging present goods for future goods. All production decisions include considerations regarding the concepts sooner and later.

We have already established that people have time preference. We prefer to have our ends met sooner rather than later. We value present goods more highly than we value the same goods in the future. The same holds for the good money. In order to agree to invest money now, capitalists require a premium to compensate them for the postponement of the satisfaction of their ends. People value present money more than they value present expectations of the same amount of money in the future. This explains why a banker offers to pay interest to people who deposit money in savings accounts. The banker knows that he must sweeten the financial pot so as to convince savers to invest their funds in the bank's savings accounts.

Because investors, like all people, value present money more highly than the same amount of money in the future, they require a positive premium for supplying money in the present in exchange for money in the future. This premium commanded by present goods over future goods is called interest. Suppose that Joe Capitalist's value scale is as shown below:

Joe's Value Scale

($1,060 a year from now)

$1,000 today

($1,050 a year from now)

($1,000 a year from now)

If Joe has the opportunity to invest $1,000 of his savings in a particular project that he thinks will allow him to receive only his initial $1,000 in one year, he will not make the investment. This is because, as time preference implies, he values the $1,000 now more than $1,000 a year from now. In order for him to voluntarily save and invest his $1,000, he will require some amount more than $1,000 a year from now.

What is the minimum amount of money he must expect to receive in one year in order to make his investment? According to Joe's value scale, he must receive at least $1,060. Joe's value scale indicates that his time preference is such that even a $1,050 total return a year from now would not be enough to induce him to invest his money in the present. He would, of course, be happy to receive any amount greater than $1,060 a year from now. He would be willing to invest his money now to receive $1,070, $2,000, or $2 million a year from now. The minimum he requires, however, is $1,060.

What, then, is the minimum premium required for Joe to make the investment? This equals the net revenue he would receive on an investment returning him his minimum required future return. In Joe's case the premium required is $60, or $1,060 minus $1,000. Because the $60 is the premium required for Joe to part with his present money, this $60 is called interest.

THE INTEREST RATE

The interest rate is closely connected to the amount of interest earned on an investment. The interest rate can be defined as the rate of return required as a payment for supplying present goods in exchange for future goods. As such, it is not a price, but a ratio of prices of present goods to the prices of future goods.

One thing to be gleaned from this definition is that the interest rate is a rate of return. The rate of return on an investment is very helpful to capitalists and entrepreneurs because it can be used to compare the value of different investments over a period of time. When deciding to invest in the production of hot dogs or barbecue beef sandwiches, for example, investors can compare the expected rate of return in order to make the investment that is best for them.

The rate of return is a number calculated using a mathematical formula comparing net revenue to the initial investment. You can calculate the rate of return on an investment by using the following formula:

$$\text{RATE OF RETURN} = \text{NR} \div \text{INV PER TIME PERIOD}$$

where NR = net revenue and INV is the capitalist's initial investment. The phrase "per time period" highlights the fact that the money returned on an investment comes in the future, after a period time has passed. It further implies that a rate of return can be calculated for different periods of time. Typically, interest rates are calculated for a time period of one year, or annually.

An example will assist us in understanding the rate of return. Suppose that Joe Capitalist invests $1,000 in some project in the present and in one year's time the project brings in $1,300 in total revenue. Joe's net revenue would be $1,300 - $1,000 = $300. His rate of return then would be 300/1000 or three tenths. This equals a 30% rate of return.

THE FINAL STATE OF REST
AND THE EVENLY ROTATING ECONOMY

Entrepreneurs and investors are always on the lookout for the best investment that enables them to achieve more of their ends. Just as people tend to prefer higher incomes to lower, other things equal, investors prefer higher rates of return to lower. When faced with different rates of return from different investments, capitalists will decrease their investment in those lines of production that reap lower rates of return and increase investment in those lines of production that reap higher rates of return.

If there were no changes in people's values, natural resources, population, and technology, the actions of investors would set in motion a process that would bring all rates of return from all investment into equality with one another. Again, suppose that Joe Capitalist has a value scale as illustrated earlier:

<u>Joe's Value Scale</u>

($1,060 a year from now)

$1,000 today

($1,050 a year from now)

If he has two opportunities in which he can invest $1,000, one returning $1,040 a year from now and the other returning $1,070 a year from now, which would he choose? He would not choose any that returns $1,050 or less, because his time preference will not allow that. Consequently, the investment opportunity returning $1,040 a year from now is out. He would be willing to invest his money for $1,060 but he prefers $1,070 even more. Other investors are similar in that they prefer greater returns to lesser.

Suppose for simplicity's sake that capitalists have two investments in which they can invest their savings. They could invest their funds in the production of golden flax seed and earn a 7% rate of return. On the other hand, they could invest their funds in the production of steel and earn a 4% rate of return. We can gain a better understanding of the market adjustment process by examining what will happen in each industry one at a time.

If investment in golden flax seed production reaps a rate of return greater relative to alternative investments, investors will act so that, over time, the rate of return in this industry will tend to decline, other things equal. Once investors recognize that savings invested in flax seed production reaps a greater rate of return, these capitalists will direct their savings out of steel production and into golden flax seed production.

Investment in golden flax seed will increase. Investment is the direction of saved resources toward obtaining factors of production used to make a particular good that will be sold in the future. If there is increased investment in golden flax seed production, this means that there would be an increased demand for all of the factors used in producing golden flax seed. There would be a greater demand for land, farm implements, labor, water, fertilizer, and other goods used to make the flax seed. This would, other things equal, result in higher prices for all of those factors of production.

Over time, these factors would be put to work producing golden flax seed, so that producers would have more to sell, increasing the supply of the final product, golden flax seed. A greater supply would result in a lower market price for the flax seed. As the price of factors of production increases, this increases the monetary expenditures necessary to produce a given quantity of golden flax seed. At the same time, every unit of flax seed would be sold for a lower price, because of the increase in market supply. When expenditures on factors of production increase and the price of the final product decreases, this tends to decrease net revenue. As net revenue declines, this will, other things equal, reduce the rate of return on this investment. Consequently, as investors recognize that they can receive a greater rate of return in a certain investment, such as golden flax seed production, they will act to set in motion a process by which the rate of return on that investment will decline.

While investment in golden flax seed production would increase, investment in steel production would decrease. Investors would send their savings elsewhere. As this occurs, the market demand for factors used to produce steel will necessarily decrease. As the demand for factors such as land, factories, iron ore, furnaces, and labor used in steel plants decreases, the price of all these factors will be bid down by the more eager buyers.

Because fewer factors of production would be purchased or hired, less steel would be produced, resulting in a decrease in the market supply of steel. A decrease in supply would result in frustrated buyers if the market price remained at its prevailing level, but in fact, the more eager buyers would bid up the price until everyone who wants to buy can.

The end result is that the rate of return on steel making would tend to increase over time. Those still investing in steel production would find that the savings necessary to spend making a given quantity of steel would decrease. At the same time, the increased price of steel means that the price brought in by a given weight of steel increases. When expenditures on factors used to produce a unit of steel decreases and the price of the product produced with those factors increases, this results in an increase in net revenue. As net revenue increases, other things equal, the rate of return in that industry increases. Consequently, as investors recognize that they can receive a lower rate of return in a certain investment, such as steel production, they will act to set in motion a process by which the rate of return on that investment will increase.

We see, then, that whenever investors have different investment opportunities with different expected rates of return, they will act so as to set in motion a process by which the rates of return in all investments become equal. In those industries where the rate of return on investment is relatively high, capitalists will increase their investment, resulting in a decline in rate of return over time. In those industries where the rate of return is relatively low, investors will decrease their investment, resulting in an increase in the rate of return over time.

This might seem paradoxical, for why would someone want to invest in the production of golden flax seed if he has taken an economics class and he knows that as more investment occurs the rate of return will tend to decline. In fact, the reason why capitalists will still invest in a line of production, even if its rate of return might, over time, decline, is that it still might be higher than the alternative. Suppose, for example, that golden flax seed investments reap a 7% rate of return and investment in steel production yields a 4% rate of return. Investment in golden flax seed production will increase and investment in steel production will tend to decline. We would expect that the rate of return in the golden flax seed industry would decrease. Suppose that, as investment increases, the rate of return on golden flax seed production decreases to 6%. Likewise, suppose that, as economic theory teaches us, as investment in the production of steel decreases, the rate of return increases to 5%. Where will capitalists direct their savings now? To the same place: golden flax seed. Even though the rate of return for golden flax seed production decreased to 6% and that for steel production increased to 5%, 6% is still greater than 5%. Capitalists would still be better off investing in golden flax seed production. An incentive to increase golden flax seed investment at the expense of steel investment would not exist forever, however. At some point, the rate of return for each investment would become equal. At this point, capitalists have no further incentive to either increase investment in golden flax seed production or decrease investment in steel production.

What is true for investment in the production of goods at the same stage of production is also true for investment in the production of goods at different stages in the structure of production. If, for example, the rate of return on golden flax seed is greater than that for investment in the production of tractors, investment in golden flax seed production will increase and investment in tractor production will decrease. Consequently the rate of return on flax seed investment will decrease and the rate of return on tractor production will increase. This process will continue until the rate of return for both activities will be equal. When this occurs there is no further incentive for capitalists to either increase their investment in golden flax seed production or to decrease their investment in tractor production.

We see, then, that there is a general tendency in the free market for the rate of return on all investments at every stage of production to become equal to one another. If people's subjective values, population, quantity and quality of natural resources, and the level of technology were to remain constant, we would reach the point where the rate of return on all possible investments did become equal. The economy would reach a final state of rest. All of the markets for all of the final products and factors of production would be in equilibrium. All of the difference between the prices of the final products and the sum of the prices of the factors of production would be due to the positive time preference of everyone in the economy.

Consequently, the rate of return toward which all investments tend and which would obtain in the final state of rest if reached would be reflective only of the time preferences of members of society. There would not be any uncertainty, because we have assumed that those things like subjective preferences or population would be constant. Because we know that the resulting single rate of return earned in the entire economy would be determined only by time preference, this rate of return is an interest rate. Because this interest rate is only reflective of social time preferences, it is called the *pure interest rate*. It has also been referred to as the *natural interest rate* and the *originary interest rate*. It is called the originary interest rate because time preference is the root of all interest. It is the phenomenon from which all interest rates originate.

DETERMINATION OF THE INTEREST RATE

We have seen that the pure rate of interest is the result only of people's time preference. We have not yet, however, discussed exactly what determines its precise level. We begin our investigation into that very question by remembering that interest is the result of an exchange of present goods for future goods. Interest is the payment capitalists require to invest present money in exchange for future money. Investors give up money in the present in return for a larger sum of money in the future. The interest rate is the discount on the value of future goods as compared to that of present goods. It is a ratio between the price of the final product sold in the future and the sum of the prices of factors of production purchased in the present.

The nature of interest must be understood in order to forestall a number of errors into which even some of the most eminent economists have stumbled, or in some cases leapt with both feet. Some economists have characterized the interest rate as merely the price for loanable funds. It is easy to see why people would think of interest as the price for loanable funds. Banks do charge interest on loans after all. The reason why defining interest as the price for loanable funds is an error is that such a definition is not general enough.

It is true that people do charge interest if they are loaning money to other people and that this interest is required because of time preference, but the loanable funds market is not the only market where we find interest. If, for example, Fanny Farmer decided to produce flourless chocolate cakes and bought the bittersweet chocolate, butter, eggs, mixers, springform pans, ovens, kitchen, land, and hired labor all with only her own savings, so that she did not have to borrow a dime, she would still receive interest. She still made an investment in which she obtained the services of factors of production with money in the present in exchange for monetary revenue in the future. She still exchanged present money for future money, so she still required a price for her flourless chocolate cake greater than the sum of prices of the factors of production. We see then, that interest is a return on all investments, not just those involving loaning money. Defining interest merely as the price of loanable funds is an error because it does not tell the whole story.

Another, more blatant error is defining interest as the price, or cost, of obtaining money. Sometimes economists who think along these lines, such as the celebrated John Maynard Keynes, have referred to interest as the cost of liquidity. Liquidity here refers to the ease with which a good can be turned into ready cash. The easier it is to sell an asset for cash, the more liquid that asset is. Because of the active stock market in the United States, shares of stock are usually thought of as being more liquid than real estate, because finding someone willing to buy a plot of land is often harder than finding a ready buyer for shares of stock. The most liquid good of all, of course, is money, because it is itself money, the most marketable good.

Again one can understand why one might be drawn into the error that interest is the price of money. After all, if you have money in your purse or wallet you could have instead deposited that money in a savings account or interest bearing checking account. It is then argued that, because the cost of something is what you sacrifice to have it, the cost of holding money is the foregone interest. Therefore, it is claimed, interest is the price of holding money, or the price of liquidity.

We can see the error of this view of interest, however, by recalling that money is the general medium of exchange. There is a multitude of things that we do with our money. We can indeed deposit it in a savings account. However, we could also give it to our church for missions, buy a set of Sibelius symphonies and tone poems conducted by Sir Colin Davis, take our sweetheart on a tour of a Victorian mansion, buy our child a toy, or merely buy an IBC root beer. Suppose that Natty Bumpo has $25 in his pocket and a value scale as indicated below:

<u>Natty Bumpo's Value Scale</u>
$27
$26
(Leatherstocking Tales)
$25

According to his value scale, Bumpo's option is between holding his $25 in cash and spending it on a used set of James Fenimore Cooper's *Leatherstocking Tales* in two volumes. Remember that the cost of doing anything is the value of the alternative given up for an action. The price Bumpo pays to hold his $25, his cost, then, is the value of the alternative he forgoes to hold onto his money. In this case, he passes by a used set of *Leatherstocking Tales*. Consequently, the cost to Natty Bumpo of maintaining his liquidity, of holding on to his money, is not foregone interest at all, but the value of those used books. Likewise the person who owns the used book store can obtain $25 merely by finding someone willing to pay the price. For the owner of the books, the price of obtaining $25 has nothing to do with interest, but everything to do with exchanging books in the present for money in the present. We cannot, therefore, correctly define interest as the price or cost of money, because the cost of anything is the value of the alternative a person has to give up in order to obtain it, and obtaining money often has nothing to do with interest. The interest rate is not merely the price of loanable funds and it is not the price of money. It is the ratio of prices of present goods compared to prices of future goods. The key issue is time.

In a modern economy all transactions are made through the medium of exchange, so in a modern economy such as ours, the present and future goods that are being exchanged are both money. In our modern economy, the time market is the market of present money in exchange for future money. Because all goods are traded for money in a modern economy, we may correctly define the interest rate as the rate of return required for the supply of present money in exchange for future money.

The interest rate, then, is determined by the market of present money in exchange for future money. Because of the importance of the concepts of present and future in this market, it can be characterized as the time market. In order to understand the precise determinants of the interest rate, it is necessary to properly identify all the time market entails.

There are two components of the time market: the loanable funds market and the production structure. It is in both of these two components that present goods are exchanged for future goods.

The loanable funds market is rather straightforward. Money is demanded by borrowers in the present and paid back with interest to lenders in the future. Today, most college students participate in the loanable funds market by taking out student loans. Students borrow money in the present in order to pay their tuition. In doing so, they agree to pay the money back with interest at some point of time in the future. These students are buying present money in exchange for future money. The price they agree to pay is the interest rate. Likewise, people who borrow money to pay for a house or automobile also borrow money in the present in exchange for the money that they will pay back in the future.

Understanding the production structure component of the time market is a bit more complex. This is because of the nature of the production process. We must never forget that production is an action that takes time.

Therefore, the entire act of producing and selling a good also involves the exchange of present money for future money. Producers pay money to the owners of factors of production in the present. These factors are turned into a final product that is sold for money in the future. Therefore, the production structure is also a very important component of the time market. A graph of the time market can be seen in figure 8.1.

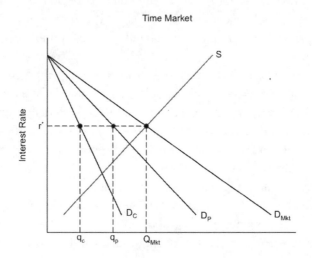

Figure 8.1. A graph of the time market. The interest rate is determined by the supply of and market demand for present money in exchange for future money.

Like all markets, the participants in the time market can be divided into two groups of people: sellers and buyers, suppliers and demanders. The simpler of the two to grasp is the supply side. The suppliers in the time market are those that supply present money in exchange for future money. These are the people who lend present money to people who will pay it plus interest back in the future. Some lend money directly, while others do so indirectly through financial institutions like banks. People deposit money in a bank for a fixed period of time and receive a certificate of deposit. Banks then lend this money to other people who want it to buy cars, houses, or expand their businesses. Such depositors are, in effect, lending money to the people through the intermediary of the bank. Others invest present money directly into the production process. They invest money in the present and produce a product that they sell for money in the future.

The common denominator between both the lender and those who invest directly in production is the *source* of the present money. Both the person who invests present money in a certificate of deposit and the person who invests present money in obtaining the use of factors of production must fund their investments out of savings. We conclude, consequently, that the supply of present goods in exchange for future goods comes from everyone's savings.

Like all supply schedules and curves, the supply of present money in exchange for future money is a manifestation of subjective human value scales. As such they exhibit the law of marginal utility. As people supply a greater quantity of present money in the time market, the marginal utility of the money they still hold in their cash balances increases. As its marginal utility increases, people require a higher price (in this case a higher interest rate) in order to part with more present money. Consequently, we conclude that the supply curve in the time market is sloping upward to the right similar to the supply curves for all other goods.

The demand side of the time market is a bit more complicated, because of the two components of the time market. Much confusion can be avoided by simply remembering that demand for present money in exchange for future money is made by those receiving the present money. This can be seen most clearly in the loanable funds market. One component of the demand for present money in exchange for future money is what we call consumption demand. This is the present money demanded by borrowing consumers. As mentioned earlier, examples of this are money borrowed in the present to pay for an automobile or a house.

Consumption demand is represented on the time market graph by the curve marked D_C. Like all demand curves it is a representation of the subjective value scales of the demanders. In this case the demand curve is a manifestation of the preferences of borrowing consumers. As such, consumers' demand conforms to the law of marginal utility. As people borrow more money, their stock of present money increases and the marginal utility of that present money decreases. Borrowers value the marginal unit of present money less and will be willing to borrow more only if they can pay a lower rate of interest. Consequently, there is an inverse relationship between the quantity of present money demanded by borrowers in exchange for future money and the interest rate. The higher the hypothetical interest rate, the smaller the quantity of present money consumers would be willing to borrow. The lower the hypothetical interest rate is, the larger the quantity of present money consumers would be willing to borrow.

The second, more complicated part of the demand side of the time market is the demand for present money in exchange for future money found in the production structure. This is the demand by those selling the services of their property in the advancing of production. These demanders are the landowners, laborers, and owners of capital goods. They are, in other words, the owners of factors of production. Hence, we call this component production demand.

It is here that you can easily fall into confusion. Above, we said that consumption demand is the demand for present money by consumers. Production demand, however, is not demand for present money by producers, but by the owners of the factors of production. In the production structure, it is the owners of land, labor, and capital goods who demand money in the present in exchange for money in the future in the form of the services of their factors. When entrepreneurs obtain use of factors of production, they do so by paying the owners of the factors present money. To the entrepreneur, the services of the factors represent to them future money, because they use the factors to produce final products that they sell for money in the future. Therefore, throughout the production structure, the owners of factors of production demand present money from entrepreneurs and supply to them future money in the form of the services of their factors.

In the graph of the time market in figure 8.1, production demand is represented by the demand curve marked D_p. Like other demand curves, it slopes downward to the right. In the production structure the present money exchanged for future money is money spent on business investment. If there is an increase of present money received by the owners of the factors of production, this must be the result of increased investment on the part of the capitalists. Increased investment results in an increased demand for factors of production, which in turn results in increased prices for these factors. As has been already seen, increased prices of the factors of production that result from increased investment due to decreased time preferences lower the interest rate. Consequently, we see an inverse relationship between the interest rate and the quantity of present money demanded by the owners of factors of production in return for future money.

Consumption demand and production demand are two components of the same time market. To derive the market demand for present money in exchange for future money we simply sum the consumption and production demands across every interest rate. This summation results in a market demand shown on the graph of the time market as D_{Mkt}. Because both the consumption demand and production demand curves slope downward to the right, so does the market demand curve.

We have determined that the interest rate is an exchange ratio—a price. Like all other prices, it is determined by the voluntary decisions on the part of people to exchange or not exchange present money for future money. Like all markets, there is a tendency for a single interest rate to emerge at which everyone who wants to buy present money in exchange for future money can buy and everyone who wants to sell present money in exchange for future money can sell. This interest rate is the rate that will clear the time market. Like all other markets, this equilibrium interest rate is illustrated on the time market graph as that interest rate that obtains where the market supply and demand curves intersect. In figure 8.1 it is identified as r^*.

Just as in any market for economic goods, where there is one price that is charged for all goods in any market, there will be one interest rate earned and paid whenever present money is exchanged for future money. Why is this? Suppose, for whatever reason, the actual interest rate happened to be above the market interest rate. In the production structure, if the rate of return is above the market rate of interest, investment would increase, so that the rate of return would decrease down to the market clearing rate. There would be an excess supply of present money in exchange for future money in the loanable funds market. The most eager lenders would bid down the interest rate they offer borrowing consumers until everyone who wants to lend can lend at the prevailing interest rate.

On the other hand, if the rate of interest would be below the market rate for any reason, there would be an excess demand for present money in exchange for future money. The most eager borrowing consumers would bid up the interest rate they are willing to pay for their consumer credit. In the production structure, investment would decrease, because the rate of return would not be high enough to cover investor time costs. This general decrease in investment would set in motion a process by which the rate of return for all investments would increase toward the market interest rate.

So, there is one market interest rate that will apply to every participant in the time market. This means that all borrowing consumers will pay the same interest rate. The interest rate earned by all investors throughout the structure of production will also be the same. Additionally, the interest rate earned by lenders and by investors in production will also be equal. If the rate of return on investments in the production structure is lower than the market interest rate in the loanable funds market, there is an incentive for capitalists to shift their savings out of investment in production and into the loanable funds market. As this occurs, the rate of return on investment in production will increase and the market interest rate received for loans will decrease until both rates become equal. The quantity of present money borrowed by consumers is denoted on the time market graph as q_c. This is the quantity of present money consumers would borrow at the interest rate r^*. Likewise, the quantity of present money invested in production is identified as q_p. This is the quantity of present money exchanged for future money by capitalists throughout the production structure. In other words q_p is the quantity of business investment there is if the interest rate is r^*. Summing q_c and q_p results in the total quantity of present money exchanged for future money in the time market, identified as Q_{Mkt}, which will be the quantity indicated where the market supply and demand curves intersect.

We see, then, that the interest rate is determined by the voluntary actions of demanders for and suppliers of present money in exchange for future money. Their actions are determined by their value scales. Because the goods being traded are present and future, the overriding preference that determines the quantity of the good exchanged and at what price is time preference. It is time preference that ultimately determines whether people will be a borrowing consumer or a net saver and investor. It is the time preferences of people therefore, that will determine the interest rate that will clear the time market as well as the quantity of consumer borrowing and business investment that takes place at that interest rate.

CHANGES IN THE INTEREST RATE

Because the pure interest rate is determined by everyone's time preferences, it is reasonable to conclude that changes in the pure interest rate will be the result of a change in people's time preference. This is undoubtedly the case. In terms of changes, there are only two things that time preferences can do: increase or decrease.

If a person's time preference increases, this person becomes more present-oriented. His preference for present goods over future goods becomes more intense. A person who becomes more present-oriented tends to increase consumption and, hence, decrease saving. A decrease in savings can be shown on the graph rather simply. Recall that everyone's savings is the source of the supply of present money in exchange for future money in the time market.

An increase in time preference by members of society will result, consequently, in a decrease in the supply of present money, which is illustrated in figure 8.2 by a leftward shift of the supply curve in the time market graph.

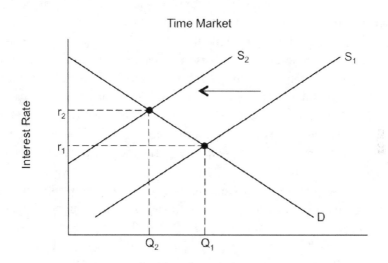

Figure 8.2. A decrease in the supply of present money in exchange for future money will result in an increase in the interest rate.

It is easy to see from the graph that a decrease in savings resulting from people developing higher rates of time preference will result in an increase in the interest rate. At a higher interest rate, there will be less consumer borrowing and business investment. We can conclude, then, that there is direct relationship between people's time preference and the interest rate. Such changes in the interest rate will be accompanied by changes in the loanable funds market and the production structure. To understand just why this is so, consider the two components of the time market.

If people save less because of their higher time preferences, this will have an impact on the loanable funds market. If people save less, banks have less money to lend to borrowing consumers. At the prevailing interest rate, there would be frustrated demanders. The most eager demanders would bid up the interest rate, the price of present money in exchange for future money.

Additionally, less saving has a definite impact on what happens throughout the production structure. A decline in saving will necessarily result in a decrease in business investment. Entrepreneurs will spend less on factors of production. The drop in the demand for these factors will result in their prices following suit. Decreases in the prices of factors of production will tend to lead to higher net revenues and rates of return. Because all of these changes are the product of a general increase in time preference, they are not happening in only one market or industry, but economy-wide. All rates of return are moving toward a new, higher, pure interest rate.

The exact opposite will tend to occur if there is a general decrease in people's time preference. A decrease in time preference indicates that people are less present oriented. Their time preference becomes less intense. While they still value present goods more highly than future goods, they do not do so as earnestly. If their rate of time preference decreases, people are more likely to put off present gratification. They are likely to consume less and save more. A decrease in time preference, therefore, results in an increase in saving.

We see the effect of savings on the time market graph as an increase in the supply of present money in exchange for future money. This is illustrated in figure 8.3 by a rightward shift in the supply curve.

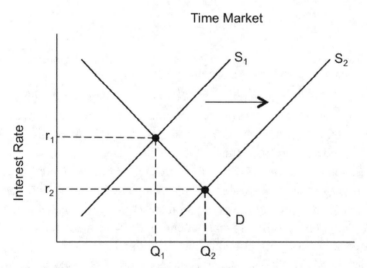

Figure 8.3. A decrease in social time preference will result in an increase in the supply of present money in exchange for future money. This in turn will result in a lower interest rate.

An increase in the supply of present goods in exchange for future goods leads to a lower pure interest rate. Following an increase in supply, there would be more suppliers of present money than demanders at the prevailing interest rate. There would be frustrated suppliers. The most eager sellers of present money would bid down the interest rate. At this lower interest rate, consumers would borrow more and there would be more business investment.

Banks would lower the rate at which they would lend money to borrowing consumers. Because of lower time preferences more people would deposit more savings in banks. These banks would have more money with which to lend. In order to entice consumers to borrow from them instead of their competitors, they would bid down the interest rate at which they would lend them money.

More savings also impacts production through an increase in investment. As savers save more, they have more to invest. Increased investment results in more spending on the factors of production. An increase in the demand for factors of production will result in an increase in the price of factors of production. This results in a decrease in the differential between the price of the final products and the sum of the prices of the factors of production. If the prices that entrepreneurs have to pay for the services of their factors increase, this will lead to decreases in their net revenues and rate of return. Consequently, the pure rate of interest toward which all rates of return are tending will decrease. We again conclude that there is a direct relationship between people's time preference and the pure interest rate. If there is a general decrease in people's rate of time preference, people will set in motion a process by which the pure rate of interest decreases.

PROFIT

Jesus asks what, perhaps, is the most profound question ever asked in human history. "For what will it profit a man if he gains the whole world, and loses his own soul?"(Mark 8:36). The obvious answer to Christ's question is *nothing*. If a man loses his soul in order to gain the world, Jesus is telling us that the man has made a bad trade—a terrible trade. In the end we have no profit, but an everlasting loss. Interestingly, Christ casts this question in economic terms, using the word profit. One of the benefits of understanding economics is that it can actually help us better understand Scripture. Too many people assume that to think in terms of economic concepts is to be stained with worldliness, yet Jesus used such concepts to drive home a strong theological point.

There are two senses in which the word profit is used. The first is the most general and it is the sense in which Christ seems to use it. Profit in the broad sense means simply net gain derived from action. The Greek word translated as *profit* in Matthew 8:36 means *benefit*. Profit in this sense is the difference between the higher satisfaction of achieving our end and the lower satisfaction of the end we give up. It is the result of the greater benefit over our cost. Christ is telling us that if we choose the whole world and lose our soul, then in the end we will not receive a net gain. We would, in fact, lose our soul eternally in hell.

This broadly defined concept of profit is a subjective phenomenon. Because it is the difference between the satisfactions of the end we achieve and of the end we sacrifice, it is determined by where these ends rank on our value scales, which are subjective. As such, we cannot measure profit. We cannot measure the value we place on either the end we achieve or the end we give up. Therefore, we certainly cannot measure profit in this broad sense.

In the market economy, however, the term profit takes on a more precise meaning. In our modern monetary market economy, all things are bought and sold for money. In order to persuade the nice folks at the snack bar to give you a cookies and cream malt (heavy on the malt), it is necessary to give them some money. The snack bar, in turn must pay for the facilities, ice cream, milk, malt, mixer, land, and labor necessary to make a cookies and cream malt with money. Because all things are traded for money, all of these goods have money prices.

Consequently, in a market economy, profit can be expressed in definite amounts of money. Mathematically, we can express profit as part of an entrepreneur's net income. Remember from chapter 7 that we indicated that a producer's net income is made up of three categories: interest, profit, and wages for his labor. His net revenue is equal to his total revenue minus his total expenses:

$$\text{Net Income} = \text{Total Revenue} - \text{Total Expenses}$$

His net income is equal to the sum of the interest on his investment, his entrepreneurial profit, and entrepreneurial wages. The interest on his investment is the positive return that compensates him for supplying his present savings in a production project in exchange for revenue in the future. Entrepreneurial wages compensate for the labor the entrepreneur personally expends carrying out the production project. If he personally manages his production process, he cannot at the same time labor elsewhere, so some of his net income must compensate him for these foregone wages. Therefore entrepreneurial profit is equal to net income minus interest and entrepreneurial wages. One can express the entrepreneur's profit mathematically using the following equation:

Entrepreneurial Profit = Total Revenue – Total Expenses – Interest – Entrepreneurial Wages

If the total revenue is greater than the expenses on factors of production, foregone interest, and forgone wages, the entrepreneur reaps a profit. If the sum of an entrepreneur's expenses, interest, and foregone wages are greater than total revenue, then he suffers a loss.

Because entrepreneurial profit is a monetary amount, this concept of profit is not a statement about the entrepreneur's happiness or subjective satisfaction. Profit, in this sense, is ultimately a statement about the entrepreneur's contribution to societal effort as appraised by other members of society. In the free market, the wants and desires of the people in society are manifested in their demand curves. These will ultimately determine the demand for and prices of factors of production. Remember that value flows up the structure of production. People value factors of production because they value the consumer goods those factors can be used to produce. If demand for a particular consumer good increases, this will result in a higher price for that good. Because the price of that good increases, producers will strive to make more, increasing their demand for the factors used to product that good. As the demand for those factors rise, so do their prices. Therefore, it is the demand for all of the consumer goods that ultimately determine the demand for and price of all the factors of production. We will examine these relationships later in the text, but it will suffice for now merely to be acquainted with them. We will see shortly just how and why profit indicates how well an entrepreneur is serving society.

THE TASK OF THE ENTREPRENEUR

The economic role of the entrepreneur is to forecast future market conditions and act accordingly. They advance money to the owners of factors of production in return for the use of those factors in making a product that can be sold in the future for money. It is the entrepreneur who directs production in a market economy. Entrepreneurs decide which factors are used to produce which products. In this act of production, however, entrepreneurs are guided by the wishes of consumers as they seek to earn the greatest profit possible. Entrepreneurs will maximize profit as they reap the largest rate of return possible.

In order to gain a better understanding of the nature of profit and the task of entrepreneurs, it is helpful to contemplate a world where there are no profits and no need of entrepreneurs. Suppose the economy made all of the adjustments necessary to reach the final state of rest. For this to occur, there could be no further changes in people's subjective values, natural resources at their disposal, technology, or population. After the final state of rest is reached, the economy would become an evenly rotating economy (ERE). If there were no further changes in any of the things mentioned above, people would continue to do the same things day after day. Producers would make the same products and consumers would demand those same products with the same intensities. The prices of all of the consumer goods and factors of production would be the same day after day. All factors would be allocated to their most valuable and productive use. In the evenly rotating economy, the future rates of return would equal the pure interest rate and would be the same in all industries.

No changes would come along that could upset the economic apple cart. Consequently, there would be no uncertainty. Because all rates of return would equal the interest rate, there is no benefit to be gained from investing less in one line of production and more in another. There would be no profit. All that would be earned is income for laborers and landowners and interest for capitalists. Because everything remains the same in the evenly rotating economy, generating the same incomes and interest, there is no need for people with insight regarding the states of future market. There is no need for producers to be people with insight regarding future market demand, because the goods demanded in the future would be exactly the same as those demanded today. In other words, there would be no need for the entrepreneur, and in any case there would be no profit for entrepreneurs to earn.

The evenly rotating economy is obviously an imaginary construction. It surely does not take a PhD in economics for anyone to recognize that our economy does not merely produce the same goods, incomes, and interest day in and day out. Values change, technology changes, population levels change, and the availability of natural resources changes.

There is a whole lot of changin' going on. If the evenly rotating economy is purely imaginary, why then spend any time discussing it at all? Because elucidating the type of activity that takes place and the forms of income that arise in such an imaginary evenly rotating economy can help bring the market phenomena in the real economy into bold relief. We can learn some principles about the economy of the real world in which we live by comparing it to an imaginary construction.

The primary difference between the evenly rotating economy and our own is, as noted above, change. Because things never remain perfectly constant from one period of time to the next, it becomes clear that tomorrow will not be exactly like today. Next year will not be exactly like this year. The next century will not be exactly like this century. Because we have no way of foreseeing perfectly what tomorrow, next year, or the next century will hold, we are brought to a firm conclusion: the future is uncertain. It is the uncertainty of the future that makes all actions speculative to a certain degree. This is no less the case for acts of production.

In the real and changing world, differences in the ratios between the prices of the final products produced and the sum of the prices of the factors of production emerge again and again. Therefore it is not enough for producers to know what could be produced and sold for a positive return today; they must forecast which products will be in sufficient demand in the future to justify certain expenditures on the factors used to produce those products today. They must be entrepreneurs.

No matter how fabulously successful an investment has been in the past, the future can always be different. As investment prospectuses often state, "past gains are no guarantee of future performance." History is flush with examples of such changes in the state of economic affairs. Investors in the motion picture industry in 1948 might be excused for expecting that business was good and it was only getting better based on past industry performance.

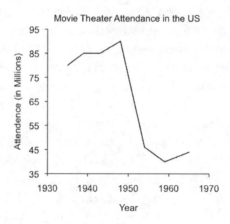

Figure 8.4. Motion Picture Attendance in the United States from 1935–1965.

In 1948 theater attendance tied an annual record of 90 million in paid attendance. However, anyone who made a substantial investment that year based on performance up to that time saw their positive returns eliminated like so many extraneous scenes on the cutting room floor. The advent of television and the U.S. Justice Department's destruction of the Hollywood studio system, helped bring on a decline in motion picture attendance, such that in only six years, ticket sales were down nearly 49%.

More recently, ask our friends at McDonalds how secure future returns are based on past success. One would be hard pressed to find a better success story in the history of free enterprise than McDonald's, which reaped profits continually since its origination in the 1950s. Investors in McDonald's may have thought there are financial constants after all. Such investors were in for a rude awakening during early 2003, when McDonald's Corporation announced its first quarterly loss ever, to the tune of $344 million.

The point is that no matter what has happened in the past, the future is uncertain. People might decide they want a change. Population might decrease. The make-up of the population might change. Natural resources used to produce what was successful in the past might become in short supply so as to make the price of those resources prohibitively expensive. Consequently, in order for producers to be successful over time, they must be successful entrepreneurs.

The main function of the entrepreneur is to forecast future market conditions. Because the future is uncertain, these future market conditions must be estimated by entrepreneurs. Production requires the entrepreneur to advance present money to factor owners in the speculation that the final product completed in the future can be sold for more money that it cost to produce.

As they anticipate future market conditions, entrepreneurs are always on the alert for investments in which they can earn rates of return greater than the going rate of interest. Suppose for instance, that Alexander Bullock is a previously successful entrepreneur whose most recent financial fortunes have turned rather thin. He is itching for another success, thereby avoiding bankruptcy. Bullock notices that the going rate of interest, the rate at which most investments are bringing in, is 6%. Bullock is not merely a capitalist; he does not want to invest his money in just any investment that reaps a positive return equal to the interest rate. He is looking to maximize his rate of return.

Based on a tip from his faithful butler Godfrey Parke, Bullock estimates that he can invest $1,000,000 now in the purchase of factors of production he can use, as general contractor, to make an office building that he can sell for $1,150,000 a year from now. The net revenue on his venture would be his total revenue $1,150,000 minus his initial investment of $1,000,000, or $150,000. His rate of return, then, would be 15%. 150,000 divided by 1,000,000 is equal to the fraction 15/100 or 15%. Bullock's rate of return is 15%. But what about his profit?

The profit reaped on any investment is equal to the difference between the entrepreneur's actual net return and the sum of his interest and entrepreneurial wages. This is because from the entrepreneur's perspective, interest and foregone wages are opportunity costs. Remember that the opportunity cost of any action is the highest valued alternative that must be sacrificed to do that action. If Bullock decides to invest in building production in the hopes of earning a 15% rate of return, he must forego the opportunity to earn 6% from another other investment. If the going rate of interest was 6%, he could take his $1,000,000 and deposit in a certificate of deposit or buy the average bond or purchase shares in the average stock and get a 6% return on his investment. If he decides to invest in buying factors of production used to make a building, Bullock cannot at the same time use that money to deposit at a bank. In an attempt to reap a 15% return he must give up his opportunity to reap 6%.

Therefore, the opportunity cost of any investment includes the foregone interest he could have received if he invested elsewhere.

Even the rest of his net income is not profit, however. Some of it compensates him for his labor as general contractor. As general contractor Bullock supervised the entire project, making sure materials are where they were supposed to be and when. He hired and oversaw the work of the specialists such as electricians. To serve as a general contractor for his building, Bullock had to forego the opportunity to work anywhere else. Suppose that he could have worked for a year for a local construction firm as a general contractor at a salary of $70,000. By undertaking his own building project, Bullock had to forgo the opportunity to earn $70,000 as a contractor in another occupation. Therefore, $70,000 of the net income earned in his own building project compensates him for his labor.

Consequently, in order to calculate an entrepreneur's profit, we must subtract the foregone interest and entrepreneurial wages from net revenue. Bullock did not reap a profit equal to $150,000. Out of that net income $60,000 was interest and $70,000 was wages compensating him for his labor. The profit that pays him specifically for his entrepreneurial foresight is equal to his net income minus interest and entrepreneurial wages. $150,000 - $60,000 - $70,000 = $20,000. Bullock's actual entrepreneurial profit is $20,000. This is the sum of his net income above his payment for time and payment for labor. This $20,000 is the money that he receives for providing the productive service of correctly forecasting the future building market and acting accordingly.

Bullock made a $20,000 profit, because he invested in the production of a good for which factors of production were underpriced from the point of view of future market conditions. Remember that the pure interest rate is determined by the time preferences of members of society. If that interest rate is 6%, then the marginal investors would be just willing to make an investment that earns them a 6% rate of return. Of course, they would prefer an investment that would return 10%. If more entrepreneurs would recognize that the future demand for office buildings would be such that investments earn 10%, there would be an increase in investment in building construction and, as we have already seen, the rate of return on such production would decrease. Investment would continue to increase and the rate of return would continue to decrease until it becomes equal to those for all other investments and the pure interest rate. Helping to bring about this result is the fact that increased investment leads to an increased demand for factors used in office building construction. Increased demand for these factors results in higher prices for them.

Bullock makes a profit precisely because he recognizes that the factors of production are underpriced compared to what future market conditions indicate. Other entrepreneurs were underestimating the value of the factors of production, while Bullock was able to see this better than his competitors and acted on his insight. Consequently, we see that profit is the *ex post* fulfilment of the entrepreneur's *ex ante* forecast.

As implied in the above paragraphs, actions of successful entrepreneurs actually work toward the elimination of profits. If investments are earning profits in an industry, investment in that industry will increase. Other entrepreneurs want a piece of the pie. This is why when one company makes a product that becomes a big seller, we often see many different versions of the same good produced by several different manufacturers. In 2002 for example, a big seller for Land's End was their All Weather Moc, a slip on shoe that protected one's feet from the elements as if wearing a boot. It was not long before consumers began seeing what was essentially the same footwear marketed by different sellers. In the L. L. Bean summer 2003 catalogue, the Comfort Moc appeared, selling at the same price as its counterpart from Land's End.

Increased investment in those industries where profits exist will set in motion a process by which the profits disappear. Increased investment is equivalent to an increase in the demand for factors of production. Increases in the demand for factors used to produce profitable final products will result in an increase in the price of those factors. As more final products are produced due to the increased investment, the supply of the final products will increase, so that their price will decrease. Higher prices for the factors of production and lower prices for the final product will tend to decrease net revenues and hence profits.

One might ask why entrepreneurs would want to invest in any line of production if they know that in so doing, they will begin a process that will result in their profits disappearing. Would they not rather want to invest in those projects for which the rate of profit would increase? Entrepreneurs are always on the lookout for the most profitable investment at a given point in time, other things equal. Remember our discussion of how the rates of return in every industry tend toward the pure interest rate that would obtain in the final state of rest. We saw that even though increases in investment in the production of golden flax seed led to a decrease in the rate of return in that industry, investment in that industry continued to increase as long as the rate of return on investment in golden flax seed exceeded that of steel production.

The same principle applies to a profitable investment in the changing world in which we live. Even though entrepreneurial activity will work to reduce profits wherever successful investment occurs, entrepreneurs will continue to invest as long as the rate of return reaped on that investment is the best alternative. Of course, successful entrepreneurs do not usually build a career out of investing in one and only one line of production their entire lives. They are always seeking to anticipate how future market conditions will be different than those at present and change their invest-ments accordingly. Because of the nature of our changing world, profit-able opportunities emerge again and again; those entrepreneurs insightful enough to anticipate them reap the profits.

What happens, however, if an entrepreneur is not successful in an-ticipating future demand? What about those investors who anticipated that future market demand for a particular good was going to greater than it actually turns out to be? What happens, in other words, if an en-trepreneur makes an error? When entrepreneurs are wrong about their market forecasts, they earn losses instead of profits. Economic losses are the result of entrepreneurial error.

Such errors occur when entrepreneurs make poor estimates of future selling prices of their final products and hence their revenues. Think again of our friend Alexander Bullock. Based on advice from his butler, Godfrey Parke, he invested $1,000,000 of his savings to buy factors he used to con-struct an office building he thought he could sell for $1,150,000. Suppose, however, that Godfrey, and hence, Bullock got it wrong. Suppose that the actual demand for Bullock's office building, once he was finished, was not as great as anticipated.

Suppose that in order to sell his building, he had to lower his price such that his total revenue was only $900,000. He invested $1,000,000 in the present and reaped only $900,000 a year from the time of the invest-ment. In this case, Bullock's net revenue would be as follows:

$$NR = \$900,000 - \$1,000,000 = -\$100,000$$

Bullock's net income is not even positive, but a negative $100,000. Because he misforecasted future market demand in the office buildings, he invested more heavily than he should have in the production of those buildings. So much for my man Godfrey.

Suppose, alternatively, that his net revenue, while not what he had hoped for, was still positive. Suppose that after investing $1,000,000 in the present, Bullock was able to sell his building at a price that yielded total revenue of $1,050,000, so that his net income was $50,000. This is positive net revenue, generating a positive rate of return equal to 5%. Many readers might be tempted to assume that Bullock would, in this case be reaping a profit because his rate of return is positive. Readers who assume this are forgetting about interest and entrepreneurial wages. The interest rate was 6% and foregone interest is an opportunity cost of investment. If Bullock's rate of return is 5% and the going interest rate is 6%, it is true that he earned a positive return, but he still reaped an economic loss. He reaped positive net revenue of $50,000, but an economic loss of $80,000. $50,000 minus his foregone interest of $60,000 minus his entrepreneurial wages of $70,000 equals - $80,000.

What economic principles can we glean from all this talk of winners and losers? For one thing, in order for the economic category profit to exist, we must be living not in the imaginary evenly rotating economy, but in the real, changing world where the future is uncertain. It is the uncertainty of the future that causes profits and losses to re-emerge over and over again.

Additionally, which entrepreneurs are profitable is determined by how well they forecast future market conditions. If an entrepreneur forecasts the future state of the market more correctly than anyone else, he will reap a profit. Every entrepreneur expects to make a profit as a result of his investment. No one hopes to lose his capital. Every entrepreneur expects that the market has underpriced the factors used to produce the good he thinks will turn a profit in the future. If the entrepreneur is correct, he earns a profit; if he is incorrect he reaps a loss.

However, it is not enough to merely forecast the future correctly. Entrepreneurs reap profits from forecasting *more* correctly than competitors. If everyone correctly anticipated future market conditions, then it turns out that no profit or loss would emerge. If everyone, for example, recognized that the future demand for recordings of symphonies by Sibelius would be much higher than at present, the prices of the factors needed to produce such recordings would immediately adjust to the level they would obtain in the final state of rest. The rate of return on Sibelius recordings would immediately equal the pure interest rate and profits on Sibelius recordings would evaporate. If all entrepreneurs forecast the future correctly, the prices in all industries would adjust quickly and the rate of return on all investments would be the same everywhere. There would be no profit or loss earned in any investment. We would be in an evenly rotating economy.

SOCIETY'S BENEFACTOR

We have seen that the task of the entrepreneur is to direct production by obtaining those factors used to produce goods based on speculations of future market conditions. We have yet to examine the socially beneficial nature of entrepreneurship.

While entrepreneurs are the ones who direct production, in order for them to actually achieve their goal of reaping profits, they must produce those goods that consumers want and they must make them less expensively than anyone else. Frequently one reads in the popular press that certain people—Corporate CEO's, Hollywood Stars, Professional Basketball Players, or owners of computer software companies—take in way too much money. All too often we read or hear that wealthy entrepreneurs get rich by stepping over the little people and therefore, justice demands that they "give something back" to the community. What this line of thinking ignores is that the only way entrepreneurs can continually reap profits over time in a free market is to give the community what it desires better than anyone else. Any entrepreneur who has earned a profit has already given something to the community: exactly what it wants. In a free market, entrepreneurs reap profits not by cheating people, but by serving consumers better by anticipating where factors of production are the most valuable. The factors of production have greater value because, as a result of the actions of insightful entrepreneurs, they are being used where they can best satisfy consumer demand. All sellers ultimately are dependent on consumers. The sellers of consumer goods are in direct contact with consumers. The desires of the consumers are transmitted from the sellers of consumer goods to the producers of higher order goods.

In our discussion of the production structure we noted that the value of higher order goods is derived from the value of lower order goods. If there is an increased demand for pizza, for example, this will increase the price of pizza, making pizza production more profitable. Entrepreneurs will tend to increase their investment in this line of production. The demand for flour, pepperoni, tomato sauce, mozzarella cheese, ovens, labor, land, and everything else used to make pizza will also increase. As a result of the increased demand, the prices on all of these factors will increase as well. Those entrepreneurs who can supply these factors at the lowest possible price will be the ones chosen by the pizza producers.

If the increased demand for pizza just posited would be due to people getting tired of hamburgers, the demand for hamburgers would decrease concurrently with the change in the pizza market. The decrease in demand for hamburgers would result in a decrease in price of hamburgers, making selling hamburgers less profitable. Producers would demand fewer buns, pickles, grills and less ground beef, ketchup, labor, land, and everything else required to make hamburgers. The prices of all these factors would decrease. Those entrepreneurs who cannot make a profit selling at the lower prices will leave that market and invest elsewhere, perhaps in pizza production.

While the entrepreneur is the one directing production, it is the consumers who determine which entrepreneur is successful and which one is not. It is ultimately the subjective judgements of consumers that determine not only the prices of consumer goods, but also the prices of all the factors of production. Therefore, when entrepreneurs reap profits, it is a sign that mismatches between how factors of production were being used and how consumers wish them to be used are being resolved. Profits are an indication that entrepreneurs are directing resources into uses that best satisfy the wants of members of society. The greater a person's profit, the greater he has been at reducing maladjustments and producing what society wants. On the other hand, entrepreneurial losses are a sign that those bearing the loss wasted capital from the point of view of society.

Capital tends to accumulate in the hands of efficient entrepreneurs. Those who are best at serving the wants of society, and therefore reap profits, have more income to save and invest. Consequently, they are able to accumulate capital. Those who make imprudent investments earn losses and their capital funds diminish. If Bullock invests $1,000,000 in office building production and reaps net revenue of only $900,000, he has less to invest next year. If unwise entrepreneurs continue to earn losses, eventually they will go bankrupt, run out of capital funds, and be forced to leave the ranks of the entrepreneur and enter the ranks of the laborers working for other entrepreneurs.

THE SUPREME IMPORTANCE OF MARKET PRICES

In the free market, prices function to transmit changes in supply and demand throughout the entire economy. Entrepreneurs use money prices to calculate expected profits and losses when comparing alternative investments. Without money prices entrepreneurs would have no way to compare the market value of different factors of production with the market value of different goods they could be used to produce.

To see how entrepreneurs use market prices to calculate expected profit and loss, let us examine the case of Mr. Blandings, an entrepreneurial construction contractor in the business of building people their dream houses. Among the many questions with which he is faced, one is the following: should he use concrete for building the structure of the house or should he use wood?

For simplicity sake, suppose there are only two uses for concrete: to build houses and to generate electricity in the form of a hydroelectric dam. Suppose that his various investment possibilities are as documented below:

Investment Opportunity	Rate of Return	Interest	Profit
Houses	5%	3%	2%
Dams	5%	3%	2%

Suppose Mr. Blandings must spend $2 million on factors used in producing a house. Half of that amount, $1 million, is the price of the concrete necessary to build one house. Suppose that he expects he can sell the dream house for $2.1 million in a year's time. His rate of return is 5%. If the interest rate is 3%, he earns a profit.

Likewise, it appears that investment in the production of electricity via hydroelectric dams is 5%. It costs producers $20 million in factors of production, half of which is concrete, to make a dam that will reap $21 million in electricity revenue in a year's time. The interest rate is the same 3%, meaning that producers earn a rate of profit equal to that for Mr. Blandings.

Now suppose, and this is where the fun starts, that the demand for electricity increases. Such a demand increase will result in an increase in the price of electricity, so more entrepreneurs will pour more investment into hydroelectric dam production. Consequently, the demand for concrete used to produce dams increases, resulting in an increase in the price of concrete.

As a result of this increase in the demand for electricity, less concrete will be used for house construction. Because of the increased market price of concrete, Mr. Blandings must now spend more of his savings on concrete necessary to build the same size house. Instead of obtaining enough concrete to build for $1 million, suppose it now costs him, $1.2 million, increasing his total investment to $2.2 million.

An increase in the price of concrete does not necessarily affect the demand for dream houses at all. Therefore, even though it now costs him $2.2 million to build a dream house, he is able to sell it in a year for the same price of $2.1 million. If Mr. Blandings continues to use concrete, he will reap losses and eventually go bankrupt. Those entrepreneurs who correctly foresee the increased demand for electricity and, therefore, invest in hydroelectric dam production will reap profits. Those, such as Mr. Blandings, who continue to build houses, will begin to use other materials such as wood or metal beams to frame their dream houses. Their actions free up concrete to be used in the construction of dams, which is exactly where the concrete is shown to be more valuable by consumers.

The beauty of free market prices, however, is not just that entrepreneurs can use market prices to calculate profit and loss, although that is indeed a wonderful thing. The beauty of it is that these same prices that allow them to make rational entrepreneurial decisions transmit the desires of consumers throughout the economy and serve as incentives for entrepreneurs to do precisely as consumers wish. Market prices provide for the coordination of productive activity toward the wishes of the consumers. Higher prices that entrepreneurs receive for their product allow them to better cover their costs of production, and encourage them to produce more of what people want.

Higher wages paid by entrepreneurs provide the incentive for more people to work at a particular job. In the late 1980s and early 1990s, one of the largest industries found in Omaha, Nebraska was telemarketing. There were firms that specialized in outbound calling, referring to when the telemarketer calls an unsuspecting resident. There were others who specialized in inbound telemarketing, in which buyers call a toll-free number to place an order, apply for a credit card, or something similar. There were, of course, some firms that did both. In the early 1990s, the going wage rate for inbound telemarketing was $5.00 per hour. It was relatively easy and much less stressful than outbound.

One firm, with which I was to have more than a mere casual acquaintance, contracted with a bank in the Midwest to accept telephone credit card applications for them. The response to the television advertisements broadcast throughout five states was much more than the telemarketing firm expected. They were swamped. The people they had working the phones were busy with wall-to-wall callers from the moment their shift started to the moment it ended with not a nanosecond of downtime. The company was in desperate need of more telemarketers. They needed to do something to encourage more people to come to work for them. To accomplish this, they placed large ads in the local newspaper promoting that they had openings paying $7.50 an hour. This was an unheard of rate for inbound telemarketing.

As one might expect, they had a huge response. Novices applied. Veteran telemarketers left their other employers to work for them. Those wanting to apply came out of the woodwork. It was clear that all of those people filling out their job applications and those waiting in line to apply were responding to the incentive of the higher price for their labor. Prices provide people incentives to do what consumers want them to.

Not only do market prices allow for entrepreneurs to calculate profits and losses so as to direct production out of some industries and into others, market prices are also useful in allowing managers to make the same sort of decisions in large companies. Monetary prices are used for effective management in the free market system.

Double-entry bookkeeping allows the entrepreneur to evaluate the profitability of the entire firm and also of each division within a firm. Double-entry bookkeeping is the form of bookkeeping that keeps track not only of how much money was spent, but also assigns each amount spent to the particular thing on which it was spent. A check book is basically a single-entry system. When you write a check and list its number, amount, and person to whom it was written, you make a single entry.

A double-entry system, however, is different. Suppose a firm uses $1,000 in cash to purchase equipment. There will be two entries showing the transaction. First, $1,000 will be taken out of the cash account, similar to what would show up in a person's check book. However, there will be a $1,000 increase in the office furniture account. This type of accounting allows the firm to not only keep track of how much it has spent, but also the value of all its assets, such as desks, computers, plant, and equipment, as well as the value of its liabilities and capital funds.

The double-entry system allows entrepreneurs to divide a firm's activities into different divisions and evaluate the profitability of each one. Each division can have a balance sheet, which can be assessed as if the division was its own firm. In this way, the entrepreneur can evaluate how each division is contributing to the success of an enterprise.

Each division is assumed to have ownership over the capital at its disposal. Each buys and sells the services from other sections. Each has its own expenses and revenues. Each has its own profit or loss. In a multi-divisional company, each division has its own manager that has a relatively large amount of independence. The only directive is to make as much profit as possible, while adhering to company policy. Such a system allows large companies to continue without an owner having to be immersed in all of the minute details. Branches that are evaluated as not profitable are eliminated while those that are profitable are retained and expanded.

Finally, it should be clear that merely because entrepreneurs have been successful in the past is no guarantee that they will be reapers of profits in the future. Every day requires another speculation. Every day brings with it the opportunity for profitable investment to be sure, but also for unwise investments that lead to loss.

Consequently, the possession of capital does not necessarily generate profit. Having the funds to invest, by itself, does not in any way guarantee profit. It is necessary to wisely invest savings. Profits are not the result of merely being thrifty; they are the results of investing funds in those projects that serve consumers better than anyone else.

SUGGESTED READING

Mises, *Human Action,* 286–91, 521–34. Mises' sections explaining entrepreneurial profit and loss and the rate of interest.

———, "Profit and Loss," 108–50. This is the quintessential essay explaining the nature and implications of profit and loss.

Rothbard, *Man, Economy, and State,* 367–451, 509–16. Rothbard's exhaustive exposition of the interest rate and entrepreneurial profit and loss.

9

Prices of Factors of Production

THE PREVIOUS CHAPTER WAS devoted to two categories of income received by different participants in the market economy. The first was interest income earned by the capitalist who saves a part of his income and invests it in some project that will yield a positive return. Interest income is, of course, earned by capitalists in the market economy, but would also be earned in the evenly rotating economy in a world without uncertainty.

The second category of income discussed was profit. Profit is the positive sum reaped by entrepreneurs according to their ability to forecast future market conditions better than anyone else. Those who make entrepreneurial errors earn losses. Because the existence of profit and loss is due to the uncertainty of the future, this category of income is earned only in the real market economy, and would never be reaped in the evenly rotating economy.

The present chapter focuses on the only other categories of income earned in the free market: payments to the owners of factors of production. Those who exchange their labor services or the use of their land for money are supplying future money (in the form of their factors) in exchange for money. Wages are income received in exchange for the provision of labor services. Likewise, monetary income for the rent or sale of land is just that: monetary income received by landowners in exchange for the services of their land. The principles of factor price determination also apply to the prices of capital goods.

In the tenth chapter of the Gospel of Luke, the beloved physician relates the account when Christ sent out seventy followers two by two to minister to the surrounding region. Christ told them to enter the homes of those who will have them and receive their meals from their hosts because, "the laborer is worthy of his hire." The word *hire* here means wages. Our Lord is teaching that those who labor are worthy of their wages. The question that economics helps us to answer in this context is: what determines the wages that a worker is worthy to be paid?

An important step toward answering this question is taken by noticing that wages are a specific amount of money paid to workers for a particular labor service. In other words, the wage rate is the price of labor. If someone is paid a wage of $28 per hour for work as a computer programmer, then the price of his labor is $28 per hour. Likewise, if the owner of a coffeehouse rents land and a building for $7,500 per month, then the price of that land and building is $7,500 per month. Once we understand that the wage rate and the rental price for land or capital goods are indeed prices, then how such payments to the owners of land, labor, and capital goods are determined becomes much less mysterious.

As it turns out, the prices of factors of production are determined like the prices of consumer goods: through the bargaining of suppliers and demanders. Now, you may be asking yourself, if the prices of factors of production are determined by supply and demand, just like those of consumer goods are, why have a separate chapter devoted to the determination of factor prices? After all, there are already two entire chapters devoted to price determination and price changes. This is a reasonable question. It is true that landowners, laborers, and capital goods producers make and sell their goods for the same reason as do makers of consumer goods. They hope to reap a profit. Additionally, the supply curve for sellers of factors of production tells us the same thing as for marketers of consumer goods: the quantity of their product they are willing to sell at any hypothetical price for their product.

Likewise, the demand curve for buyers of factors of production communicates the same information as the demand curve of consumers: the maximum quantity of the good buyers are willing to buy at any hypothetical price of the good. We must not forget, however, that there is a reason that we distinguish between consumer and producer goods. Consumer goods, you will remember, provide utility directly upon their use. They are demanded because they satisfy the buyer directly. Nothing has to be added to the consumer good for it to satisfy the user.

Producer goods are different. They do not yield utility directly, but only satisfy the user because they help him produce another good that he can either consume or sell for money that he can then spend on consumer goods he desires. For those of us who enjoy the occasional flourless chocolate cake, the butter, eggs, and semisweet chocolate used to make them do not provide satisfaction from consuming them directly. They satisfy us indirectly because we can use them to make a flourless chocolate cake that does provide us with an almost unspeakably ultimate chocolate experience. Therefore, while consumer goods are demanded for directly satisfying the wants of consumers, land, labor, and capital goods are demanded only because they help entrepreneurs reach their goal of producing a good that can be sold for a profit.

Additionally, even supply considerations are different between a worker supplying labor and an entrepreneur supplying a consumer good. The worker's opportunity cost is either the value of his labor in personal use or the value of leisure. The entrepreneur's opportunity cost of supplying a good is the foregone exchange value he could receive by selling it to someone else. Thus it makes sense to contemplate the determination of prices of producer goods separately from those of consumer goods. As has been the practice in this text when examining the determination of prices, we will look at supply and demand one at a time.

THE SUPPLY OF FACTORS OF PRODUCTION

Both the sellers of consumer goods as well as sellers of land, labor, or capital goods compare the value of the goods they own with different amounts of money that they could receive in exchange for each good. The law of supply applies to factors of production just as surely as it applies to consumer goods. If, at the point of sale, the price offered for a factor of production is higher, sellers would be more willing to sell additional quantities of the factor. Likewise, if the price offered for a factor is lower, sellers would be less willing to sell additional units of the factor. As the law of supply indicates, in general, there is a positive relationship between the hypothetical price of a factor of production and the quantity of that factor that sellers are willing to sell.

If we were to graph the supply curve for any factor of production, it would look basically the same as that for a consumer good. Supply for any factor of production can be represented by a curve that slopes upward to the right as shown in figure 9.1.

Factor of Production

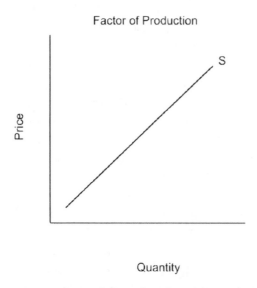

Figure 9.1. The supply curve for a general factor of production.

DEMAND FOR FACTORS OF PRODUCTION

What truly makes the process of determining factor prices distinct from that of consumer goods is the reason the factors are demanded. All goods are demanded because buyers think that the goods will help them achieve their ends. Factors of production are demanded by producers only because they think that these factors will help them earn a profit. What do entrepreneurs have to do to earn a profit? We have already seen that the important factor is how well the entrepreneur forecasts future market conditions. If he forecasts the future market better than his fellow entrepreneurs, he will reap a profit. If he forecasts worse than the others, he will reap losses and may even go bankrupt if he makes errors that are large enough or frequent enough.

Of course, merely forecasting the future state of the market better than anyone else will not be enough to actually bring him a profit. The entrepreneur must act on his forecast. If, in the early 1980s, Bill Gates had been able to perfectly foresee the world of computing and then leaned back in his La-Z-Boy waiting for the checks and letters to start pouring in, he would not have made one dime. In order to profit by a correct forecast, the entrepreneur must produce those goods that will actually be demanded by buyers in the future. Bill Gates became a very rich man because he successfully *acted* on his correct forecast.

Factors of production, then, help entrepreneurs achieve their end of earning a profit by allowing them to physically produce those goods that entrepreneurs think that they can sell in the future, at a price high enough to reap a profit. Consequently, the maximum price entrepreneurs are willing to pay for each hypothetical quantity of any factor of production is determined not by the direct utility it provides, but instead according to its contribution to the entrepreneurs' revenues.

This bit of knowledge, however, points to another puzzle. In our modern economy, most production processes are relatively complex, utilizing a large number of different factors. How do entrepreneurs figure out how each of the different factors affects their revenues? The problem is even greater than indicated by the question. Entrepreneurs do not decide whether to purchase or rent merely classes of land, labor, or capital goods. They do not even make decisions on whether to buy the services of particular types of land, labor, and capital goods. In order for an entrepreneur to make a profit, he must produce only those goods that he can sell at a profit. He must produce a precise number of goods. Therefore he will want to use a specific number of factors. Entrepreneurs are regularly faced with the decision of whether to rent or purchase specific quantities of factors of particular types. How does a computer company decide that forty programmers is the right quantity to hire instead of thirty-five or fifty? Our first step in discovering the economic principles of factor price determination is to revisit the concept of derived demand.

DERIVED DEMAND

Producers demand factors of production because they expect that by selling the goods that they make, they will be able to maximize their income. They therefore value factors of production to the extent that they help them in producing goods they think they can readily sell for a profit. This points us toward the principle of derived demand.

The demand for factors of production is derived from the demand for the product they are used to produce. Suppose, for example, that a new article in the *New England Journal of Medicine* reports on a link between consumption of semi-sweet chocolate and reduction in various forms of cancer. This would likely encourage people to consume semi-sweet chocolate and economic theory tells us that this would tend to increase demand for semi-sweet chocolate desserts that feature it—like flourless chocolate cake. If this is so, dessert makers would find the demand for flourless chocolate cake increasing, making it more profitable to produce. Producers would wish to crank up production and perhaps even new producers would enter the dessert market hoping to cash in on the new flourless chocolate cake craze.

What would happen to the demands for semi-sweet chocolate, butter, eggs, springform pans and all of the other factors used to produce flourless chocolate cake? They would increase as well. Because the demand for the consumer good increased, the demands for all of factors used to produce the consumer good increased as well.

On the other hand, suppose that for whatever reason, the demand for bread falls to zero. What would happen to the demands for wheat flour, yeast, bread pans and all of the other factors used to make bread? They would fall as well. If people do not want bread, bakers cannot reap profits by making and selling it. They could make it, but it would languish on the shelves, a moldy memorial to a grave entrepreneurial error. Producers would not willingly make a product that they knew would reap negative net income. Therefore, they would not pay good money for factors used to make a product that will not sell. If the demand for bread fell to zero, the demand for flour, yeast and everything else used to make bread would fall drastically as well.

We conclude, then, that as the demand for the final product goes, so goes the demand for the factors of production. This is what economists mean when they say that value is imputed up the structure of production. As people value consumer goods, producers of those goods value higher order goods. As those higher order goods are valued, so the makers of those higher order goods value the even higher order goods used to make them. There is a positive relationship between the value of the final product and the value of the factors used to produce the final product.

Because the demand for factors of production is derived from the demand for consumer goods, the prices of all factors of production are ultimately determined by the subjective preferences of all consumers in society. What we want to examine in this chapter, however, is the interconnection of the prices of lower order and higher order goods. The maximum amount entrepreneurs are willing to pay for the service of specific units of factors of production is determined through a process of appraisement. Entrepreneurs do not so much impute subjective value to factors, but appraise the largest amount of money they are willing to pay for the services of the factors they are considering using. Entrepreneurs begin the appraisement process by estimating the economic contribution specific units of factors make to their production operation.

THE MARGINAL REVENUE PRODUCT

The demand for a factor of production is positively related to the value it contributes to production. Remember, however, people never employ classes of goods to achieve their ends, but particular units of goods. They only take into account those units relevant for action, or the marginal units. The same is true of factors of production. When deciding whether to hire more workers, the producer does not contemplate using all the labor in the world, but a specific number of workers. Likewise, the producer does not consider renting land in general, but only specific plots of land with specific characteristics. Therefore the important consideration for a producer is how the marginal unit of each factor of production contributes toward achieving his end.

The value contributed by a relevant unit of a factor of production is that factor's marginal revenue product (MRP). The MRP is defined as the revenue attributed to the employment of the marginal unit of a factor. Because the marginal revenue product of a factor is equal to the revenue contributed by using that factor, producers will appraise specific units of specific factors of production according to each factor's MRP.

A factor's MRP can be calculated in two ways. If an enterprising woman Cecily Isabella produces Hawaiian shirts and is considering expanding her output capacity, she knows she will need access to another sewing machine. Buying another machine outright is not a financial commitment she wishes to make at this time, so she is considering renting one on an annual basis. The MRP of sewing machines is the extra revenue she thinks she can bring in with the additional sewing machine. On the other hand, if she already rents the sewing machine, the MRP of the sewing machine is the revenue she would lose if she had to do without the marginal machine.

Note that both ways to consider marginal revenue product involve revenue considerations. Not surprisingly, then, our understanding of how to calculate revenue holds the key to understanding how to calculate a factor's marginal revenue product. Producers use the expected demand for their product and what we call the production function to estimate each factor's marginal revenue product.

The production function is a mathematical formula showing the various ways different factors of production can be arranged to produce a given quantity of goods. Each good will have a different production function, depending on the technology needed to produce it. Some production functions will be very complex, involving many different factors of production. Some will be relatively simple.

Let us look at the production of Cecily Isabella who specializes in producing goods that evoke late-1950s suburban popular culture. Part of her enterprise includes a line of Hawaiian shirts. Suppose that our friend Cecily has the production function as seen in the following equation:

$$1L + 9N + 6C = 2,400 \text{ shirts}$$
$$1L + 9N + 7C = 3,000 \text{ shirts}$$

where L = acres of land, N = laborers, and C = capital goods. Her production function tells us that if she uses one acre of land, nine workers, and six capital goods, she is able to produce 2,400 Hawaiian shirts. Of the nine people working for her full-time, six of them operate sewing machines, one is a runner who prepares fabric and other materials for the sewing machine operators, and two work maintenance and clean up. Cecily, like all other entrepreneurs, does not merely want to make Hawaiian shirts just for fun. She is not interested in using them to decorate her home. She wants to make shirts because she thinks she can sell them in the future for a profit. What concerns Cecily is the revenue she will take in by producing and selling those 2,400 shirts. This is where her forecast of future market conditions comes in. Suppose that Cecily estimates that she can sell each shirt for $10 each. The total revenue that she will take in with her operation is $24,000.

Suppose Cecily finds that she really only needs one person for maintenance and clean-up. She estimates that if she had use of another sewing machine, she could take one of the clean-up people and assign them to run a sewing machine and therefore, increase her output of shirts. To determine how much she would be willing to pay to rent another sewing machine, she must estimate how much revenue the new machine will contribute to her shirt operation. She must again turn to her production function and demand for her shirts. Her production function shows that if Cecily adds another sewing machine to her operations, she is able to increase her output by 600 shirts. This additional quantity of output is the marginal physical product (MPP) of sewing machines in Cecily's firm. The marginal physical product is defined as the output attributed to the employment of the marginal unit of a variable factor. In this case, the variable factor we are considering is a capital good and the marginal unit is the seventh sewing machine. The total quantity of shirts Cecily can produce with seven machines and her other factors is 3,000. Recall that she can sell each of them for $10. This means that by employing a seventh machine, her total revenue increases to $30,000. The amount by which her total revenue increases is the marginal revenue product of sewing machines.

There are two ways for Cecily to arrive at her MRP for sewing machines. The first way is to simply subtract the total revenue earned without the marginal machine from the total revenue earned when using the marginal machine. In other words, she subtracts $24,000 from $30,000 and calculates that the MRP of sewing machines is $6,000.

Another way is for her to multiply the MPP times the price of her output. If, by obtaining the use of the marginal sewing machine (in this case the seventh machine), she is able to increase her output of shirts from 2,400 to 3,000, the MPP of sewing machines for Cecily is 600. This is because that is how many additional shirts she could produce by employing the seventh machine.

The reason her revenue increases with more capital goods is that she can produce more shirts. Therefore, if she multiplies the number of additional shirts she can make with the marginal sewing machine, 600, times the market price she can receive for each shirt, $10, she arrives at the additional revenue brought in as a result of her use of the marginal unit, or her MRP for sewing machines. If she sells 600 shirts for $10 each, she brings in $6,000 more. Again we see that Cecily's MRP for sewing machines is $6,000. Not surprisingly, this is the same amount we arrived at when we subtracted the total revenue reaped without the marginal machine from the total revenue earned when using the marginal machine. Both methods of calculating the MRP result in the same number. If you use both methods to calculate MRP and arrive at two different results, you have not made an economic breakthrough worthy of the Nobel Prize. You have made an error.

MARGINAL REVENUE PRODUCT
AND THE LAW OF RETURNS

For most entrepreneurs, their factor demand schedules include more than one unit of each factor. On their value scales they compare different amounts of money and different marginal units of land, labor, and capital goods. In order to more fully understand the factor price determination process, we need to proceed to a discussion regarding how entrepreneurs value not just one unit of a factor, but each marginal unit. Just as people in general do not value each unit of a homogenous consumer good equally, entrepreneurs do not value each unit of a factor equally, even if every unit is of the same quality.

Graphically, we can represent the MRP of different units of a factor of production as a negatively sloped curve. This is because, holding the quantity of other factors constant, there is an inverse relationship between the quantity of a factor and that factor's MRP. This inverse relationship exists for two reasons.

One reason is the law of returns. Recall from chapter 3 that if we vary the quantity of one factor that we use, while holding the quantity of other complementary factors constant, there is an optimal quantity of the variable factor to be used in order to achieve our end. This is the result of the physical limitations of finite goods. As the quantity of the variable factor used increases, while the quantity of complementary factors remains constant, a point is reached where the MPP of the variable factor begins to decrease. Remember that in order to allocate different factors so that none are wasted, entrepreneurs should use them so that their MPP is decreasing, but not negative. Because each unit of a variable factor will have a different MPP according the law of returns, each unit will also have a different MRP.

Suppose that Cecily's production function is such that her MPP schedule for sewing machines is as in table 9.1.

Table 9.1

MPP of Sewing Machines		
Machines	Shirts	MPP
0	0	---
1	600	600
2	1500	900
3	2400	800
4	3000	600
5	3400	400

Just as a demand schedule can be illustrated with a demand curve, so can the MPP of a factor of production be derived by plotting different factor unit-MPP combinations as points on a graph. Doing so, as is done in figure 9.2, allows us to see clearly the law of returns in action.

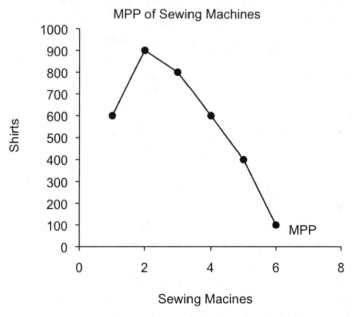

Figure 9.2. The marginal physical product for a variable factor.

Notice that if Cecily uses more than two sewing machines, the MPP decreases for each additional unit.

If a factor's MPP decreases as the quantity used increases, holding the other factors of production constant, that factor's MRP will decrease as well. Remember that the MRP of a factor is equal to that factor's MPP multiplied by the price of the final product. If each successive unit of a factor increases output by a smaller amount, then the revenue that factor brings into the firm decreases as well. If Cecily only uses three sewing machines, the marginal machine increases her output by 800 shirts. If she can sell those shirts at the market price of $10 each, then using the third machine increases her revenue by $8,000. If she uses a fourth sewing machine, this new marginal machine will increase her output by 600 so her revenue further increases by $6,000. Note that, as she uses more sewing machines, the size of the revenue increase, the MRP of the machines, is smaller. Although her revenue increases, it increases by a smaller amount.

Therefore, as the quantity of a factor increases, the revenue attributed to the marginal unit of that factor falls.

Additionally, we know that every actor will allocate each unit of means to the most highly valued end not already satisfied. As the quantity of factors of production increases, each successive unit of a factor will then be used to satisfy a less valued end. Therefore, the value attached to each successive unit will decrease. This is merely an application of the law of marginal utility. We can readily conclude that, holding the quantity of other complementary factors constant, as the quantity of a factor of production increases, that factor's MRP decreases, which can be shown as a negatively sloped curve as illustrated in Figure 9.3.

Figure 9.3. Marginal revenue product curve.

THE DISCOUNTED MARGINAL REVENUE PRODUCT

If Cecily is considering how much she is willing to pay for a seventh sewing machine and that machine has an MRP of $6,000, you may think that she would be willing to rent the machine for a maximum of $6,000. This is a natural assumption, but it would be incorrect. Do not forget that production takes time. The MRP of this sewing machine is an amount of money that Cecily expects to receive as a result of the production process. If the production of her shirts was instantaneous, then she would be willing to rent an additional sewing machine for up to $6,000, but, in fact, the MRP is always a future good. When Cecily has to make the decision whether to rent the machine, she has not seen a dime of the MRP. This is money that she will receive only *after* she has rented the sewing machine and used it during the period of production to make the shirts. Cecily rents the sewing machine in advance and only later will she receive that capital good's MRP.

What has this to do with how much she will pay for a sewing machine? Time preference. It is a universal condition of mankind that we value goods in the present more highly than we value those same goods in the future. The necessary implication is that people value future goods less highly than they value those same goods in the present. Cecily would not be willing to give up $6,000 in the present in order to receive $6,000 a month from now, after the shirts are made and sold. This is because she does not value both sums of $6,000 equally. The rent she has to pay is a present good and the MRP she receives after production is a future good. Consequently, she values the $6,000 she would have to pay in the present for the use of the sewing machine more highly than the $6,000 MRP she will receive in a month. Conversely, if she expects to receive a $6,000 MRP from using the marginal sewing machine, she will only be willing to rent it for some amount less than $6,000. The maximum for which she will be willing to rent the sewing machine will be a discount of the MRP. Hence we arrive at our conclusion regarding the most an entrepreneur will be willing to pay for each unit of a factor of production. The maximum price that an entrepreneur will be willing to pay for a unit of a factor of production is that factor's discounted marginal revenue product (DMRP).

The question remains, how much will entrepreneurs discount the MRP in question? Entrepreneurs discount the MRP because of the difference in value between present and future goods. Recall that the price of present goods in exchange for future goods, the minimum rate of return investors demand to cover their cost of time is the interest rate. Consequently, in order to arrive at a factor's DMRP, entrepreneurs discount that factor's MRP by the interest rate.

Suppose, for simplicity's sake, that the social rate of time preference is such that the annual interest rate is 12%, so that the rate of discount amounts to 1% a month. With this interest rate, the amount by which the MRP for the marginal sewing machine is discounted is $60, or the $6,000 MRP multiplied by 1%. The DMRP for Cecily's capital good, then, is equal to $6,000 minus $60 which equals $5,940. Consequently, the maximum amount Cecily will pay in the present for renting a sewing machine that will yield a MRP of $6,000 in one month is $5,940.

Because the DMRP of a factor is merely that factor's MRP discounted by the interest rate, the DMRP curve also has slope similar to the MRP curve. It is downward sloping as the quantity of the factor increases. Additionally, because the DMRP of a factor is always that factor's MRP minus the interest discount, the DMRP of a factor will always be less than the factor's MRP and the DMRP curve for a factor will always be below that factor's MRP curve.

With all of this talk of MPPs, MRPs, and DMRPs, it might be easy to forget the point of this discussion. We are trying to explain the principles related to the determination of factor prices. We need to find the conceptual link between all that we have been talking about and the price of a factor of production. This link comes from our discussion of the demand curve many chapters back.

Remember that the demand curve represents the maximum price buyers will pay for any given quantity of a good. We have seen that the maximum price a producer would be willing to pay for the use of a specific quantity of a factor is that factor's DMRP. Consequently, a factor's DMRP curve can be viewed as the demand curve for that factor. We find, then, that an individual producer's demand for the service of a factor is determined by three things: The marginal productivity of the factor or its MPP, the price for which the producer can sell her output, and social time preference which determines the interest rate.

DEMAND FOR FACTORS OF PRODUCTION: FROM INDIVIDUALS TO THE MARKET

Recall from chapter 4 that the market demand for a good is the sum of the quantities demanded by all buyers at each hypothetical price. Likewise, the market supply of a good is the sum of the quantities all sellers are willing to sell at each hypothetical price. For factors of production that are not specific to a production process, meaning they can be used in other lines of production, there are many entrepreneurs that can make use of their services. For example, many more than one firm or even industry make use of janitors or administrative assistants or assembly line workers. Every firm that uses their services is a demander of their labor. Of course, the many, many people who are willing to work as janitors, administrative assistants, assembly line workers, or college professors are suppliers of such labor.

In order to derive the market demand for a particular factor of production, we need to consider the other producers besides Cecily who want to make use of sewing machines in their operations. These other producers use their respective production functions to determine the MPP of machines for them. They also use their MPP calculations in conjunction with the estimated market price of the good they are selling to derive the MRP of sewing machines for them. They likewise discount their MRP for machines by the interest rate in deriving the DMRP of sewing machines in their operations. Their MPP, MRP, and DMRP schedules and curve manifest all of the same properties as they did for Cecily.

The market demand schedule for a particular good is the sum of all the individual demand schedules for that good. What is true for goods in general is true for specific factors of production. The market demand schedule for sewing machines is derived by summing up the individual DMRP schedules individual producers have for those machines. Therefore, because all of the individual DMRP schedules and curves manifest an inverse relationship between the quantity of a factor of production and its DMRP, the market DMRP schedule and curve will also show such an inverse relationship. Throughout the market, as the quantity of a particular factor of production increases, its DMRP decreases, hence the lower the hypothetical price producers would be willing to pay.

Once it is understood that the market DMRP curve for a factor of production is the demand curve for that factor, understanding the price determination process becomes much easier. As with the price of any good, the price of the factor is determined by the interaction between buyers and sellers. Just as with any economic good, a market clearing price will be arrived at through the bargaining of suppliers and demanders. For every factor of production, the market price will be that price at which the quantity demanded is equal to the quantity supplied.

The culmination of the process can be illustrated in figure 9.4.

Factor Market

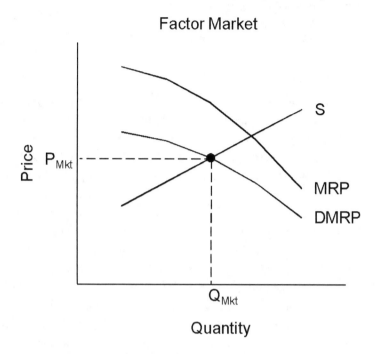

Figure 9.4. The market for a general factor of production.

If the price factor owners tried to obtain was above the market clearing price, there would be an excess supply of the factor as the quantity of the factor supplied would be greater than the quantity demanded. This would result in some frustrated factor owners who would be willing to supply the services of their factors at that price but could not find any buyers. In order to ensure that they would be the ones receiving income, the more eager factor sellers would bid down the price until everyone who wanted to sell could sell, to the price at which quantity demanded equals quantity supplied.

On the other hand, if the price producers tried to pay was below the market clearing price, there would be excess demand for the factor. The quantity of the factor demanded by the producers would be greater than the quantity supplied by the factor owners. This would result in some frustrated producers because they would be willing to buy the services of the factor at the prevailing price, but factor owners would not be willing to sell enough to meet demand. In order to ensure that they are able to obtain the services of the factors of production that they desire, the more eager producers would bid up the price of the factor until everyone who wants to buy can buy. In the market for every factor of production, the price that clears the market prevails.

CHANGES IN PRICES FOR FACTORS OF PRODUCTION

Just as changes in the supply of and demand for consumer goods will yield different market prices for those goods, changes in the value scales of buyers and sellers of factors of production will result in different market clearing prices for those producer goods. In an earlier chapter we identified a long list of factors that could lead to changes in the price of consumer goods.

Because of the nature of factor markets, we can simplify the analysis and focus on four things that result in changes in the market price of producer goods. The market price for factors is determined by demanders and suppliers. Therefore a change in the price of a factor of production will be the result in a change in demand or supply. Three determinants of the demand for factors of production have already been identified: the marginal productivity of the factor, the price of the good produced, and social time preference. If any of these three factors change, there will be a change in the factor's DMRP, its demand schedule, and hence, its market price.

The marginal productivity of factors is determined by technology and the quantity of complementary factors. A complementary factor is a factor that is used in conjunction with others to produce a particular good. A hot dog bun, frankfurter, ketchup, mustard, warmer, electricity, labor, and land are all complementary factors assisting the production of the consumer good: a ready-to-eat hot dog. Likewise, a farmer uses seed as well as fertilizer and land to grow a crop.

If a producer increases her stock of complementary factors, the productivity of the factor used with them will tend to increase. If for example, Cecily buys more fabric and hires more labor, she can produce more shirts with every sewing machine she rents. An increase in her use of goods that are complementary to the sewing machine increases the MPP for every unit of the machines. Because every sewing machine can produce more shirts, more revenue is attributed to each machine, increasing the MRP for every unit of sewing machine. Because the MRP for every sewing machine increases, the DMRP for every unit increases as shown in figure

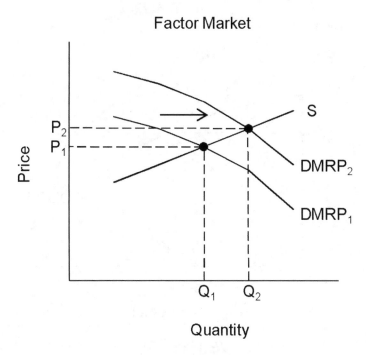

Figure 9.5. An increase in the DMRP of a factor of production results in an increase in its market price and quantity sold.

This increase in the DMRP schedule is equivalent to an increased demand, so the market clearing price of the factor will increase and more of the factor will be employed.

On the other hand, if for any reason producers find their stock of complementary factors decreasing, this will decrease the productivity of the factor used with them. If Cecily loses the lease on her land and if some of her workers quit without her being able to find replacements, she will not be able to produce as much with each sewing machine. The decrease in the stock of complementary land and laborers reduces the MPP for every machine. Because each sewing machine is now less productive, it will bring in less revenue, so the MRP for every machine decreases as well. Likewise, if the MRP for sewing machines decreases, its DMRP will follow suit, as illustrated in figure 9.6.

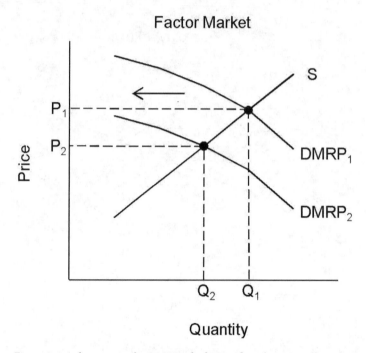

Figure 9.6. A decrease in the DMRP of a factor of production will result in a decrease in its market price and quantity sold.

As indicated in the graph, a decrease in the DMRP of sewing machines will result in a lower market rental price for sewing machines and fewer machines utilized in production.

Another determinant of the demand for a factor of production is the demand for the product it is used to produce. Suppose that people grow weary of Hawaiian shirts and the demand for the shirts decreases. This will result in a lower market clearing price for the shirts. A lower price for the shirts will result in all of Cecily's factors of production being worth less to her in that line of manufacturing. Even if all of her land, labor, and capital goods are just as productive as they always were, Cecily will bring in less revenue with each unit.

Suppose she is still considering renting the seventh sewing machine. It had an MRP of $6,000 because the output attributed to it was 600 shirts she could sell for the market price of $10 each. Now suppose that, due to the decreased demand, the market price falls to $9 per shirt. Even though the MRP of sewing machines is still 600 shirts, the revenue that the seventh unit brings in is no longer $6,000 but only $5,400. A decrease in the price of the final product will, therefore, result in a decreased MRP. As the MRP for a factor decreases, so does its DMRP. As we saw above in the graph in figure 9.6, if the DMRP for all sewing machines falls, the more eager machine owners will bid down their price, lowering the market price and reducing the quantity of machines rented.

If there is a general increase in the demand for Hawaiian shirts, however, there will be an increase in the market price for the factors used to produce the shirt. If people demand more Hawaiian shirts the market price for these shirts will increase. This increases the MRP for sewing machines that Cecily uses, because, while the machines are no more productive than ever, she can charge a higher price for each shirt that she produces.

Because the MRP for sewing machines increases, the DMRP for the same increases, meaning that Cecily finds the marginal machine more valuable. The factor price changes as illustrated in Figure 9.5. An increase in DMRP for sewing machines is synonymous with an increase in the demand for those machines, which will result in an excess demand at the original market clearing price. The more eager producers using sewing machines to produce Hawaiian shirts will bid up the price of the machines until everyone who wants to rent one can do so. An increase in the demand for the final product leads to an increased DMRP schedule for all factors used to produce that product. The market clearing price for those factors will rise and the quantity of them employed in production will increase as well.

Finally, changes in time preference affect factor prices through changes in the interest rate. The interest rate, you will remember, is the rate at which producers will discount a factor's MRP when determining the maximum amount he is willing to pay for its services in the present. The higher the interest rate, the greater is the discount, because the more valuable money in the present is compared to money in the future. We have already seen that when social time preferences increase, people are less likely to supply present money in exchange for future money, resulting in a higher interest rate that must be paid for such intertemporal exchanges. This increases the rate at which producers discount their factor's MRPs, which results in lower DMRPs for all of their factors.

Again consider that seventh sewing machine Cecily is contemplating renting. Using it for a month would reap an MRP of $6,000 a month from the present. The interest rate was 12%, so her monthly rate of discount was 1%. Consequently, the most she was willing to pay in the present for renting the sewing machine, the machine's DMRP, was $5,940. Now suppose, due to increased time preference that the annual interest rate increases to 16%. The monthly rate of discount would then increase to

1.33%. The size of the discount is larger. Instead of discounting the $6,000 MRP by $60, Cecily discounts it by $80, so the DMRP of the seventh machine used by her is $5,920 not $5,940.

What is true for the seventh sewing machine is true for every other machine Cecily could employ. An increase in the interest rate lowers the DMRP schedule, increasing the gap between the MRP and DMRP curves. If social time preferences increase, the DMRP schedules for all factors decrease. The results of such a change are shown in figure 9.7.

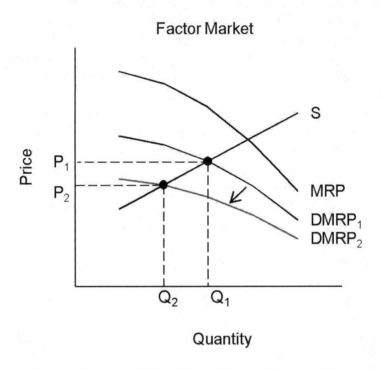

Figure 9.7. An increase in social time preference will increase the interest rate and cause a decrease in the factor's DMRP, which results in a decrease in its market price and quantity sold.

A decreased DMRP schedule results in excess supply of factors at the original price because the quantity supplied is greater than the quantity demanded by producers. Therefore, there would be frustrated factor owners if the price remained at that level. The more eager factor owners would bid down the rent of their factors until everyone who wanted to sell could sell. There would be a decrease in the market clearing price for the factors and fewer factors would be employed in production. This makes sense because if people have higher time preferences they are by definition not as willing to invest in production for future consumption. The services of fewer factors would be purchased at lower prices.

Of course social time preferences could decrease. Lower time preferences result in a lower interest rate, which decreases the rate at which future MRPs are discounted by producers. Even if productivity of factors and the price of final products do not change, so that the MRPs of factors of production do not change, a decreased interest rate will increase a factor's DMRP. If instead of the annual interest rate increasing to 16%, it decreased to 8%, this would lower the monthly rate of discount to .67%. The amount by which Cecily would discount her monthly $6,000 MRP on the seventh sewing machine decreases from $60 to about $40. The DMRP of that seventh machine would increase from $5,940 to $5,960.

As the rate of time preference decreases, the higher the maximum price producers like Cecily are willing to pay to obtain the services of factors of production. Therefore, decreases in time preference result in increased DMRP curves as seen in Figure 9.8.

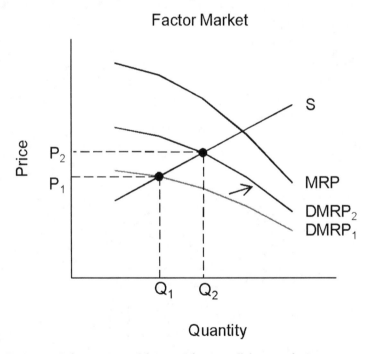

Figure 9.8. A decrease in social time preference will decrease the interest rate and cause an increase in the factor's DMRP, which results in an increase in its market price and quantity sold.

The gap between the factor's MRP and DMRP curves decrease. The increased DMRP results in what we would expect from an increase in demand for any economic good. If the price remained at the original price there would be excess demand for the factor, because the quantity demanded by producers would be greater than the quantity willingly supplied by factor owners. There would be frustrated producers, so the more eager producers would bid up the price offered to the owners of the factors. As the price of the factors is bid up, some producers who would have liked to purchase services of producer goods at lower prices will decide not to buy and drop out of the market. The price will continue to get bid up until every producer who wants to buy can. At this new higher market clearing price, the quantity of factors sold increases as well.

We have found that there are three variables that have a sure effect on the demand for factors of production. For any particular factor of production, its demand will be positively related to the stock of complementary factors. The demand for a factor will also be positively related to the demand for the final product it is used to produce. The demand for a factor will be inversely related to social time preferences.

Of course, the market for factors of production will also change if there is a change in supply. An increase in the supply of a factor will have the same result as an increase in the supply of any other good. Suppose there is an increase in the supply of sewing machines as illustrated in figure 9.9:

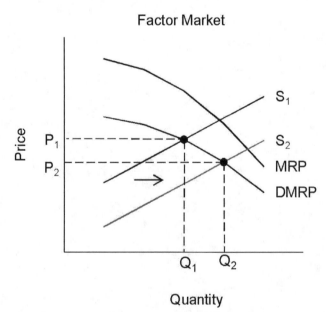

Figure 9.9. An increase in the supply of a factor will result in a lower market price and larger quantity sold.

In this case, the supply of sewing machines would increase. At the original price there is an excess supply because there is now a larger quantity for sale than there is demanded by producers. The more eager sewing machine owners will bid down the price of their machines until everyone who wants to sell can. As a result of the increased supply, more sewing machines are rented at a lower market price.

Conversely, if the supply of sewing machines decreases, the price all producers must pay for renting one will increase. This is illustrated in Figure 9.10.

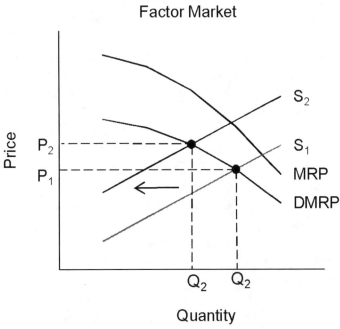

Figure 9.10. A decrease in the supply of a factor of production will result in decrease in its market price and larger quantity sold.

If the supply decreased, at the original price there would be excess demand because the quantity of sewing machines demanded would be greater than the quantity supplied. Some entrepreneurs would be frustrated. In order to ensure that they would have access to the machines they need, the more eager entrepreneurs would bid up the price for the sewing machines until everyone who wanted to rent could. Because of the decreased supply, fewer machines would be rented at higher prices.

While most of the above analysis has used the example of a capital good in deriving the principles of factor price determination, the same principles apply to land and labor. For both land and labor, the maximum price entrepreneurs will be willing to pay for specific units will be the DMRP of each factor. In a free market, the prices of land and labor will tend to be the prices at which the quantity of each factor demanded equals the quantity of their respective supply.

THE ECONOMICS OF FACTOR PRICES

When Christ sent out seventy of his followers to minister to the surrounding area, he told them "the laborer is worthy of his hire." Economics helps us understand what determines the wages that a worker is worthy to be paid, as well as what determines the price of land and capital goods.

Like the prices of consumer goods, the prices for factors of production are determined by suppliers and demanders. Factors of production are demanded by entrepreneurs because they provide productive services. Land, labor, and capital goods are demanded by entrepreneurs because those factors contribute value to the production process. They help entrepreneurs produce goods that they will sell for income in the future. Consequently, the demand for factors of production has three determinants: the productivity of the factor, the demand for the good being produced, and time preference. Entrepreneurs negotiate with factor owners supplying the use of their land, labor, and capital goods to agree upon a mutually beneficial price for their services. In a free market, the price system directs factors of production toward their most highly valued use. An unhampered market is crucial for scarce factors of production to be used wisely in the fostering of social prosperity instead of being wasted producing things people do not want.

SUGGESTED READING

Menger, *Principles of Economics*, 149–74. The seminal explanation of the value of factors of production from the fountainhead of Austrian economics.

Mises, *Human Action*, 330–36. Mises' exposition of the principles of factor price determination.

Rothbard, *Man, Economy, and State*, 453–507. Rothbard's outstanding chapter explaining the general pricing of the factors of production.

10

Competition and the Number of Sellers

THROUGHOUT THE FIRST NINE chapters we have discovered some of the first principles of human action; we have then derived economic principles regarding the mutually beneficial nature of exchange, the division of labor, the determination of prices, income maximization, and the sources of incomes in a free market economy. Most of the principles regarding price and income determination were derived while considering the case where there are many buyers and many sellers of the economic good in question. A good number of people, however, think that the economic principles considered so far do not similarly apply to an economy in which there are only a small number of sellers. They tend to think that in a market in which there is only one seller or a small group of sellers, some people are needlessly deprived of beneficial exchanges. These people argue that such arrangements harm consumers, and therefore, we should consider government regulation. Such was the rationale for the U.S. Justice Department's prosecution of Microsoft. This chapter examines what happens in a market when different sellers join together to form a cartel and also considers the issue of monopolies.

WHO IS SOVEREIGN?

Much of the concern people have regarding cartels and monopolies stem from their concept of consumer sovereignty. The word sovereignty implies authority coupled with power. The Bible reveals that one of the characteristics that God possesses is sovereignty. He is the creator and sustainer of all there is. He directs all things that come to pass according to his eternal decree. As such, God is the supreme ruler of the universe. He is sovereign.

The free market has been characterized as a system where the consumer is sovereign. This is understandable from a descriptive standpoint. If an entrepreneur wants to make a profit, and he does, he must supply what consumers want better than anyone else. He is not free to merely supply those goods that he would like to sell. If radio station owner, Gummo, for instance, persists in broadcasting Bach, Beethoven, Brahms, and Bruckner over the airwaves while the public at large has the disappointing taste to prefer music sung by the likes of Justin Timberlake and Britney Spears, Gummo will not reap a profit merely because he has better taste. If other entrepreneurs are more successful than he is in providing vapid music that consumers want at a lower possible price, Gummo will earn a loss. For Gummo to reap a profit, he must do what the consumers want. Therefore we can say that in a free market, the consumers are sovereign.

Unfortunately some people—many economists included—have sought to take the descriptive concept of consumer sovereignty and make it into an ethical imperative. They have moved from understanding that if entrepreneurs wish to reap profits they must serve consumers better than anyone else, to insisting that serving only the wishes of consumers is a moral requirement. We have seen earlier, however, that our sovereign God ordained private property. This implies that, with regard to our social relationships, neither the consumer nor the producer is sovereign in all economic affairs. Instead, each person in his stewardship is sovereign over his own actions. The right to property is the right for each property owner to use and dispose of his property as he sees fit. This is not to say that we are all sovereign over our lives in any theological or moral sense. Every person has a moral duty to live a life that is righteous before God. When we say that each person is sovereign over his actions, we are merely stating that no other mere human has a moral claim on his property.

On the other hand, people often adopt producer sovereignty as an ethical concept, especially if they think that the fortunes of particular producers are vital for local employment and a thriving economy. Thus, if people fear outsourcing some or all of particular production processes, they may call for protectionist tariffs and quotas on imports. Such policies protect domestic producers from foreign competition. However, tariffs and quotas also harm domestic consumers because they are forced to pay higher prices for fewer goods. In this case domestic producers are assumed sovereign over consumers. Producer sovereignty is also behind the use of eminent domain laws to seize someone's property in order to give it to private producers who want to use it to put up a new plant or retail store. In the case of such eminent domain seizures, producers are assumed sovereign over homeowners.

From an ethical point of view, the free market is a system that mandates neither consumer nor producer sovereignty, but personal sovereignty. Producers have the same right to use their property as the consumers have to use theirs. Both buyers and sellers are free to pursue their own ends. This means that a seller has the right to either sell the good he has produced or keep it. The consumer has no right to forcibly demand that the producer sell his good to him. At the same time, the seller has no right to force people to give up their property or to force people to buy from him.

The distinction between consumer sovereignty as a practical observation in the free market and personal sovereignty as an ethical concept is very important. Much confusion in economic policy regarding forms of business organization has occurred because the powers that be have not maintained this proper distinction. Keeping in mind these issues of sovereignty, however, will help avoid needless prejudice while examining the economic theory of cartels and investigating the question of monopoly.

CARTEL THEORY

In chapter 2 it was established that people act in order to achieve ends they most prefer. Entrepreneurs act to reap profits. Forming a cartel is one way sellers sometime seek to increase their profit. If we are going to examine economic principles that are applicable to cartels, it is helpful to define just what a cartel is. A cartel is a particular form of business organization in which a group of sellers agrees to act in concert to jointly maximize profits. I used to tell my students that a cartel was a group of sellers that acted in concert to maximize joint profits, until it occurred to me that this definition implied that all cartels were in the narcotics trade.

A key word in the above definition is *concert*. To act in concert means to act together. Sometimes different churches will hold what they call a concert of prayer. This is an event where different people come to pray together usually about something specific. Likewise, when a symphony orchestra plays a concert of Beethoven symphonies, each member of the orchestra, from the cellists to the oboists, join together to each perform their part in fulfilling their end of producing and participating in an exquisite musical experience.

A cartel, then, is formed when different sellers agree to act together. Specifically, the businesses involved agree to act as if they were one firm. Their goal is to reduce competition between themselves. By doing so, they hope to take advantage of a more inelastic demand, as illustrated in figure 10.1, allowing them to raise the price for the goods they sell and hence increase their total revenue and profit.

Figure 10.1. What a difference a cartel makes. Because members of a cartel face less competition from other sellers, each seller in the cartel faces a less elastic demand curve.

Why do they hope to face a less elastic demand after forming a cartel? Remember that the key determinant of demand elasticity is the availability of close substitutes. If the sellers agree to act together and treat each other as different divisions of one firm, the number of substitutes available to the consumers decreases. Suppose that there are three barbers in town, Figaro, Monsieur Beaucaire, and Floyd, each competing against each other and charging the market clearing price of $11 for one haircut. If only Figaro tried to increase his profits by raising the price of haircuts to $15, he would most likely be left with little more than a fist full of talcum powder. This is because consumers of haircuts perceive that the services of Beaucaire and Floyd are very close substitutes that cost 27%

less. If Figaro increased the price of his haircuts to $15 he could expect the quantity demanded by consumers to fall drastically. Such a move would be disastrous to his total revenue.

If Figaro would convince Monsieur Beaucaire and Floyd to work together with him, however, the three are no longer competing against each other, but working together to jointly increase their profits. If the three work together as if they were each a division of a larger barber business, the number of sellers effectively decreases from three to one. The number of available substitutes to haircuts from Figaro, Beaucaire, and Floyd decreases. Therefore demand is more inelastic. If their demand becomes sufficiently inelastic, they can agree to decrease their output and raise the price of their haircuts in order to increase their total revenue and profits.

Suppose the market for haircuts is such that, without a cartel, the market clearing price is $11 per cut. At that price 10,000 haircuts would purchased over the course of a year as indicted in table 10.1. Total expenditures on haircuts in this market would then equal $110,000.

As indicated in table 10.1, consumer demand is such, that of the ten thousand haircuts sold, Figaro and Monsieur Beaucaire would sell three thousand each, so their total revenue for the year would equal $33,000 each.

Table 10.1

Forming a Cartel Can Result in Greater Revenues											
Seller Revenues with No Cartel:											
Figaro			Beaucaire			Floyd			Market		
P	Q	TR	P	Q	TR	P	Q	TR	P	Q	TR
11	3,000	$33,000	11	3,000	$33,000	11	4,000	$44,000	11	10,000	$110,000
Inv = 30,000			Inv = 30,000			Inv = 40,000					
RR = 10%			RR = 10%			RR = 10%					
Seller Revenues with a Cartel:											
Figaro			Beaucaire			Floyd			Market		
P	Q	TR	P	Q	TR	P	Q	TR	P	Q	TR
15	2,700	$40,500	15	2,700	$40,500	15	3,600	$54,000	15	9,000	$135,000
Inv = 27,000			Inv = 27,000			Inv = 36,000					
RR = 50%			RR = 50%			RR = 50%					

At the market price Floyd would sell four thousand, making his total revenue for the year $44,000. If the investment necessary to produce haircuts equals $10 per haircut, Figaro and Beaucaire must invest a total of $30,000 each so both of their net revenue would be $3,000 resulting in a 10% rate of return (RR). Similarly, Floyd, to produce his 4,000 haircuts must invest a total of $40,000 so that he would reap net revenue of $4,000, also resulting in a 10% rate of return (RR).

Now, suppose that the three entrepreneurs decided to quit splitting hairs and work together forming a cartel. Suppose further that consumer demand does become less elastic as indicated in Figure 10.1. In that case, Figaro, Beaucaire, and Floyd could agree to raise the price of their product to $15 and, while the quantity demanded would decrease, the decrease is small relative to the increase in price. At a price of $15 consumers would buy nine thousand haircuts, increasing total expenditure in the market to $135,000. This larger total revenue can now be divided up by the same number of barbers.

If all three barbers agreed to cut their output by 10% (the same percentage drop for the market as a whole), Figaro and Beaucaire would reduce their number of haircuts to 2,700 and Floyd would reduce the number of customers serviced to 3,600. This means that, at the cartel price of $15, Figaro and Beaucaire would each bring in total revenue of $40,500. Both of their total expenditures on the 2,700 haircuts would be $27,000. Therefore, as a result of participating in the cartel, their net revenue would be $13,500, an increase of $10,500 for both Figaro and Beaucaire. The rate of return on their investment would be 50%. For Floyd, the cartel appears even sweeter. At the cartel price his total revenue would increase to $54,000. His necessary investment to provide his 3,600 haircuts would be $36,000. His resulting net revenue would be $18,000. Floyd's rate of return would be 50% as well.

It is easy to see why many buyers do not like cartels. In this case it means that in order to get a haircut in this economy, one has to pay $4 more than before the cartel existed. Again, however, we want to remember that, no matter how much consumers may dislike paying higher prices, the three barbers are not stealing from anyone. They have ownership over their factors of production. They own their shops, their chairs, their mirrors, their scissors and clippers, their combs, their land, their labor, and even the grooming products they use. No one has a moral right to a haircut. There is no canon of justice that demands haircuts be provided for only $11 each. Likewise, no one is forced to buy haircuts from the cartel. Consumers are free to look elsewhere or make do with fewer haircuts.

Nevertheless, people need not be concerned over the prospect of having their incomes bled dry by purchases from cartels. In a free market, there are two serious problems that all cartels face, making it very difficult for any cartel to continue for long. Both of these problems have to do with competition.

One challenge facing a cartel is competition from outside the cartel. If a cartel is successful and earning profits, this will attract other entrepreneurs into the cartelized industry. Suppose that, recognizing the tremendous rate of return earned by the cartel cutting hair, Sweeny Todd decides to try to get in on the action. Because this new industry entrant is not part of the cartel, Todd feels no obligation to play ball with its members. In an effort to draw customers from the cartel, entrants usually try to sell at a lower price. Todd may offer to cut hair for $11 per head. As more entrepreneurs follow suit, the total supply of haircuts increases, driving down the selling price available to the consumer. The cartel is then faced with the choice of either reducing its price and consequently its rate of return, or losing all of its customers to the new competitors. At some point, the cartel will tend to break down and adopt an every-man-for-himself strategy, which results in the sort of competition with which we are already familiar.

Another problem faced by a cartel is competition from one of its own members. This is what many economists call *cheating*. One member seizes an opportunity he sees to cheat on his fellow cartel members. Such competition could arise because of disagreements over production quotas. Those firms who already have a large output would like to keep it. Those who were expanding would like to continue. Those members who have a greater capacity for output due to larger factories will tend to demand larger quotas. The more efficient firms will be eager to expand and not be constrained by subsidizing the less efficient producers.

Perhaps Figaro, for example, does not think that he should have to cut his output back to 2,700 haircuts. He finds that he can do a lot better for himself than even the cartel outcome, if he cheats a little on his colleagues. Suppose he drops his price to $14, a dollar below the cartel price.

Because the only other barbers in the cartel customers can turn to are charging $15, Figaro's demand would most likely be very elastic. He would most likely find customers flocking to him for the removal of their locks. As shown in table 10.2, the quantity of haircuts he could provide would increase greatly along with his revenue and rate of return.

Table 10.2

Figaro's Revenue If He Cheats		
2,700 cuts @ $15/cut	=	$40,500
3,000 cuts @ $14/cut	=	$42,000
3,300 cuts @ $14/cut	=	$46,200
3,500 cuts @ $14/cut	=	$49,000

Once Beaucaire and Floyd detect a noticeable drop in their sales and discover the cause, they will not sit idly by as their customers leave for Figaro. They will tend to respond by lowering their prices and once again, the look-out-for-number-one strategy takes hold and the cartel disintegrates as each barber cuts prices to ensure that customers patronize them, resulting in competition without a cartel.

Nevertheless, it is, of course, possible that a particular cartel might find a way to make it work. If it turns out that cartels are able to earn profits while keeping outside competitors and cheaters at bay, it is still likely that the cartel will disappear. If it turns out that pooling resources in a cartel is indeed profitable for an extended period of time, then the cartel has an incentive to fully merge into one company. By merging we mean that instead of merely co-operating, the cartel members bring all of their resources into one firm. Figaro, Beaucaire, and Floyd, for instance, may decide that, because things are going so well, because there did not seem to be much of a threat from outside entrepreneurs for whatever reason, and because they did not destroy their cartel from within, it is in their interest to go ahead and form a partnership. This, of course, removes the cartel as a cartel, because there are no longer separate firms co-operating, but one firm with three divisions. We reach the conclusion that, because of the problems of competition from within and without the firm, and the incentive to merge if the cartel appears successful, in the free market cartels are fragile and fleeting.

Of course, the specter of cartel members merging to form one company will cause many to merely shift the focus of their worry from the operations of a cartel to concerns about the one large firm. This they view as something even worse: a monopoly.

THE QUESTION OF MONOPOLY

During the twentieth century much ink was spilt on economic analysis of monopolies. Part of the reason for so much discussion is that the concept of monopoly itself is not easy to definitively pin down. When analyzing something, it is helpful to know just what that thing is and have a good understanding of the nature of that thing. It turns out that the term monopoly is notoriously ambiguous. There are many ways to define monopoly, all of which will result in different analytical conclusions.

One approach some use to define terms is the linguistic approach. One can trace the etymological roots of a word. For instance, monopoly is of Greek derivation. It is from two words: mono, meaning *one* and polein, meaning *seller*. Literally, then, monopoly means *one seller*. This is not to be confused with the literal meaning of the word politics, which, rumor has it, may be derived from the Greek poly meaning *many* and tics meaning *blood-sucking parasites.*

The literal definition of monopoly, however, is not overly helpful if we want to undertake meaningful economic analysis. Although *one seller* seems exact enough, in reality it opens the door for a plethora of contrasting conclusions. For example, if one defines a product narrowly enough, then everyone is a monopolist. Very few goods are exactly alike. Everyone's labor is at least of slightly different quality than others'. Different products have different qualities that make each unique. Although some are more unique than others, none are exactly the same. Consequently, no seller is selling exactly the same product as other sellers. This makes each seller the only seller of a unique product. However, this makes every seller a monopolist, which makes analysis of monopoly rather meaningless. If every seller has a monopoly, the term monopoly ceases to be a distinct concept requiring economic analysis. All there would be is analysis of the market.

On the other hand, if the product is defined broadly enough, then no one is a monopolist. Most goods have substitutes that can be used in their stead. Veal can be eaten in place of beef. Apples can be eaten in place of mangoes. Additionally, from a very broad economic standpoint, every good is a substitute for every other, because every good can be used to achieve some end. If I forego purchasing a mango, I can spend my money on beef instead. If my end is to increase my state of satisfaction, I can substitute beef for mangoes. In this sense, every seller competes for the consumer's dollar with every other seller. This implies that no one is a monopolist. There are no firms that are the only sellers because every seller is a marketer of a good that can increase satisfaction. If there are no monopolies, it would be rather silly attempting to analyze such a fiction.

Additionally, even if there is a seller that can be identified as the only seller of a particular good, it is unclear that this fact alone makes the case worthy of special analysis. Just because one is the only seller of something does not guarantee anyone profit. If no one demands the product, no one will buy it even if there is only one seller.

The issue of buyer demand brings us back to the important truth that it is ultimately the consumer who determines which seller is the only seller of a particular good. Whether there are many substitutes for a certain product or no close substitutes is determined by the subjective values of buyers. Indeed, some items that have the same physical make up will be perceived as different units of the same good by some people and as completely different goods by other people. For example, some people like to drink colas, but think they all taste alike, so to them Coke, Pepsi, and RC are different sellers of the same product. Other people, however, feel very definitely that RC is clearly superior to the others and that the maker of RC is one seller of a unique good. Consequently, the subjective nature of economic value itself makes the concept of a monopolist as the only seller of a good ambiguous at best, and hence, not very useful for economic analysis.

Another definition gets us closer to a meaningful description of monopoly. It is also the original definition. When the word *monopoly* was first used, it meant a grant of special privilege by the state. This can be seen by taking a look at the most popular economics textbook written by an American author in the mid-1850s. During the early to mid-nineteenth century, there were three very popular economics texts. One was the famous *Wealth of Nations*, another was *Treatise on Political Economy* by the Frenchman, Jean Baptiste Say. The third was by the Calvinist Northern Baptist and President of Brown University, Francis Wayland. In his best-selling text, he defines monopoly as "an exclusive right granted to a man, or to a company of men, to employ their labor or capital in some particular manner."[1]

There have been a number of various types of government privilege that the state has granted certain sellers over the years. One is a govern-

1. Wayland, *The Elements of Political Economy*, 116.

ment license. A government license is a legal right granted by the state to enter an occupation or industry. Examples would be teacher certification, cosmetology licenses, and taxi medallions (signifying that the holder has the legal right to sell taxi services). Some states even have laws requiring shampooers in hair salons to be certified by the state. A government license gives only those who possess the license the legal right to sell their product.

If one does not have the state license, he cannot legally provide the good. For example, in Arizona a seventeen year old entrepreneur started a successful business of rat proofing homes. For $30 each he would climb onto the roof of his client's house and cover any openings with wire mesh. Because demand for his services was so great, he hired three of his friends to help. However, the day after his business was publicized in a news story in the *Arizona Republic*, the state told him he had to cease and desist his pest control business. He was guilty of doing structural pest control on buildings without a license. Because this teen did not have a government license, he was prohibited from serving customers.

Another form of government-granted special privilege is the patent. A patent is a monopoly right granted by the state to an inventor of a product or process. The patent gives only the inventor the right to produce and sell the good for a certain number of years. As such, the patent prevents other investors from producing and marketing similar products. In the United States, the monopoly right to sell a product for which a producer has a patent lasts twenty years. Similar to patents are copyrights, which give authors, publishers, or composers the monopoly right to market their creative product.

A third form of special privilege granted by the state is the public franchise. This is a business' exclusive state-granted right to produce and sell a commodity or service. This type of state-granted monopoly is similar to the government license, because only those given the privilege by the state are allowed to produce and sell their goods. It is different, however, in that the recipient of a public franchise is the *only* seller who can legally sell the product in question. The best example of the public franchise in the United States is the Postal Service. The Postal Service is the only business in the United States that can legally deliver first class mail. There are a number of transport services such as UPS and Federal Express that can compete against the U. S. Postal Service in the delivery of overnight letters and parcel packages. However, no one is allowed to deliver first class mail besides the Postal Service. Another example would be cable television. In most cities, there is only one supplier of cable television who has the right to sell based on a public franchise granted by the city council.

In every case, grants of special privilege by the state keep some entrepreneurs from entering the industry in which the monopoly privileges are being passed out. Because such actions keep entrepreneurs from entering a profitable industry, such special privileges are called barriers to entry. The government license, patent, and public franchise all prohibit some entrepreneurs from entering the monopolistic markets.

The economic effect of all of these state-granted special privileges is two-fold. In the first place, because some sellers are kept out of the protected markets, the supply of the product will tend to be less than it otherwise would be. Additionally, because such monopoly grants reduce the number of competitors, it also reduces the number of close substitutes for the product sold by the monopolist. Consequently, grants of special privilege result in demand for the goods being less elastic then they would be without the monopoly grant.

Although the definition of monopoly as a grant of special privilege given by the state is consistent and makes good sense, it should be clear that such grants could never be made in a free market. A free market is a market in which every entrepreneur can use his property as he sees fit, including entering a profitable industry if he deems it wise. Because grants of special privilege are the result of government intervention, we will delay detailed analysis of such grants until later in the text. Additionally, when most economists use the word monopoly, this is not the case they are thinking of.

The third, and most popular (if not most correct) definition of monopoly is a firm that has *market power*. You might naturally be asking: what is market power? It is the power to charge a monopoly price. What is a monopoly price? It is the price that is charged by those with market power. The theory is a bit more straightforward than that. The claim is that if a seller's demand is inelastic at the competitive price, then the monopolist can restrict output and increase the price of the good in order to increase total revenue and profits. The monopolist can reap monopoly profits in perpetuity, as long as demand is inelastic.

There are several problems with such a definition. In the first place, no matter how inelastic demand is, a literal monopolist is not removed from competition. If a seller continues to raise his price, at some point the demand for his good will become elastic, so that he could no longer raise prices without decreasing total revenue. Additionally, as the market develops, goods become more specific, there are more substitutes for a seller's good, and demand for it becomes more elastic.

Another area of concern regarding firms with market power is what has been called *cutthroat competition*. This occurs when a seller sells his product at a very low price, so that he drives the competition out of business and then has a monopoly so he can, and will, raise prices. However, we must remember that the failure of any entrepreneur is due to forecasting the future incorrectly. Business failure helps direct factors of production to where they will be used more productively. Selling goods at a very low price is great for the consumer. When Microsoft was charged with violating Federal anti-trust laws, it was mainly due to the fact that they were giving away their web browser, Internet Explorer, along with their Windows operating system. They were charging the very low price of zero. The criminal complaint brought against Microsoft was not brought by computer consumers. It was brought by Microsoft's competitors. Consumers were happy to benefit from Microsoft's giving away its web browser for free.

Additionally, if a seller is able to charge a high and profitable "monopoly price," other entrepreneurs will be drawn into the industry, as long as it is a free market. As soon as this happens, of course, the supply of goods increases and the demand curve for the monopolist's product becomes more elastic. Finally, it is hard to tell whether a seller is engaging in cutthroat competition merely by looking at the price being charged. Instead of trying to drive competitors out of business, an entrepreneur could be selling at low prices merely to liquidate his stock following an unwise investment.

Given the above analysis, it appears that, like the case of the cartel, consumers do not need to be overwrought at the prospect of monopolists hiding under every bush ready to spring out with another monopoly price. In a free market, there will be entrepreneurs entering a market whenever profits are being earned. Such entry always puts downward pressure on the market price for the good being sold.

Further analysis shows, however, that not only should we be unconcerned about monopoly prices, but the concept of a monopoly price might itself be an illusion. This is because the concept of a monopoly price only makes sense in contrast to a competitive price, but in a free market there is no identifiable competitive price that we can distinguish from an alleged monopoly price. Remember that the most common contemporary definition of a monopoly price is the higher price that a seller charges after taking advantage of an inelastic demand. He keeps the price high by restricting his output. It seems, then, that there are two tests we can use to identify a monopoly price: is the seller increasing his price in the face of an inelastic demand and is the seller reducing his output?

Neither of these tests, however, will necessarily tell us what we want to know. We have already seen that after an investment in production is made, every seller will want to charge a price that earns him the highest total revenue. He does this by increasing the price he will accept for his good as long as demand is inelastic. He will stop raising his selling price as soon as demand turns elastic. Now if this is the practice of every seller, one can hardly identify every entrepreneur who acts accordingly as a monopolist. This would make everyone a monopolist, which again makes the concept itself irrelevant.

Because of this problem, some have sought to use restriction of production as a test for monopoly price. As indicated above, however, this cannot be used as a scientific indicator, because there are additional reasons a seller might decrease his output besides attempting to charge a higher price. A producer reducing his output may be doing so because he has not sold all of his output in the past. He might be attempting to cut his losses. In fact, this is exactly what we have already seen tends to happen when entrepreneurs begin to earn losses in a particular line of production. Investment leaves that industry, so that the supply of goods in that market decreases, resulting in an increased price. None of this activity resulting in lower outputs and higher prices is the result of monopolistic behavior. It is the result of entrepreneurs leaving an industry earning losses. This is hardly the picture of a seller rolling in monopoly profits.

Our analysis leaves us with the interesting conclusion that in a free market not only is it not unjust for a seller to charge a so-called monopoly price, but the concept itself is rather meaningless. Monopoly gains only make sense in an environment where the state is erecting barriers to entry by giving some firms grants of special privilege.

ECONOMICS OF LABOR UNIONS

One more form of commercial organization that is of much interest is similar to a cartel because it features suppliers acting in concert to increase their income. However, the suppliers in question are not capitalists or producers, but sellers of labor. The organization of which we are speaking is the labor union.

Labor unions are organizations of workers who come together to bargain collectively with their employer. Instead of each individual worker negotiating personally with the owner of the firm, the workers band together to negotiate with the firm and arrive at a labor contract that applies to every worker employed by the firm. It is hoped by the workers, that, by banding and bargaining together, they will be able to negotiate a better deal, resulting in increased compensation. Often this means higher wages, and in the analysis that follows we will examine what will be called the union wage.

Understand that there are other forms of compensation that firms can provide to their employees. One is health insurance coverage. Another is vacation and another is sick leave. Workers also try to achieve some measure of job security so that they will not be as worried about losing their job in the near future. Additionally, a firm can simply try to make the working environment better. Consequently, when we use the term *wage* in the following analysis, we are speaking about compensation in general and not merely the literal money price for labor. We are speaking of the full monetary value of compensation for labor that includes health care, vacation pay, sick pay, and expenses incurred for job security and improving the workplace.

Cartels are often accused of engaging in monopolistic practices because, as you remember, they attempt to increase their total revenue and profit by restricting output in order to increase their selling price. Labor unions also tend to bargain for higher wages and are often accused of trying to enforce monopoly wages. This is incorrect, however, because a cartel or alleged monopoly increases its selling price by restricting its own output. As we shall see, a labor union does not increase the wages of its members by decreasing the supply of labor offered up by its members. In a free market each person owns himself and does not withhold his labor if he is working at the union wage.

Instead union wages are better characterized as *restrictionist* wages. How so? A union is successful only to the extent that it can convince the employer of its members to pay higher wages than the firm would otherwise. The union, in effect, negotiates a voluntary minimum wage. The effect of such a wage is seen below in Figure 10.2.

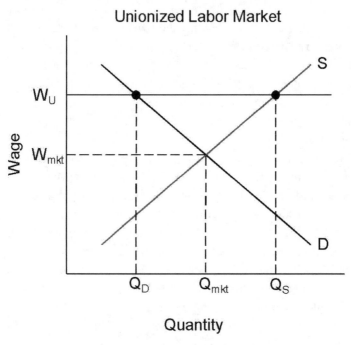

Figure 10.2. A unionized labor market. At a union wage above a non-union market wage, the quantity of labor hired in the unionized industries will be less than the quantity of labor supplied.

Without a successful union, the wage rate will be what we would expect. The market wage, W_{mkt} would be that wage at which quantity supplied and quantity demanded are equal. This market quantity Q_{mkt} is the quantity of labor hired in this labor market.

If the union is successful in negotiating increased compensation, the employer is contractually obligated to pay a wage that is no lower than the union wage, W_U. Because this wage becomes the minimum wage necessary to induce workers to supply their labor, the supply of labor curve becomes, in effect, a horizontal line at the union wage. It is horizontal until quantity Q_S. In order to hire more than this quantity, workers' value scales are such that the firm would have to offer an even higher wage. Therefore, whenever a union wage is enforced, the effectual supply of labor curve looks like a hockey stick lying on its side.

Once we have identified the relevant supply curve in a union environment, we are able to analyze the results of a union successfully negotiating increased compensation. We are also able to see why a union wage is a restrictionist wage and not a monopoly wage. If a union successfully negotiates wage rate W_U, two things become clear: there will be fewer workers hired in this industry, but those who remain employed will received increased compensation. Note, however, that the higher wage earned by union members is not the result of their limiting their own supply, it comes by restricting competition by other workers who are willing to do the same job for a lower wage. A union wage is made possible by members restricting the labor of other workers.

This brings us to an important point often overlooked in discussions regarding labor unions and labor-management relations. It is often thought that labor and management are adversaries, that employee and employers are the ones competing against each other. This is not correct. In the first place, both groups—those who traditionally have been classified as *labor*, such as assembly-line workers, and those who traditionally have been classified as *management*, the workers' immediate supervisors—are laborers. Assembly line workers and their foremen, teachers and their principals, coal miners and their managers, are all employees of the firm's owner, the entrepreneur. In an economic sense, both labor and management are in the same boat. Both are hired by the firm's owners to help achieve a profit. Both are, indeed, laborers.

More importantly, in order to achieve production and receive incomes, the employee and the firm must cooperate. They must work together and engage in mutually beneficial exchange in order to increase both of their levels of satisfaction. A worker who is hoping to land a job at an acceptable wage is not competing against his employer, but against other workers. It is the existence of other workers that threaten the position of union members earning higher union wages. In the free market, laborers, landowners, capitalists, and entrepreneurs cooperate with each other. Workers compete against other workers. Entrepreneurs compete against other entrepreneurs. Capitalists compete against other capitalists.

As a consequence of successful collective bargaining on the part of the union, those working in unionized industries and firms earn higher wages and increased benefits. They do so, however, at the expense of other workers. As a result of the higher wage, fewer workers are employed in the unionized industry. Firms in the industry that have just agreed to the increased union wage cannot afford to hire the same number of workers. Therefore, some will be laid off. Those people must look elsewhere for work. They will find it only in those areas where they are less suited, where their DMRP is lower. In a free market, if their DMRP would have been higher in the other industry, they would already have been working in that industry. The fact that they were not indicates that this move is a move to a place to which they are less suited. Because their DMRP is lower in this other non-unionized industry, they will earn lower wages than they were at their job before they were laid off.

However, those laid off in newly unionized industries are not the only workers whose wages will be lower. As indicated in figure 10.3, wages for *everyone* employed in non-union industries will tend to decrease as well.

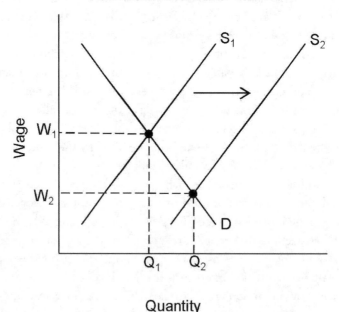

Non-Unionized Labor Market

Figure 10.3. As workers are restricted from being hired in the unionized industry due to an above-market union wage, the supply of workers in non-unionized industries increase. This results in lower wages received by workers in non-unionized firms.

If there are fewer workers hired in unionized firms, some of which are laid-off, where will these workers try to find gainful employment? Non-unionized industries, naturally. The supply of labor in those industries will, consequently, increase, resulting in lower wages for all workers in those industries.

We see, then, that when a union is successful in bargaining for increased compensation for its members, its members benefit at the expense of other workers. The costs of union wages, however, run even deeper. Unions also benefit their members at the expense of society as a whole. How so? In the first place, the greater the scope of unionism, the fewer non-union jobs there are, so the harder it is for non-members to find a job even at lower wages. Consequently, the more pervasive unionism is, the more permanent mass unemployment will be. Additionally, successful union activity is akin to retarding the division of labor. Because union wages result in allocating labor toward less productive uses, total output is reduced and societal wealth and general standards of living are not as high as they would be otherwise.

Because of these benefit-cost trade-offs and because union wages are higher than market wages, several economists have argued that union activity is inherently opposed to the principles of the free market. This is not true. Labor unions and their collective bargaining are not necessarily incompatible with the free market. As long as membership in the union is voluntary and as long as unions do not violate anyone's right to their property, they are compatible with the free market. Within a Christian private property ethic, it is legitimate for any group of workers to organize and attempt to bargain collectively with their employer. In a free market, the employer would also be able to negotiate with the union collectively if he so chooses or only with each worker individually if he desires.

With these free market characteristics in mind, let us investigate the methods labor unions use to achieve their ends of increased compensation for their members. The most well-known method unions use is the strike. A strike occurs when union members walk off their jobs until their demands are met. It is hoped by the union, that the mere threat of the strike will be enough for the owner to agree to the desires of the union and increase the workers' compensation accordingly. In a free market, the strike can be successful if there are not very many people willing to cross the labor picket lines and work for the firm, replacing striking workers. Historically, many workers have adopted a commandingly high place on their value scales for the goal of not undercutting union wage rates. One can identify periods when people really believed in the sanctity of the picket line, when crossing the line and going to work for an entrepreneur was a moral wrong. Social pressure is also applied to those who replace striking workers. They are often subjected to verbal abuse. The common pejorative slung at such workers is "scab," as if people eager to work in order to support their family are no better than a piece of scratchy dried blood covering a wound. Such behaviour is definitely not living out loving one's neighbor as oneself.

While the strike can be a legitimate course of action for a labor union to undertake in a free market, in today's environment throughout the Western world, labor unions also greatly benefit from government intervention in the market. Union shop rules are laws requiring a firm to deal only with the union as the sole bargaining agent for all of the workers at the firm. If there is a union that is designated as the sole bargaining unit for the workers, even those laborers who are not union members and who wish to negotiate their own contract cannot do so, but must be represented by the union and have the terms of their labor contract stipulated by the union agreement.

Another tool that unions use to achieve their goals is lobbying for increases in the national minimum wage. Although union members almost never earn the minimum wage (union wages are usually much higher) it is in the union's interest to have a relatively high minimum wage because it helps shield them from lower wage competition, increasing their bargaining position. If for instance, owners of a factory in Ohio are being hurt financially due to union labor costs and are thinking of moving the plant to an area in the south where wages are lower because of fewer labor unions and less capital investment, they are less likely to do so if the minimum wage they would have to pay in the South is higher due to a federal law. That helps keep the factory in Ohio and makes it more likely that the labor union can bargain from a position of strength.

In sum, labor unions can operate legitimately in the free market as long as they do not violate anyone's private property rights. If they are successful in their endeavor to increase compensation for their members, the higher union wage will benefit union members at the expense of all other workers and of society in general. To the extent that labor unions and their activities are propped up by legal privilege through labor regulations, they are party to aggressing against the property of entrepreneurs.

ECONOMICS AND THE NATURE OF COMPETITION

One of the benefits of economics is that it helps us understand the nature of competition and to make sense of various types of business organization such as cartels, monopoly, and labor unions. In the free market, people only have power over themselves. They cannot force anyone to do anything. Buyers cannot force sellers to sell them products if the sellers do not want to. Likewise sellers cannot force buyers to buy their products if they do not want to. Nevertheless, once we realize that the end of most entrepreneurs is to reap a monetary profit, it is clear that, in order to do so, entrepreneurs must serve their customers better than anyone else. Therefore, the market has been likened to a daily plebiscite in which everyone votes with his dollars. In order for entrepreneurial Rosalee to make profits she must convince enough buyers to spend their dollars buying her products. In order for her to continue to reap a profit, she must convince these buyers to buy her products every day. She cannot rest on the fact that she made profits yesterday. Reaping profits in the past definitely does not guarantee that she will reap profits in the future. Buyers can change their minds and better entrepreneurs can enter the scene.

In a free market, sellers compete against each other to best satisfy the demands of buyers. Producers compete against each other to satisfy consumers. Laborers and landowners compete against each other to satisfy entrepreneurs. Competition is a real dynamic market process that encourages the production of the most goods that society wants in the least costly way. Competition should not be understood in an artificial way that presumes a certain large number of sellers who have perfect information about the market and who never make mistakes. As long as a market is free, meaning as long as private property rights are secure, an economy will manifest true competition.

In light of these facts, economics teaches us that we do not need to fear the specter of cartels and monopoly in a free market. If a group of sellers form a cartel they face competition from both within and without the cartel, which places severe limitations on how successful a cartel can be in restricting output and raising prices. Similarly monopolists are not immune from consumer demand and competition. In a free market, they are constrained by what buyers are willing to buy and pay. If they do charge a price that allows them to reap large profits, other entrepreneurs will be attracted to that industry, resulting in an increased supply of goods and lower prices.

Economic analysis also helps in understanding the nature and economic consequences of labor unions. While unions may be able to assist their members in receiving more compensation for their work, unions can never benefit all laborers in general. Unions benefit their members by negotiating a restrictionist wage. Those not employed in the unionized industry will receive lower wages and society in general will have fewer goods to use in achieving their ends.

SUGGESTED READING

Armentano, *Antitrust and Monopoly*, 13–48. An excellent analysis and critique of conventional monopoly theory in light of market competition.

Mises, *Human Action*, 354–84; 771–73. Mises' detailed discussion of the theory of monopoly prices and his explanation of the economics of labor unions.

North, Gary, "A Christian View of Labor Unions," 1–2. A concise, yet substantial discussion of the economics and ethics of labor unions.

Rothbard, *Man, Economy, and State*, 629–754. This chapter is Rothbard's definitive exposition of competition, monopoly, cartels, and labor unions.

11

The Production Structure and the Social Economy

Economics is a social science that begins by analyzing individual human action. The early chapters of this text acquainted you with the first principles of human action, culminating in the derivation of the law of marginal utility. Those principles were then applied toward the establishment of certain economic principles relating to exchange and price determination. We further applied those principles to discover how interest is determined, how and when entrepreneurs earn profits, and what determines the prices of factors of production.

Some issues concern more than individual people, households, or firms. Phenomena such as recession, increases in overall prices (commonly called inflation), and economic expansion and development are all occurrences that sound economics should be able to explain, but that involve more than any one person, market or industry. Because these issues affect the social economy as a whole, many refer to them as macroeconomic in nature. In the next few chapters we will be applying principles of economics to explain inflation, recession, and economic expansion and development. These issues can only be understood if we understand how every individual market is integrated with each other into the broader social economy.

Until the middle of the 1930s, most economics texts built up an integrated body of thought about the working of market economies and often culminated with discussions of the purchasing power of money and perhaps a chapter on business cycles. Unfortunately, with the coming of John Maynard Keynes, that all changed. After the widespread zealous acceptance of his *General Theory*, economics was cut into two hermetically sealed halves: macroeconomics and microeconomics, with the former given the greater honor. Macroeconomics became the glamorous half. It dealt with hip and swinging topics such as economic depression and recovery, inflation, and unemployment. Microeconomics became the realm of economic nerds, focusing on rather mundane topics such as price determination, interest theory, and economic profit.

Such a clear bifurcation of economic theory is unfortunate because economics does not, in fact, come in two neatly wrapped containers of microeconomics and macroeconomics. They are not independent halves. For instance, in order to understand the network of markets referred to as the national economy, a subject for macroeconomics, one must first understand the workings of individual markets, and before one can understand movements in overall prices, one must first understand how individual prices are determined. Nevertheless, since the rise of the Keynesians and Paul Samuelson's extremely influential 1948 text, *Principles of Economics*, the economic profession has followed his lead and has kept the rather hard distinction.

There is, however, no real justification for such a theoretical dichotomy. At most any division between *microeconomics* and *macroeconomics* is topical. The same principles of human action we use to develop principles of exchange and price determination are the same we use to examine the determination of the purchasing power of money and cause of recession.

In this chapter you will see how the actions of people and firms in different markets at different stages of production integrate with each other into the general social economy. As a result you can understand the nature of the interconnectedness of the entire market economy. How is it that the vast array of markets for particular goods are coordinated so that consumers have their material needs and wants most efficiently met without society degenerating into chaos? What determines how incomes are allocated in a complex extensive social economy?

To arrive at a satisfactory answer to this question, attention must be turned to the economic activity of everyone in the economy. There are two things that integrate all economic activity of the entire social economy: the production structure and money. Both will be examined in succession. We will consider money and its purchasing power in chapter 12. The focus in this chapter will be on the production structure.

The production of every consumer good in the economy is supported by a structure of production. The existence of a production structure for every consumer good has a very important implication for our analysis of the overall economy. Our analysis of the overall social economy must include the entire capital structure. Because all production takes time, if we leave capital out of our analysis, we will be ignoring a key aspect of economic theory that is necessary to explain how production occurs, what leads to economic expansion, and what causes recessions. In short, it is absolutely necessary to understand the production structure in order to understand the workings of the social economy as a whole.

THE FREE MARKET REVISITED

Before getting into the nuts and bolts of the production structure, it will be helpful to revisit some of the key characteristics of the market economy and of the human action that takes place in such an economy. The market is where buyers and sellers come together to make voluntary exchanges in order to achieve more preferred ends. Until now the focus throughout this book has been on individual markets for homogenous goods, such as the gasoline market, the DVD market, the education market, the automobile market, and the pork belly market. A lot of time has been spent discussing the market for Beethoven records. Each of these markets involved the buyers and sellers for the particular good identified. Participants in the gasoline market are buyers and sellers of gasoline. Someone seeking to reap a profit peddling pork bellies is not going to spend a lot of time in detailed analysis of the Beethoven market, because the goods are so different that what happens in the Beethoven market has little if any noticeable effect on the market for pork bellies.

The institution of the free market, however, is broad and incorporates all of the individual markets that exist. The free market is the entire network of voluntary exchange that takes place. All markets are integrated into a complex structure of production. It was implied above that the pork belly market is unrelated to the market for Beethoven records. It is not true, however, that they are entirely unrelated. If investment is directed toward the production of pork bellies, it is funded by savings that cannot be directed toward the production of Beethoven CDs. It is very possible that the CEO of a meat packing plant and the CEO of a factory producing compact disks may be competing for the same blue collar workers.

As has already been seen in chapter 2, the structure of production culminating in any given consumer good is sort of like that good's family tree. The production structure included the entire quantity of every producer good at every production stage that is necessary to bring a consumer good into existence. In order to produce a flourless chocolate cake, it is first necessary to have butter. In order to have butter, the butter maker must obtain cream. In order for the dairy producer to make cream, he must first obtain the services of dairy cows and so on. Throughout the structure of production for flourless chocolate cake, the higher order goods at each stage of production are used to advance people toward the next stage in their ultimate goal of consumption of the quintessential chocolate dessert.

As consumers decide to buy or abstain from buying different goods, producer goods throughout the entire structure of production are directed accordingly. Suppose that people, having discovered the beauty of flourless chocolate cake, begin to prefer that as their dessert of choice instead of their previous standby, apple pie. There would be an increase in demand for flourless chocolate cake corresponding to a decrease in demand for apple pie. The increased demand for flourless chocolate cake will result in an increase in the demand for factors of production throughout the entire production structure. To meet the increased demand, dessert makers will increase their demand for semi-sweet chocolate, butter, eggs, baking equipment, land, labor, and everything else they need to make the cake. To meet the increased demand for chocolate, butter, eggs, and baking equipment, their producers will increase their demand for the factors necessary to produce each of these capital goods.

At the same time, the decreased demand for apple pie results in pie makers decreasing their demand for apples, cinnamon, flour, shortening, ovens, and everything else needed to make apple pies. Decreased demand for a consumer good will generate a decreased demand for factors of production throughout the structure of production of that consumer good.

Each consumer good has its own structure of production that is integrated into the larger social economy through the money nexus. As has been explained in chapter 5, money is, by definition, the general medium of exchange used to buy all goods. In a monetary economy, all goods are bought with and sold for money.

Remember the many important benefits gained from the emergence of money. Monetary exchange allows for the division of production into many very specialized operations with each producer selling his product for money and using that money to buy goods he wants. At any stage of production, producers use land, labor, and capital goods, exchanging their money for the uses of the services of these factors. They then sell the good they produce with these factors for money to producers in the next stage of production. This process continues until, in the lowest stage of production, the consumer good is made and sold to consumers for money.

Because goods are bought and sold for money at each stage of production, it is money prices that entrepreneurs use to calculate expected profit and loss at each stage, and hence, use to direct factors of production toward best satisfying consumer wants. If, following a decrease in demand for apple pie and an increase in demand for flourless chocolate cake, entrepreneurs directed more factors of production toward the manufacture of more flour for pie crusts when those factors could have been used to produce chocolate and eggs, these resources would be wasted from the point of view of society. We see, then, that all of the activity that occurs throughout the production structure is coordinated with the use of market prices.

Consumers obtain the money they use to buy consumer goods by selling goods they own. They can sell durable consumer goods, such as used cars, stereos, computers, and houses. Most consumers receive their income by selling the productive services of factors of production they own. Most net income earned in the current economy in the United States comes from people selling their labor. They can also sell the use of their land or capital goods that they own.

We can sum up our observations regarding monetary transactions as follows: nearly all exchanges in a modern economy are made using money. In their role of producers, people sell consumer goods and producer goods in exchange for money. Entrepreneurs buy producer goods with money. In their role of consumers, people buy their consumer goods with money and receive the money that they spend by selling either durable consumer goods or the productive services of factors of production. The universality of monetary transactions which allows for the use of money prices to calculate profit and loss also allows for the intensive and extensive division of labor. This division of labor, we have seen, allows for a great increase in productivity as each person is able to engage in that production in which he has a comparative advantage.

THE PRODUCTION STRUCTURE

Before the culmination of our investigation regarding the aggregate production structure for the entire economy is reached, it will be helpful to explore in further detail some principles related to production structures in general, beginning with a review of some key characteristics of the structure of production.

Production effort always runs *down* the structure of production. In order to produce a consumer good, the higher order goods must first be produced. Before flourless chocolate cake can be made, semi-sweet chocolate must first exist. Before semi-sweet chocolate can be produced, cocoa powder must be made. Before cocoa powder can be made, cocoa beans must be harvested. The order of productive effort, then, begins at the highest stages of production. Land, labor, and cocoa seeds are combined to produce cocoa plants, from which cocoa pods are harvested. The resulting seeds are then ground to make cocoa powder, which is then used to make semi-sweet chocolate. Only after all of this effort has been expended at the higher stages of production, can the flourless chocolate be baked. Productive effort begins at the highest stage and runs down the structure of production.

Income, on the other hand, runs *up* the structure of production. It is the prospect of income earned from selling flourless chocolate cakes that stimulates the dessert maker to spend income on semi-sweet chocolate. The chocolate maker reaps income from selling chocolate and spends money on cocoa powder. The potential income reaped from making cocoa powder motivates the cocoa producer to spend money buying cocoa seeds. Monetary income, then, is transmitted from the consumers to the producers at the lowest stage of production, up the various stages of production, culminating in monetary payments to those at the highest stage of production. Monetary income moves up the structure of production.

The allocation of this monetary income is a key problem in production theory. At the lowest stage of production, sellers of a consumer good receive money from consumers. How is this money allocated? It will be spent on three sets of suppliers of productive services. A good portion of it will be paid to owners of land and labor. Because land and labor are original factors of production in that neither of them must be produced, gross income earned by both workers and landowners equals their net income.

Another portion of the income received by the maker of the consumer good is paid to the owners of capital goods. The gross income received by the producers of these capital goods is not equal to their net income, because these second stage producers had to make their capital goods before they could sell them to the producer of the consumer good. Therefore, the makers of the capital goods must spend much of their income in obtaining factors of production. Owners of capital goods receive income, but must also expend money. They both earn revenue and make expenditures in the production process.

We must also not forget production is like every other type of action in that production takes time. Because it takes time, producers do not receive their income until after their good is produced. However, they obviously need to obtain the services of the factors they use to make their product before they can sell it. Consequently, producers must spend money on factors of production first, then make their product, then receive their income from selling it.

Therefore, in order to reach their ultimate goal of receiving money by selling their product, capitalists must save. They must restrict their consumption spending and with these saved funds purchase factors of production. As mentioned earlier in our text, the role of the capitalist is to supply present money to the owners of factors of production in return for the services of those factors needed to produce goods that will be sold in the future. Capitalists provide the service, then, of satisfying the higher time preferences of owners of land and labor.

Because of their positive time preference, capitalists demand a positive return on their investment. They demand a premium for supplying present money in exchange for future money. This would even be the case in the evenly rotating economy (ERE). It is true that in the ERE there is no uncertainty and, hence, no profit. However, the production process still would take time, so capitalists still would be supplying present money in exchange for future money. No entrepreneur will bid prices of factors used to make a capital good up to the point so that their sum is equal to the capital good's selling price. The entrepreneur at the next lower stage is willing to pay the selling price, instead of the lower buying prices of the factors used to make the capital good, because doing so advances him in time toward his goal.

Capital is not an independently productive factor. The existence of all consumer goods is ultimately traced to land, labor, and time. The incomes earned throughout the production structure, then, are earned from the services of the property of landowners, laborers, and owners of capital goods. The service provided by the owners of capital goods is one of time. They restrict consumption so they can advance money in the present to the owners of land, labor, and higher order capital goods. They use the services of those factors of production to make a good that will be sold in the future. In doing so, they are contributing to the process by which a future good is converted into a present good.

As discussed above, the entire production process culminating in consumer goods is divided up into many stages. At each stage, the factors of production work and advance the process from higher to lower stages of production. At each stage, the product is sold for money to another capitalist. A steel maker sells its steel to a bolt manufacturer. The bolt producer sells his bolt to Kitchen Aid who makes fine stand-alone mixers. Kitchen Aid sells its mixer to Just Desserts, a maker of world-class flourless chocolate cakes. Just Desserts sells its product to the consumer.

At each stage of production, income is earned from each productive participant. Wages are earned by laborers and rent is earned by landowners. Capitalists, remember, must be compensated for their waiting time. This payment for time is *interest*. Therefore, the capitalist acquires interest income at each stage of production.

THE AGGREGATE PRODUCTION STRUCTURE OF THE SOCIAL ECONOMY

The vast network of all voluntary exchanges in which we participate rightly can be called the social economy. The social economy gets its name from the fact that it is in this network of voluntary exchange that people come together to form society. They participate in voluntary exchange and the division of labor, thereby both benefiting and receiving benefits from each other. The social economy includes all voluntary exchanges in all markets for all consumer and producer goods.

The same economic principles that apply to the production structure for a single consumer good can be brought to bear as we begin analyzing principles that relate to the social economy as a whole. This is because the general macroeconomy is an integrated aggregate production structure that supports the production of all consumer goods. Throughout the entire social economy, money moves from consumer goods industries up through the higher stages of production. At the same time, production effort in the form of goods flows from the higher stages to the lowest order of production, finally culminating in the sale of consumer goods.

At any given moment in time there is productive activity occurring simultaneously at the various stages in the production structure. At the lowest stage producers of consumers goods exchange their products for money. Money moves from consumers to suppliers of consumer goods. These purchases are *not* time transactions. They feature exchange of present money for consumer goods. In other words, exchange of present goods for present goods.

To engage in production, however, sellers of consumer goods do engage in the exchange of present money for future money. In order to invest in production of consumer goods, they must spend present money on land, labor, and capital goods. They invest in the services of factors of production to make consumer goods. Their investment consists in the purchasing of land, labor, and higher order capital goods. These are time transactions. They involve supplying present money in exchange for future money. These producers supply money in the present to owners of factors of production in order to use those factors to produce a good they will sell for money in the future.

Likewise, higher order capital goods producers such as bakers, clothing makers, and wholesalers supply capital goods to sellers of consumer goods such as grocers, retailers and fast-food restaurants. They purchase factors of production from owners of original factors and even higher stage capital goods producers. They also must spend present money on land, labor, and capital goods. They also supply present goods and purchase future goods (in the form of factors of production). This pattern continues throughout every stage of production.

Because income flows up the structure of production, from the lowest stage to the highest stage, people can often fall into the error of believing that the secret to economic expansion is to increase consumption spending. However, our analysis of the production structure reveals that consumption does *not* drive the economy. Saving and investment does. All of the exchanging of present money for future money that takes place at every stage of production in the entire economy points to one inescapable fact: the amount of money spent at any time on capital goods is much larger than the amount spent during the same period on consumer goods. The production of consumer goods is and must be supported by a vast, complex capital structure. The entire structure of production is supported by saving and investment.

Those who work producing consumer goods are working at the stage of production that results in goods that will sustain them (i.e. consumer goods). Workers who are employed producing higher order goods, however, are farther removed from the consumption stage. It will take longer time for the processes they are working in to come to completion. Therefore, in order to participate in production processes at the higher stages of production, they must be supported by savings (restriction of consumption). In order for workers and producers at the higher states to be sustained during their production process, purchasing power must be supplied by savers and transferred to owners of original factors farther up the production structure. This saving of income and supplying of present money to higher order factor owners results in their obtaining necessary consumer goods. It allows for consumer goods to be available for land-owners and laborers whose services are employed far removed from the stage of production where consumer goods are actually made.

The main point to remember, again, is that all of this requires re-peated acts of saving. Notwithstanding the mistaken notions of many fi-nancial journalists and policy makers who are quick to distribute money in the forms of subsidies and rebates in the hopes of taxpayers going on spending binges, consumption spending does not drive the economy. As we have seen, the manufacture of all consumer goods is supported by a large structure of production, which is made possible by investment in capital goods. It is gross investment expenditure that actually drives the economy, because it is only through savings and investment that capital is accumulated and maintained. The capitalists at each stage of production have a crucial role to play in maintaining the capital structure.

If capitalists decide not to save, but instead spend all of their income on consumption, the production structure would immediately break down. For example, if first stage capitalists who receive payment from consumers spend all of their revenue on consumer goods, none of their income gets advanced to the owners of higher order capital goods or the original factors of production. There would be no higher stages of pro-duction. Without the saving and investment of capitalists, all production processes for all consumer goods would be extremely short and the econ-omy would devolve to a most primitive state. The division of labor would be destroyed and we would literally be back to living hand-to-mouth, while a large portion of the population would die due to starvation. Not a pretty picture. It is the capital structure that supports the production of all consumer goods. Therefore, it is the savings and investment decisions of capitalists that ultimately provide the engine by which the economy progresses.

In the aggregate social economy, the ratio of gross savings and investment to consumption is determined by aggregate time-market schedules, which are determined by the subjective time preferences of everyone in society. Consequently, the important consideration regarding whether the economy will develop and prosper is not whether there will be adequate consumer spending, but whether there will be adequate saving and investment. The important factor is social time preference. The lower people's time preferences are, the more saving and investment that they will engage in. The more saving and investment there is, the more production there will be and greater will be the quantity of consumer goods available to society over time. The higher people's time preference, the more present-oriented they will be. The more present-oriented people are, the more consumption and the less saving they do. Less saving means less investment and production. Consequently, the higher social time preferences are, the fewer consumer goods a society will be able to enjoy over time.

Another characteristic of the production structure that bears remembering is that it is very complex. At the many stages of production there are a myriad of heterogeneous capital goods suited to tasks that are more or less specific. Some capital goods are relatively non-specific. Hammers for instance can be used in several various forms of construction and other production projects. Some machines, however, are made to perform only one task in one specific production process. Because of the relative specificity of capital goods, it is important that all of the investment and production that takes place throughout the entire social economy is properly coordinated somehow. If we are to be good stewards with the goods that God has given us, we should not waste them by using them to make goods that are neither useful nor wanted.

All of the coordination that is necessary in the market division of labor occurs via the market price system. Investment is coordinated horizontally between goods at the same stage of production via profit and loss calculations using prices. Investment is coordinated vertically between goods at different stages of production via calculations of the ratio of market prices of goods at different stages. Remember that this ratio is the interest rate. The interest rate is equal to the ratio between the price of the product and the sum of the prices of the factors of production at each stage. The larger the difference between the two, the higher the rate of return.

In our earlier discussion of interest we found that there is a tendency for the rate of return for all investments to move toward the same rate of return. If there were no more changes in preferences, resources, population, or technology, differences in rates of return would be eliminated as entrepreneurs sought out profitable investment until we reached the ERE in which the rate of return earned from all investments would be due solely to time preference and would be the pure interest rate. Not only is this true for different investments that occur in the same stage of production, but it is true for all investments in different stages of production as well.

Suppose, for example, that different producers at different stages of the production structure earn rates of returns as indicated in table 11.1.

Table 11.1

Rates of Return in Different Stages of Production			
Higher Stage	Medium High	Medium Low	Lowest Stage
Bolt	Mixer	Baker	Grocer
6%	5%	4%	5%

Grocers—those who sell consumer goods such as bread—are reaping a 5% rate of return, as is the manufacturer of the bread mixer, a capital good that is sold to the baker. Producers in the bread baking industry are reaping a 4% rate of return and those in bolt-manufacturing, the highest stage shown, are reaping a 6% rate of return. Entrepreneurs who recognize such opportunities will adjust their investments accordingly. Investment will flow out of the baking industry and into the bolt industry. As explained in an earlier chapter, this will result in the rate of return falling in the bolt industry and the rate of return increasing in baking industry. Hence, both will move toward each other. The precise rate of return that all industries will tend toward will be determined by the overall social time preference of everyone in the economy.

CAPITAL AND THE STOCK MARKET

It is now time to investigate in more detail how market prices are used to coordinate all economic activity throughout the production structure for the entire social economy. Entrepreneurs use money prices to calculate profit and loss. In order to effectively direct factors of production toward their most highly valued uses, entrepreneurs and capitalists use the praxeological concept of capital. We have already used the term *capital good* to define produced means of production. The concept of capital is closely related to the stock of these goods.

Capital is the market value of the sum of capital goods. As such capital is the fundamental concept of economic calculation. Not only do entrepreneurs use market prices to estimate the expected profit or loss of various investment projects, but they use market prices to sum up the market value of their firm's capital goods available for continued production. Capital accounting allows entrepreneurs to determine the ultimate outcome of their investments by keeping track of all changes in their capital fund as a result of their production activities. Furthermore, capital accounting allows them to do this not for only the whole project upon completion, but also can provide a snapshot of the project in process at any specific point in time.

This snapshot point of view is helpful because production is a continuous process that is constantly changing. Every day capital is being used up or accumulated. At the same time, income is being distributed to the owners of the original factors of production and the capitalists. In order to understand how resources and incomes are allocated over time throughout the production structure, it is necessary, therefore, to understand the process of capitalization.

Capitalization is the process by which the capital value of productive assets and businesses are determined. This process starts with the market price of the available capital goods. The capital value of any asset is the sum of each period's discounted marginal revenue product (DMRP) the owner expects to reap in the future by the using that asset.

It was explained in chapter 9 that the unit price of the services of a factor of production tends to equal the factor's DMRP. The maximum price that an entrepreneur will pay for the services of a factor of production for a unit of time will be less than that factor's marginal revenue product (MRP). This is due to time preference. Money in the present will be valued more highly than that same amount of money in the future. Correspondingly, the present value of any future MRP will be less than the MRP, because it will be discounted. If an entrepreneur expects that the marginal unit of a factor will contribute $1000 to his firm a year from now and his time preference is such that he discounts future MRP's at an annual rate of 10%, he would only be willing to pay $900 for the factor right now. The amount by which the entrepreneur discounted the MRP is $100. If one month's labor of a worker is responsible for the production of goods that can be sold for a total of $3,000 and the entrepreneur discounts that future MRP by $500, then the monthly rent imputable to that worker is $2,500. If a printing press contributes production that can be sold for $450 in one week's time and the entrepreneur discounts that future MRP by $50, then the weekly rent of that printing press is $400. Consequently, rent is essentially the DMRP of a factor, and applies to all factors, be they land, labor, or capital goods. The concept of rent is important because it is the foundational price upon which the capital value of a factor is built.

Capitalization applies to all goods that can be bought and sold as wholes. When an entrepreneur is considering how much an acre of land or a computer or a doughnut machine is worth to him in the present, he adds up the factor's future DMRPs. This sum is the asset's capital value. It will be the price that will tend to be paid and charged for any factor bought as a whole in the ERE.

As has been implied throughout our discussion, a factor has a capital value only if it is durable. One that is not durable but is used up immediately in the production process does not have a capital value. For example, flour used in baking a loaf of Wonder Bread generates no future stream of MRPs because once it is used, it is no more. It makes a one-time contribution to the bread production process, so its price will tend to equal its rent, its DMRP. Therefore capital value only applies to durable factors of production. Because the capital value of a factor is equal to the sum of that factor's MRPs discounted by the interest rate, it should be apparent that the capital value of a factor is determined by that factor's rent, its durability, and the interest rate.

Suppose, for example, that an entrepreneur is considering buying new desktop computers for his offices. How will he decide what is their capital value? He surmises that the computers will be useful to him for three years before they become obsolete. He further surmises that their annual rent contributed to the firm, their MRP, is $1,000 and social time preferences are such that the interest rate is 10%. He uses the principle of discounting to calculate the present value of the three successive yearly $1,000 MRPs. For each year's rent, he discounts the MRP by the interest rate across the number of years until he will receive each computer's rent. The $1,000 reaped at the end of one year is discounted for a year by 10%. Consequently, the DMRP from using the computer during the first year of its usefulness is approximately $900. The DMRP generated from the second year that the computer would be used is less than that, because this $1,000 will be reaped farther in the future. It will be discounted for 2 years instead of only one. A MRP of $1,000 discounted for two years at a rate of 10% is approximately $825. The DMRP for employing the computer the third and last year of its usefulness will be even lower, because the future MRP is discounted for three years. A MRP of $1000 discounted by 10% over three years is approximately $750.

The capital value of the computer in question is the sum the DMRPs. In our example the capital value is equal to $900 + $825 + $750 = $2475. This amount is how much the computer would be worth to our entrepreneur in the present.

The bottom line is that, when calculating the capital value of the assets that they own or are considering buying, entrepreneurs sum the discounted future rents they expect to reap from using that asset. The ability to calculate the capital value for an asset in monetary terms is very helpful to the entrepreneur because he is able to make valid comparisons between different assets. Having capital values in terms of money also permits him to add them up.

Entrepreneurs can sum the capital values of different assets to determine the capital value for the entire firm. Suppose, for example, that the owner of Gerald's Barbecue wants to sell his very successful mobile barbecue business. He has sold barbecue beef and pork sandwiches, chickens, pork ribs, beef brisket, and baked beans at different grocery stores for over twenty years and decides it is time to retire. He wants to sell his business, which requires selling his assets on the market. These durable assets consist of the following:

- 4 electric warmers used to keep cooked meat warm. Each has a capital value of $400, so that the total capital value of the warmers is $1,600.

- 1 cash register that has a total capital value of $250.

- 1 hickory smoker fixed on a trailer with a total capital value of $4,000.

- 1 food service scale used to sell meat by weight. The scale has a capital value of $500.

- 1 RV-type mobile home with a capital value of $15,000.

- Assorted cutlery with a total capital value of $100.

- Miscellaneous equipment with a total capital value of $200.

The total value of Gerald's business is the sum of all of the above capital values or $21,650.

It is this capital value that Gerald will use when considering selling his business. If someone offers him $15,000 to buy the works, he will turn him down, because his business assets are worth more than that. However, if someone offers to pay him $30,000 he will probably accept because $30,000 is more than the present value of the sum of the discounted rents that he will earn in the future by keeping and using his assets. It is also this amount of capital that an investor would need to accumulate in order to enter business either in place of or in competition with Gerald. Perhaps it will cost more because of the customer good will that Gerald has built up over his twenty-plus years in the business. Such good will is also economically valuable.

It should be noted that this amount calculated to be the firm's capital value is relevant only to the extent that the capital values of each asset are rooted in actual market prices. When estimating the capital value of assets and firms, investors sometimes make the mistake of relying on the original purchase price as the value of the asset, or the original price discounted only by depreciation. It is just as possible that the capital value of an asset can change greatly because of changes in the future rent received, because of changes in future market prices resulting from changes in future supply and consumer demand. If entrepreneurs do not take these changes into account, they will end up making investment decisions that very likely will end in losses.

THE CAPITAL MARKET

Given the level of development in our modern industrial economy, it should come as no surprise that there is an active market for exchange of the capital necessary for production. The capital market is part of the time market. We have already seen that the time market has two components: the production structure and the market for loanable funds borrowed by consumers. Consumer borrowing uses what we could call *non-productive savings*. If Joe Saver deposits his savings in First Bank and First Bank lends it to Shelly Teeny-Bopper so she can pay for her t-shirt at the Gap, Joe's savings do not contribute to production in the social economy. It is only the savings that finds its way into the production structure that we could call productive savings. It is productive savings that are available to entrepreneurs as they undertake production by obtaining the services of factors of production.

There are three main ways entrepreneurs get access to the funds necessary for capital investment. They can, of course, directly invest their own personal savings. So far, as we have explained the time market and interest rate, we have focused mainly on this method of obtaining capital. Additionally, entrepreneurs can raise capital funds by merely borrowing the savings of others. We will examine the effects of producer borrowing later.

In our modern economy, another important way capital is raised is through investors pooling their savings together with others and jointly investing in the ownership of the assets of a business. The capital that is pooled together and invested in firms has given rise to what is known as the stock market. The stock market is an exchange institution in which shares of ownership of a company's assets, called shares of stock, are bought and sold.

Companies that are owned by stockholders are often called joint-stock companies or publicly held companies. They are called joint-stock companies because they are jointly owned by more than one person. They are owned together with every other person who invests capital in that company. Each investor receives a share, a certificate of ownership, certifying that he is part owner of the firm's assets. They are called publicly held because they are owned by different members of the public. In this case, *public* definitely does not mean government-owned in the sense that public schools are actually government-owned schools. Jointly-owned firms we are talking about are called publicly held, because they are owned by different people.

Such jointly-owned companies arise when various people invest their savings in one firm and jointly make decisions on the investment of their total savings. These decisions include the business purchasing land and capital goods, and hiring labor. Tracing out the chronology of their investment, we see that in the case of a joint-stock company, shareholders together successively own their pooled savings, the services of the factors those savings are used to obtain, the product that is made with the factors of production obtained with the savings, and the money received from the sale of that final product. If continually in business, at any one time a company's operation will be the combination of investment and the production and sale of output.

It is very important to note again, as we did when discussing the aggregate production structure, that there is nothing automatic or inevitable about this process of investment and production. Production can only continue if the owners choose to continue. In order for production to proceed, capital must be maintained, so further savings and investment must occur. At any point the firm's owners could decide to cease and desist. They could sell out their investment and turn to some other line. They could cash out their investment and spend all of their income on consumption. In order for a firm's productive activity to continue, the owners must decide to do so.

Should they decide to continue with their adventure in production, the assets they jointly own will be a mixture of cash to be spent on factors or factor services, factors just purchased, partially completed products, finished products ready to be sold, and money just received from the sale of their product. Assets owned by the firm at any one time will, therefore, consist of money, land, capital goods, partially completed products, and finished products ready for sale.

The market makes an evaluation of the firm by placing a monetary evaluation on the entire stock of assets owned by the firm. This monetary amount will tend to be equal to the firm's capital value. Given that the market evaluates the capital value of the firm, how do the owners allocate their shares of these valuable assets? They do so by the proportion of total investment made in the company. Suppose that there are five saver/investors who pool their savings together to form Acme Corporation and they each contribute their savings as indicated in table 11.2.

Table 11.2

Shareholder Investment		
Investor	Investment	Share
Groucho	$ 400,000	40%
Harpo	$ 200,000	20%
Chico	$ 200,000	20%
Zeppo	$ 150,000	15%
Gummo	$ 50,000	5%
Total	$1,000,000	100%

The total amount of money invested in Acme is $1 million. Each individual investor owns the same percentage of assets as the percentage of total investment he makes. We can see that Groucho who has invested $400,000 of that million has invested 40% of Acme's total capital. Therefore, Groucho owns 40% of Acme's assets. Likewise both Harpo and Chico have each invested $200,000, so they both own 20% of the firm's assets. Zeppo, with his investment of $150,000, owns 15% of Acme and Gummo owns 5% of the company's assets, because he invested $50,000, which is 5% of $1 million.

Each of the five investors receives share certificates certifying that they own their relative percentage of Acme. Groucho, Harpo, Chico, Zeppo, and Gummo could decide to issue a total of 10,000 shares with each share representing an asset value of $100. We can see from the information in table 11.2 how many shares are owned by each investor. Because Groucho invested $400,000 and because each share represents $100 of Acme capital, Groucho is issued 4,000 shares in exchange for his $400,000. Likewise, Harpo and Chico are both issued 2,000 shares. Zeppo is issued 1,500 shares and Gummo received 500 shares in exchange for his $50,000 investment. Because each investor is provided share certificates in exchange for their investment, capitalists who invest in publicly held stock companies are called shareholders.

If Acme adds to its capital by selling more stock, then the relative percentage owned by each investor will change, although the number of shares owned by each will not. Suppose that Acme wants to increase its capital by $300,000 so it sells 3,000 shares to Maggie Dumont for $100 each. This changes the ownership allocation as indicated in table 11.3.

Table 11.3

Shareholder Investment		
Investor	Investment	Share
Groucho	$ 400,000	31%
Harpo	$ 200,000	15%
Chico	$ 200,000	15%
Zeppo	$ 150,000	12%
Gummo	$ 50,000	4%
Dumont	$ 300,000	23%
Total	$1,300,000	100%

With Dumont's addition of $300,000, the total capital invested rises to $1.3 million. The original five investors find that, while the total number of shares they own has not changed, the percent of the firm that they own has decreased. Groucho, for instance, no longer owns 40% of Acme's assets. $400,000 is 40% of $1 million, the original amount invested, but only 31% of $1.3 million, the new total capital value. Harpo and Chico now own 15% of Acme's assets. Zeppo owns 12% and Gummo owns only 4%. Dumont's $300,000 investment entitles her to 23% of Acme's assets, because her investment is 23% of $1.3 million, Acme's total capital value.

Suppose that Acme is able to sell its product so that its revenue is such that its rate of return on investment is equal to the interest rate at 5%. After the sale at the end of the year, the total capital owned by the shareholders will increase by 5% to $1,950,000. This new, greater capital amount will be distributed to the shareholders in the same percentages of their investment. Groucho will own 31% of the new revenue, Harpo and Chico will each own 15%. Zeppo will own 12%, Gummo 4%, and Dumont will own 23%.

At any point in time, the capital value of a firm will be equal to the sum of the capital value of all of its productive assets. The capital value of a firm, therefore, increases any time additional investment occurs and is maintained by the reinvestment of the owners after final products are sold. If the shareholders decide not to reinvest enough of their earnings, Acme's capital will be consumed, resulting in lower productivity and ultimately losses, and perhaps even bankruptcy.

The shares of the capital assets owned by the investors are called *stock*. The market price of the stock can increase or decrease as changes in the capital value of the firm warrants. Profits may be reaped and income earned or losses can be suffered. Capital can be reinvested or removed from the company. As the capital stock changes, the price of each share will tend to change accordingly.

The stock market is the exchange institution in which shares of publicly held companies are bought and sold. It is very important because the existence of the stock market allows for a larger and steadier flow of capital from new investors. This is because the stock market makes it easier for a person with a relatively small amount of savings to invest in the ownership of a business. Additionally, the ability to easily sell shares increases the attractiveness of buying in the first place. People are more willing to risk their savings investing in a business venture if they have the ability to sell quickly if things go bad, so that they might lose only a few threads instead of their whole shirt.

Each share of stock in a publicly held company is a pro-rata share of what is called the firm's *equity*. A firm's equity is equal to the value of its assets minus its liabilities. The market value of a firm's assets is what we have defined as capital. A firm's liabilities are what it is obligated to pay someone at some point in the future. For instance, suppose that Acme, Inc. owns capital worth $100,000, but also owes $10,000 in issued corporate bonds. Acme's equity is equal not to its total capital value of $100,000, but $10,000 less than that, or $90,000. If it was a publicly traded company and went out of business, the firm's assets would be sold for $100,000 and their liabilities of $10,000 paid, so that the shareholders would split $90,000.

If a firm's liabilities remain constant, the firm's equity will increase as its capital increases. Because the price of each share is directly related to the capital value of the firm, share prices are also inversely related to the interest rate that is, of course, determined by social time preferences.

Capitalists make decisions to either save or consume based on their time preference. Therefore, the price of a share of a firm's stock will be greatly influenced by social time preferences as they are manifested in the interest rate. These time preferences determine the difference between the price of output and the sum of the prices of the factors used to produce their output. The ratio of these price spreads is the interest rate. The total capital value of a firm's assets, which is a determinate of its equity and hence its share price, will be determined by the sum of expected future earnings from the assets discounted by the interest rate. The rate of return on the stock market will tend toward equaling the rate of interest, which we have seen is the rate of return toward which all investments tend. However, at any moment, the rate of return on any particular stock investment will be different from the rate of interest by the differing amounts of entrepreneurial profits or losses earned on each investment.

Suppose, for example, that time preferences increase so that the interest rate also increases. Acme's capital value will tend to fall, because the capital value of all their assets will fall. This is because each asset's future MRP will be discounted at a higher rate. Each year's DMRP will be smaller, so the sum of the future DMRPs will be smaller as well. Therefore, as interest rates rise, the capital value of assets and firms decrease. As capital values decrease, so will the firms' equity. Share prices will also fall, because they tend to equal a company's total equity divided by the number of shares issued.

On the other hand, if social time preferences decline, so that the interest rate falls, capital values increase. The total capital value of businesses will increase because the capital value of all of their assets will increase. This is because each asset's future MRP will be discounted at a lower rate. Each year's DMRP will be greater, so the sum of the future DMRPs—the capital value—will be greater as well. Because of the lower interest rate, the capital value of assets and companies increase. As capital values increase, so does shareholder equity and, hence, stock prices.

As mentioned earlier, another way entrepreneurs gain access to capital funds is *borrowing*. They can access the producers' loan market. Entrepreneurs can borrow indirectly from lenders through the banking system or they can borrow directly from lenders by selling bonds. In its capacity as a lending institution, a bank is a financial intermediary that makes it easier for savers and borrowers to come together. A bond is a financial instrument that states that the holder of the bond is entitled to a fixed interest payment over a period of years, at the end of which the principle loan amount is also to be paid back.

Obtaining capital by borrowing is little different from obtaining capital by selling shares of stock. Suppose that Acme wants to further expand its operations and needs to raise $200,000 to do so. The company decides to borrow the money from one Helen Rosalee who now holds a $200,000 Acme bond. In this transaction Acme is exchanging a future good for present money. The future good is the promise to pay interest in the future and is supplied by the owners of Acme. The $200,000 is present money supplied by Rosalee. She is the person who has done the saving and she has supplied the new capital that is invested in future goods by the stockholders.

The rate of return on Rosalee's investment in Acme will tend to equal the interest rate. In our example, we have been assuming that social time preferences are such that the interest rate is 5%. If Acme offers to pay Rosalee 3% for her loan, she will not lend Acme the $200,000, because she could receive 5% on his money either as a stockholder in Acme or in some other investment. At the same time, Acme cannot pay Rosalee any more than 5% for her loan, because Acme's net return on its capital investment will be no more than 5%. Even if Acme earns entrepreneurial profits and therefore could pay a higher rate of interest, it need not do so because its competitors are all paying the market rate of 5%. The maximum that the firm will pay is 5% and the minimum that Rosalee will accept is 5%. Therefore, the rate of return mutually agreed to for a producers' loan such as the one in our example will tend to be equal to the interest rate that is, of course, determined by the subjective time preferences of everyone in society.

Economically, the capitalists who invest money in the producers' loan market are similar to the capitalists who invest money in the stock market. Both have saved money instead of having spent it on consumer goods. Both seek to earn interest by exchanging saved capital in the present for future money. Both of their time preferences, along with everyone else's, determine the rate of interest. The interest return on both types of investment is determined by the interest return on investment throughout the entire time market. The saver, however, is a general entrepreneur when he buys stock, because he is purchasing a share of a firm's equity based on how well he thinks that firm will satisfy future market demand. If the firm engages in unprofitable production, it will, of course, sustain losses and its capital and equity will shrink. Then the shareholder will be left holding a share that loses value. Because the future is uncertain, stock investors do not know in advance how profitable will be the actions of their firm. Consequently, stock investors are not only capitalists but entrepreneurs as well.

MARKET CHANGES AND CAPITAL VALUES

We have established the principles by which market prices and the interest rate determines DMRPs, asset and firm capital values, and thereby, stock prices. We have left to trace out the process by which changes in the market for consumer and producer goods result in changes in capital values.

As investors seek to maximize their returns, capital will be directed toward those firms that will bring in the highest future rents. Those firms are where the money will be. Because the capital values that help determine firm equity and relative stock prices are affected by DMRPs of the various assets owned by the firm, capital values are also affected by changes in the markets for a firm's output.

Suppose that, for whatever reason, there is increased demand for the goods Acme sells. Increased demand for Acme's products means buyers are willing to pay higher prices for their goods. Higher prices for their output will result in increased total and net revenue and increased profit. Additionally, higher prices for the goods that Acme produces result in higher DMRPs for all of the factors used by Acme. We have already seen that an asset's capital value is equal to the sum of its future DMRPs, so higher DMRPs will cause capital values to increase. As capital values for the assets owned by Acme increase, Acme's total capital value increases. Increases in capital value will, other things equal, increase Acme's shareholder equity, which results in higher stock prices as the demand for shares of Acme increases in the stock market.

On the other hand, if there is a decrease in the demand for a firm's product, the opposite will occur. If people demand less of Acme's output, this is the result of buyers willing to pay less for their goods. Lower prices for Acme's product lowers Acme's total and net revenue, as well as its profit. Acme might even suffer losses. Lower prices for their output also result in lower DMRPs for all of the factors Acme uses in its production. Lower DMRPs result in lower capital values, which will cause lower share prices for their stock as the demand for shares in Acme declines.

Our analysis demonstrates the link between the market prices for consumer goods, producer goods, and capital values. It also helps us to see how capital is allocated by entrepreneurs in our changing economy. As the supply and demand for various consumer goods change, these changes affect the market prices for these goods. These prices affect the DMRP of factors of production and asset capital values, which in turn determine total capital values for entire businesses. As investors evaluate such businesses, such changes in total capital values and shareholder equity will be reflected in the stock market.

ECONOMIC PROGRESS AND REGRESS

As noted earlier in this chapter, the length and size of the production structure that supports the making of consumer goods is determined by social time preferences which determine how willing people are to save and invest. It should come as no surprise, therefore, to find that changes in the level of aggregate savings generate either economic expansion or contraction. As savings and investment go, so goes the social economy.

If there is an increase in the rate of voluntary saving, what effect will this have on the social economy? Increased saving due to a lowering of social time preferences will result in an expanding economy. Lower social time preferences means that, in general, people are less present-oriented than they were, so they are more willing to restrict consumption and will save more. Lowering time preferences, then, results in additional net savings.

Because people are saving more, they have more to invest, so the ratio of gross investment to consumption increases. People decrease their demand for consumer goods. The number of stages of production in the economy increase as investment is shifted up the production structure from lower orders to higher orders. The structure of production lengthens and a lower interest rate is reaped at each stage, because increased saving results in a decreased interest rate as the price differentials between a firm's output and the factors it uses to produce its output decrease at every stage of production. The reduction of consumption and increased investment in the higher stages of production is a self-supporting process.

Increased savings will create a new disparity in profits between different productive stages. This occurs because there will be a decrease in monetary demand for consumer goods equal to the increase in savings. Entrepreneurs at stages closest to the lowest stage of production and who do not foresee such a change will reap relative losses. After a period of adjustment, the effects of losses at the lowest stages will begin to be felt in the stages closest to them. The increase in savings creates disparity between rates of return of companies at the lowest stage of production and those at the higher stages. In the lower stages of production, prices fall due to a lower consumer demand. The demand and prices for factors used in these stages will decrease. The demand for labor used in the lowest stages of production will fall, so wages in these industries will fall as well. The market for factors specific to these stages are impacted the most because they have nowhere to go. If there is some machine, like a french fry maker, that is useful only in the fast food industry and demand for fast food decreases because people are saving more and consuming less, the income received by the owner of that machine will fall by a relatively greater extent than the income received by fast food workers. Labor is relatively less specific and any fast food workers who lose their jobs will be able to find employment in the higher stages of production where there is new investment. The negative effect of savings will be weaker the father up the structure of production.

This new disparity of profits at different stages will act as a signal and incentive for entrepreneurs to allocate investment out of lower stages and into higher stages. There then will be an increase in production at higher stages. The production structure will increase in stages until a new, lower interest rate spreads uniformly throughout the entire production structure. The new production is funded entirely by either their own increased voluntary savings or by increased savings of others.

The prices of factors of production at higher stages will not necessarily increase. Increased demand for land, labor, and capital goods used in the higher stages will be offset by increases in supply of these factors as they are freed from other uses in lower stages of production. Note that for such coordination to take place, it is very important that factor markets be very flexible and allowed to adjust to new market conditions.

Often increased saving will be accompanied by a temporary slowdown in the arrival of new consumer goods to market. But this will only happen until increased productivity that results from more and better capital goods at higher stages works itself down the structure of production.

The interest rate will decrease throughout the time market. This is because the ratio of the price of the final product to sum of the prices of factors of production will shrink. The interest rate charged for loans will decrease as well. A lower interest rate will result in increased capital values of assets. The effect will be greater the more durable the good is, because there are more annual rents to be discounted by the interest rate.

Capital goods already in use will experience significant price increases. These goods will be produced in greater quantities, which ultimately will bring more production at each stage. The fall in the interest rate will make many processes for the production of capital goods profitable which until then had not been profitable. Entrepreneurs accordingly will start to produce them. Projects that result in lengthening the production structure through new, more modern stages further from consumption will be undertaken. Consequently, the capital structure is lengthened as new stages of capital goods are introduced. Changes in value of capital assets will tend to be reflected in the price of stock shares that represent the goods.

The moral of the story is that increases in voluntary saving result in economic expansion. As a result of increased saving, the interest rate at each stage will be lower. The capital structure deepens because new stages are added which did not exist before. The longer production structure will allow for increases in production of consumer goods, so that the prices of consumer goods will decrease. As saving increases, people have more funds to invest. As investment increases, capital is accumulated and the supply of capital goods will increase. Because of the increase in capital goods, the production structure can be lengthened. Longer production processes allow for greater marginal productivity of the factors of production. The end result is that, because of new net savings, more capital goods and consumer goods are produced. In the aggregate, there are more goods available for the same total quantity of money. Each dollar can buy more so real incomes increase. This results in significant increases in real wages because workers can buy more for their dollar. In the long run, people will be able to buy more goods at lower prices.

We can conclude, then, that increased saving results in an expanding economy. The structure of production lengthens. Capital is accumulated. Factors of production become more productive. More goods are produced with the same amount of labor and real incomes increase.

On the other hand, decreases in savings due to increased time preferences will result in a shrinking economy. If time preferences increase, people become more present-oriented and will save less and consume more. Overall savings and investment will decrease while spending on consumer goods will rise. The result is net dissaving and net disinvestment.

Decreases in saving result in economic regression. Less saving results in higher interest rates as the spread between the prices of output and the sum of factor prices increase throughout the production structure. If saving decreased, so would the demand for factors of production at the higher stages of production so the prices of those factors will decrease. Because of increased consumption, prices for consumer goods will rise, so there will be an increase in the demand for factors at the lower stages of production. Hence the ratio of the spread between the price of output and the price of factors of production would increase throughout the production structure. Capitalists would require higher rates of return in order to make their investments. Consequently, the interest rate would increase and lenders would require higher interest rates for their loans.

The overall standard of living will decrease due to decreased productivity. The productivity of land and labor decreases because longer, more productive processes must be abandoned due to the impatience of society's new time preferences. The capital structure would shorten because fewer stages of production could be financed. Because saving has declined, less capital is maintained, culminating in capital consumption. Aggregate output, therefore, declines due to a smaller stock of capital goods. Over time, lower productivity and fewer capital goods will result in a decrease of consumer goods.

Additionally, reduced saving generates aggregate losses in the economy, especially at the higher stages of production. When time preferences increase and investment decreases, it is the firms at the higher stages of production who first lose customers. Therefore, the assets used in those highest stages are now overcapitalized. Those who invested in them will not be able to sell output at high enough prices to break even, so they earn losses on their investments.

Over time these losses tend to disappear as firms leave these industries. In those industries that are earning losses, investment will decrease. Therefore, demand for factors of production in these industries will also decrease, lowering their prices. Decreased prices for factors will result in higher net revenues, lifting the rate of return back toward the interest rate. Lower real wage rates and land rents combined with higher interest rates result in lower capital values.

Consumption spending initially would increase. However, because fewer capital goods would be produced, the quantity of consumer goods would decrease over time. Because there would be an increased demand for consumer goods while fewer of them could be produced, over time, the prices of goods would increase. In the long run, people would be able to buy fewer goods at higher prices.

We can conclude, then, that decreased saving results in a shrinking economy. The structure of production becomes shorter. Capital is consumed. Factors of production become less productive. Fewer goods are produced with the same amount of labor. Real incomes necessarily fall. Overall prosperity declines.

GROSS DOMESTIC PRODUCT

Since the late 1930s, there have been attempts to measure the size of the social economy in order to determine whether the economy is expanding or contracting. We will examine the nature and causes of expanding and contracting economies in more detail in chapter 19. For now we want to examine the most common statistic used by economists to gauge the state of the economy, the Gross Domestic Product (GDP). We should note at the outset that many economists and politicians use GDP to measure the size of the economy, but that does not mean that the GDP is very effective at measuring the economy's size. In fact, there are many grave limitations of the GDP that leaves it wanting. However, because of its popularity with economists and politicians who use it to convince people to go along with their policy proposals, it is our duty to know of what they speak.

National income statistics, of which the GDP is the most popular and most used, are a product of state intervention in the economy. There was very little data collected and very few economic statistics calculated by the U. S. government before World War I. During and after the Great War several economists who were enamored with the economic policies of the so-called progressives gained increasing influence in the halls of power. These were economists who believed that we had entered an age where the economy could be scientifically managed in such a way to produce continuing prosperity without business cycles and increases or decreases in overall prices. With the sweeping popularity of John Maynard Keynes' *General Theory of Employment, Interest, and Money*, Keynesian attempts to manage the economy have become the order of the day since the Great Depression. It was soon recognized that attempts to manage something so that it reached certain goals requires a way to measure whether the goals are being met. Therefore, national income statistics were created in an effort to measure the production and income generated in our economy. The development of these statistics was considered very important contributions to macroeconomics, because it was thought that statistics like the GDP would allow government policy makers to more effectively do their job. Creating these statistics was considered so important that the man chiefly responsible for their development, Simon Kuznets, received the Nobel Prize in economics for his work.

As mentioned, GDP stands for Gross Domestic Product. As its name implies, the GDP seeks to measure the market value of current, final, domestic production during a specific period of time. We can unpack the meaning of this definition by exploring in more detail the implications of various words in this definition.

Notice first that the GDP is the market *value* of current, final, domestic production. Because it is an attempt to calculate the market value of production for a period of time, the government values goods at the prices actually paid in market transactions and sums them together. Consequently, GDP is calculated in dollars. There are notable exceptions to using the prices actually paid for a good, however.

Housing presents its own unique challenges for GDP statisticians. If housing is rented, rent is included in GDP calculations. Many people own their own house, however, and do not sell it regularly. Notwithstanding this, the government wants the value of owner-occupied housing accounted for in the GDP. Therefore, when calculating the value of housing in the GDP, government statisticians substitute the estimated rental value of owner-occupied housing as a proxy for its market value. The value of housing services is counted as if the owners pay rent to themselves. The value of housing services is imputed by the Department of Commerce based on the rental market in the area of the house being counted.

Additionally, any *non-market transactions* are not included in GDP. By *non-market* transaction Department of Commerce bureaucrats do not mean somehow a transaction that is not an exchange, but rather those exchanges that go unreported. When a college student is paid in cash for changing the oil in a hot rod owned by a little old lady from Pasadena there is no official record of the transaction. The government has no way of knowing the transaction took place at all; therefore it does not make it into the GDP calculation.

The GDP seeks not to be a measure of the value of any production, but of current production. This means that it only counts production made during a particular year. That is why, if you consult government national income publications, you will find tables listing the GDP numbers for various years and quarters of years. For example, the official GDP for the United States in 2007 was $13,807.5 billion; this means that the number calculated for GDP reflects expenditures on products only during 2007, not those in 2006 or 2008 or any other year for that matter. It does not represent the output for the economy for all time, but is only a chronological snapshot.

Additionally, the GDP is defined as the value of all *final* domestic production. This word *final* highlights that the only values included in the calculation of GDP are those for final products. GDP statisticians do not include intermediate goods or goods that are only partially finished in the calculation of GDP. This is done because, it is argued, the value of intermediate and partially finished goods is included in the value of final productions. The reasoning behind this practice is illustrated by table 11.4.

Table 11.4

Value Added to GDP	
Producer	Value Added
Farmer	.15 – 0 = 0.15
Miller	.20 - .15 = 0.05
Baker	.50 - .20 = 0.30
Grocer	1.00 - .50 = 0.50
Total	1.00

Suppose we want to count the contribution of the production of one loaf of bread to GDP. The grocery store that sells the bread to the consumer receives $1.00 for the loaf of bread. In order to obtain the loaf of bread, the grocer must purchase it from the baker for fifty cents. In order to make the bread, the baker must buy flour from the miller for twenty cents. In order to make flour, the miller must purchase wheat from the farmer for fifteen cents. If we were to sum up the market values of all of the intermediate goods along with the load of bread we would finish with the sum of $0.15 + $0.20 + $0.50 + $1.00 = $1.85.

Statisticians at the Commerce Department, however, are of the opinion that the value of the wheat, flour, and wholesale bread is included in the price of the bread sold by the grocer. Consequently, our $1.85 figure includes the counting of some goods multiple times. For example, in the situation above, the value of the wheat is included in the value of the flour, which is included in the value of the wholesale bread, which is included in the value of the retail bread. Therefore, the value of wheat is counted four times, not once.

To keep from over-counting, the Commerce Department sums not the market price at each stage, but the value added. As illustrated in table 11.4, the farmer's labor and land work together to produce wheat. The amount of wheat that will be transformed into the flour that will eventually make it into one loaf at the store is sold for fifteen cents. The farmer, then, generates fifteen cents in value. The miller sells the flour he produces with the wheat for twenty cents. Does this imply that the total value that the miller produces is twenty cents? According to the GDP statisticians, the answer is no. The miller produces a net value of five cents, because fifteen cents out of the twenty cents for which he sold the wheat was produced by the farmer. Likewise, the value added by the baker is the difference between fifty cents (what he sold the bread for) and twenty cents (what he paid for the flour). Therefore, the sum of the net value of productivity for all the stages is equal to $0.15 + $0.05 + $0.30 + $0.50 = $1.00. Note that this $1.00 is equal to the price of the bread at the grocery store, our final product. Therefore, in order to avoid over-counting the value added at each stage of production, the GDP includes only the market value of all final products sold during the relevant time period.

Finally, the statistic used to measure the state of the economy is the *Gross* Domestic Product. It is called gross because the depreciation of capital goods is not subtracted from the value of output.

Business inventories are treated differently depending upon what happens to them. If Roman Meal spends money on wages and other factors to make a loaf of bread during the relevant time period and the bread is unsold at the end of the time period, how is this counted? If the bread becomes moldy and rots, Roman's profits are reduced by the same amount that wages have increased, so the total expenditures and income reaped in the economy are unchanged. The only thing that changes is the distribution of income, in this case from profits to wages. Consequently, such an occurrence would have no effect on GDP.

On the other hand, if the bread does not rot, but can be put in inventory to be sold later, profit is not reduced, because Roman Meal will reap profit from the bread when it is sold. The money spent on the factors of production during the relevant time period is counted as investment, as if Roman Meal purchased the loaf of bread for its inventory. Because the increased expenditure on the factors of production raises total income, and greater inventory accumulation raises total expenditures, this occurrence increases GDP.

We have seen above that GDP is calculated by summing the net income of all producers that contribute toward the production of a final product. The GDP can also be calculated by adding expenditures instead of incomes. For every purchase someone receives income. Everyone's income, in turn, constitutes someone else's spending. Therefore, we can calculate GDP by summing up total expenditures on final products.

Calculating GDP is made a bit more complicated than our simple example above because the value of different goods must be added together. We are literally adding apples and oranges, or we should say, adding the price of apples to the price of oranges. As mentioned above, however, essentially the GDP is merely the sum of total expenditures on final products during a particular period of time. Suppose that we have a two-good Turkish-prison economy, meaning that consumers could purchase both bread and water, but that was all. If, during the fourth quarter of 2004, 1,000 loaves of bread were purchased at a price of $1.00 each and 2,000 gallons of water were purchased at $2.00 each, the GDP for the fourth quarter of 2004 for this economy would equal the price of the bread times the quantity of bread sold plus the price of the water times the quantity of water sold. In other words, GDP = ($1.00 × 1,000) + ($2.00 × 2,000) = $1,000 + $4,000 = $5,000.

There are four broad categories of expenditures that are included in the GDP, as illustrated in table 11.5.

Table 11.5

2007 GDP (in billions)	
Consumption Spending	$9,710.2
Business Investment	$2,130.4
Government Expenditures	$2,674.8
Net Exports	-$707.8
GDP	$13,807.5

The first is consumption spending which is spending by private citizens on consumer goods. This is added to business investment, which is the sum of investment in durable capital goods and business inventories. To consumption spending and business investment is added government spending, which as the term suggests, is the sum of government spending on consumer goods as well as spending on durable capital goods. Finally, net exports (exports minus imports) are included, hence the domestic part of the Gross Domestic Product. Thus, the GDP for 2007 was $13,807.5 billion, which was the sum of $9,710.2 billion in consumption spending plus $2,130.4 billion in business investment plus $2,674.8 billion in government expenditures -$707.8 billion in net exports. This number is subtracted because in 2007 we imported more goods than we exported, so we had net imports instead of net exports. Notice that, according to the official figures, the spending in the different sectors actually adds up to $13,807.6 billion, but official GDP is reported as being $13,807.5 billion. This is the result of what Bureau of Economic Analysis statisticians refer to as statistical discrepancy.

While using the GDP as a measure of overall economic productivity is wildly popular, the statistic itself is not without its problems. In fact, they are so numerous that we can conclude that GDP ultimately fails as a scientific measure of economic activity because of what it ignores, it obscures, and for which it simply cannot account.

Beware of using the GDP as a measure of human welfare. Even if the GDP were completely accurate, which it is not, it would account only for the market value of economic goods; it is only a tally of market prices for economic goods bought and sold. Prices are objective quantities of money, whereas value is subjective. Although the prices used to calculate GDP are reflections or manifestations of human values, they are not measurements of them. In fact, a good's market price most clearly reflects the value of the marginal buyer. The value of the good for all supra-marginal traders must exceed the market price. Because all value is subjective, there is no scientific way to measure human satisfaction or human welfare.

Using market prices for final products actually sold not only greatly limits the scope of GDP as an indicator of human welfare, but it can be downright misleading. For example, while people generally prefer more economic goods to fewer economic goods because more goods allow them to satisfy more of their ends, people have some ends that cannot be satisfied with goods. Ends such as the satisfaction in doing right or in having one's sins forgiven are not something that can be bought or sold or be satisfied with tradable commodities. Therefore, GDP does not include such considerations.

Some productive activity is actually ignored by GDP. Household labor is a good example. The productive work of a homemaker rearing children, cooking, cleaning, washing, shopping, and the multitude of other tasks necessary for the smoothly operating household are completely left out of the official GDP statistics.

At the same time, because GDP incorporates the prices of all final goods sold in a particular period, it includes goods bought in purchases people would rather not have to make. A hail storm that ruins your roof, for instance, will result in an increase in GDP. You will have to purchase plywood and shingles, which are included in the GDP count. While the hail storm made things worse for you, the official GDP statistics makes it appear that the economy expanded. Hurricane Charlie struck Florida in the summer of 2004 resulting in damage estimated at exceeding $20 million. The victims of the hurricane or their insurance companies are poorer by that much. However, because this money needed to be spent on newly produced building materials, most of the hurricane bill showed up as a positive contribution to GDP.

Another common mistake that people make in their understanding, and subsequent use of GDP, is viewing it as a measure of national wealth. It is true that entrepreneurs use market prices to economically calculate the market value of the goods that they are producing and the assets they own. If Wal-Mart places a monetary value of $5.95 on a DVD set of Hope and Crosby Road Pictures, *Road to Rio* and *Road to Bali*, it is because they think that they can sell them to consumers for $5.95 each. If a business says that its business is worth $2.5 million, it is because they think that at that point in time it could sell its assets for $2.5 million. Monetary evaluation requires the ability to sell the good at the point of evaluation.

It is out of the question, however, for an entire nation to liquidate all of its assets at one time. An entire nation cannot sell all of its property for money. A business can, but a nation cannot. Therefore, GDP is not a measure of national economic wealth. As indicated earlier in our calculation of GDP, essentially it is a record of net income for a particular period of time. However, even as a record of net income, it has problems.

Although the G in GDP stands for *gross*, it actually does not include true gross figures. For example, GDP only includes gross purchases of durable capital goods and consumption of capital owned by producers. It implicitly separates gross investment spending into purchases of durable and non-durable capital goods. However, from the economic point of view there is no reason to separate the two. There is no economic difference between durable capital goods and capital goods used up in the production process. By ignoring investment spending on non-durable capital goods, it ignores a large portion of the structure of production supporting the availability of consumer goods.

By ignoring investment in non-durable capital, GDP leaves the unsuspecting with the impression that consumer spending is the dominant determinant of economic performance. Many people fear that if consumption spending falls the economy will drop into depression. In fact, we do not need to worry about maintaining consumption because we always must consume at some level. It is the proportion of all investment in capital goods that is the important consideration, which means social time preference is a key to determining the rate of economic progress. As time preferences go, so goes saving and investment, and, hence, spending on capital goods. Remember, it is spending on capital goods throughout the entire structure of production that supports the manufacture of consumer goods.

Another serious limitation of GDP as even a measure of national output is the use of inaccurate data. The facts of life are that the statisticians at the Commerce Department do not always have perfect data. They are often forced to include less than accurate data if they want to publish their statistics. Two economists who were involved with the development of national income accounting statistics estimated that reliance on inaccurate data results in GDP figures with a relatively large margin for error. Nobel Prize winner Simon Kuznets estimated that imperfect data results in a GDP statistic with a margin of error of 10%. Oskar Morgenstern estimated that the margin for error is more like 20%. Kuznets thought that the margin of error was so great that it was statistically possible that the GDP figures could be telling us the opposite of what actually happened. The potential for mistake is so great that, while the GDP signals economic boom, it could be that the economy is actually in recession, and when it signals recession the economy could actually be in a recovery or boom period. This surely is not a very sturdy statistical rack upon which to hang one's policy hat.

An internal inconsistency from which the GDP suffers stems from its inclusion of government spending. At best, some things that the government purchases, such as military spending, do not directly increase human welfare and are more like intermediate goods. Yet they are included in GDP figures making them appear larger than they would be otherwise.

Worse, however, the economic impact of government expenditures is more akin to consumption rather than production. They are better viewed as consumption because they do not reflect subjective preferences of those from whom the money spent was taken via taxes, but rather reflect the personal wishes of the politicians and bureaucrats who actually spend it. In order to receive the money that it spends, the state does not produce anything. It relies on coercive taxation, borrowing, or monetary inflation. The economic effects of all these practices will be analyzed in detail in later chapters. It is important to understand that government spending is more consumptive than productive.

THE PRODUCTION STRUCTURE OF THE SOCIAL ECONOMY

The aggregate production structure is one of the things that integrates all markets that make up the social economy. The social economy includes all voluntary exchanges i3%n all markets for all consumer and producer goods. In essence, it is the material manifestation of humanity's response to God's cultural mandate. The production of every consumer good is supported by its own structure of production. The broad social economy is an integrated aggregate production structure that supports the production of *all* consumer goods. Throughout the entire social economy, money moves from consumer goods industries up through the higher stages of production. At the same time, production effort in the form of goods flows from the higher stages to the lower stages, finally culminating in the production of consumer goods.

In order for the social economy to perform as well as possible, it is important that the myriad of production plans are coordinated in a peaceful and productive way. Markets at each stage must be coordinated so as not to produce shortages or surpluses. Markets at different stages of production must be coordinated intertemporally, so that producers at each stage will have use of the factors of production they need and will produce those goods in most demand by buyers at lower stages. These facts highlight again the importance of the free market price system. Market prices allow producers to calculate profit and loss and, hence, co-ordinate production activities to best serve consumers. Furthermore, the interest rate—the ratio of prices of present goods compared to those of future goods—allows producers to make decisions about intertemporal stages of production. The interest rate will only be an accurate reflection of social time preferences if prices of all goods are allowed to freely adjust. Free capital markets also play a crucial role in allocating capital through-out the social economy.

The structure of production is one of the two things that integrates the entire social economy. The other is money, which is the focus of chap-ter 12. Chapter 13 will investigate how government intervention in the monetary system can have grave consequences for the structure of pro-duction, resulting in massive malinvestment and economic depression.

SUGGESTED READING

Huerta de Soto, *Money, Bank Credit, and Economic Cycles*, 265–346. An outstanding recent exposition of the complex capital structure supporting the production of all consumer goods, the theoretical problems of GDP accounting, and the economic causes of economic progress and regress.

Mises, *Human Action*, 292–99. Mises provides his explanation of increased per capital investment as the source of economic progress.

Osterfeld, Prosperity Versus Planning, 9–14. A masterful and brief description of the technical and political problems of GDP calculations.

Rothbard, *Man, Economy, and State*, 319–451; 517–55. These two chapters provide Rothbard's definitive and exhaustive exposition of the production structure, stock ownership, and economic change.

12

Money and Its Purchasing Power

As explained in the previous chapter, all human action in the market—the vast network of voluntary exchanges—is integrated into
a highly complex aggregated structure of production that we refer to as
the social economy. Through the structure of production, landowners,
laborers, and capitalists come together in order to benefit consumers by
producing goods that they want at prices they are willing to pay. At the
same time, the consumers benefit the owners of the factors of production and the entrepreneurs who drive the system by providing them with
income that they can use to purchase the consumer goods that they want.
Clearly the social economy is integrated via the aggregate structure of
production.

Remember, however, that the wishes of consumers are transmitted
through the structure of production through changes in the relative prices
of factors of production and final products. It is money prices that allow
entrepreneurs to calculate expected profit and loss and assess actual profits and losses after the fact. Entrepreneurs are able to use money prices
for economic calculation because in a monetary economy all consumer
goods and factors of production are exchanged for money. Consequently,
money is the one good that integrates the entire social economy and
makes possible our immense structure of production, because money is
the only good that is traded in all markets.

In chapter 5 we discovered that because of the problems associated
with the barter system, people sought out and traded to obtain more marketable goods until the entire society began using the most marketable
good as a medium of exchange. This general medium of exchange is what
we call money. We saw that the process by which a commodity becomes
money is a voluntary process. Money was voluntarily chosen, not forced
on people. The state cannot force anyone to use a specific good as money;

it has tried numerous times and failed. The most recent attempt by the U. S. government was the introduction of the Sacagawea dollar. It was a metal dollar made to look golden, at least until the outside covering wore off. The government wanted this dollar to be used by its citizens and spent millions of dollars in a promotional campaign, all to no avail. People simply did not want to use the coin as money, and instead hoarded it as a collectable. Consequently, virtually no Sacagawea dollars are used as money.

It was also explained that because money is simply the most market-able good, money is a commodity—an economic good. As such, all of the economic laws that we have derived apply to it just as much as they apply all other economic goods. This means that the value of money, a subject that was very problematic for economists for a very long time, is deter-mined in the same fashion as the value of every other good—according to people's subjective preferences.

THE PURCHASING POWER OF MONEY

Because money is a scarce good, it has a price at which it can be bought and sold. Typically, however, when we think of the market price of good, we usually think of the amount money it takes to buy that good. What about when the good for which we are trying to determine its price is money itself? What then?

In answering this question, it is helpful to remember just what is a price. The price of any good is an exchange ratio. A price is the quan-tity of one good that must be given in exchange to obtain another good. Therefore, if we must trade $4.50 for a gallon of gasoline, the price of the gallon of gas is $4.50. However, this is looking at the exchange from the point of view of the buyer of gas. What about the seller? True, he is selling his gasoline. But what is he buying? He is buying money. Specifically, he is using his gallons of gasoline to purchase money that he will sell later to buy different goods he wants. Consequently, the price of gasoline is the amount of money the buyer must pay, but it is also the amount of money that the owner of gas can purchase. We can conclude then that one can look at the price of a good, not only as what people have to give up to get that good, but also what a person can purchase with that good.

The purchasing power of money (PPM) is the quantity of goods that can be purchased with a unit of money. In the United States, the PPM is the amount of goods that can be bought with a dollar. This means, of course, that the PPM can be reflected against several different goods, because all goods can be bought with money. For the sake of simplicity, let us consider only four goods: apples, gasoline, shoes, and personal computers. Suppose the market price of each good is as reflected below in table 12.1.

Table 12.1

The Price of Goods and the PPM				
	Price	PPM₁	2P	PPM₂
Apples:	$1.00/lb.	1 lb.	$2.00/lb.	1/2 lb.
Gasoline:	$4.00/gal.	1/4 gallon	$8.00/gal.	1/8 gallon
Shoes:	$100/pair	1/100 pair	$200/pair	1/200 pair
Computers:	$800/each	1/800 computer	$1600/each	1/1600 computer

In this four-good economy, the PPM can be expressed against each of these goods. For example, if the price of apples is $1.00 per pound, then the quantity of apples that a dollar can buy—the dollar's purchasing power—is one pound. Likewise, because the price of gasoline is $4.00 a gallon, the PPM regarding gas is $\frac{1}{4}$ gallon. The price of shoes is $100 per pair, so one dollar can, in effect, purchase $\frac{1}{100}$ of a pair of shoes. You will probably have already figured out that, at a computer price of $800, the PPM can also be expressed as $\frac{1}{800}$ of a personal computer.

Now suppose that the price of each good doubles, as indicated in column 2P. The price of apples increases from $1.00 per pound to $2.00 per pound and the prices of all other goods follow suit. The important question is what happens to the PPM. If the price of apples doubles, we see in column PPM$_2$ that the quantity of apples that a dollar can purchase is cut in half. When the price of apples is $1.00 a pound, the PPM is one pound of apples. However, if the price of apples doubles to $2.00 a pound, the quantity of apples that a dollar can purchase decreases to a half a pound. Likewise, if the price of gasoline doubles to $8.00 a gallon, then the PPM decreases to $\frac{1}{8}$ gallon. If the price of shoes doubles to $200 per pair, the PPM decreases to $\frac{1}{200}$ of a pair of shoes. Finally, if the price of personal computers were to double from $800 to $1,600, the PPM would decrease to $\frac{1}{1600}$ of a computer.

By now, you should be able to identify the relationship between the money prices for goods and the purchasing power of money. As prices increase, the PPM falls and as prices decrease, each dollar can buy more goods so the PPM rises. In other words, there is an inverse relationship between the overall prices of goods and the PPM.

Knowledge of the inverse relationship between the purchasing power of money and overall prices is helpful in understanding how the PPM is established. Economic analysis reveals that the PPM is determined by the total stock of money in existence and the total demand for that money.

The total stock of money is merely the total quantity of the commodity serving as money in existence at any given time. If gold is the medium of exchange, then the stock of money is equal to the total stock of monetary gold in society. Presently the dollar is our medium of exchange, so the stock of money is equal to the total quantity of dollars available on demand. Because, at any given moment in time, there is a particular quantity of money in existence, we can represent the stock of money curve as being vertical as shown in figure 12.1.

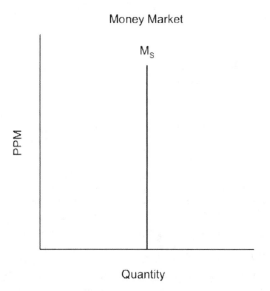

Figure 12.1. The money stock curve is vertical line illustrating the quantity of money in existence.

The PPM is also determined by the demand for money. When economists use the phrase *demand for money*, keep in mind that they are speaking of the demand to *hold* money. The demand for money is not the demand to spend money; it is the demand to hold it in personal cash balances.

As indicated earlier, because money is an economic good, all economic laws, including the law of demand, apply to money just as surely as they apply to all other goods. Therefore, just as with all other goods, there is an inverse relationship between the hypothetical price of money and the quantity of money that people demand to hold. Graphically, this means that the demand for money curve is downward sloping to the right as indicated in figure 12.2.

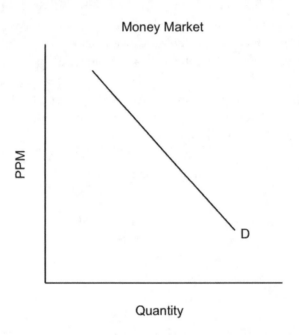

Figure 12.2. The demand for money curve slopes downward to the right, demonstrating an inverse relationship between the PPM and the quantity of money people are willing to hold.

Let us examine just exactly why this is the case. Remember that people only value and hold money because they can use it to make purchases. As the PPM, the price of money, decreases, it takes more money to make the same quantity of purchases, so people will necessarily want to hold more in anticipation of those purchases. On the other hand, as the PPM increases, it takes less money to make the same amount of purchases, so people do not need to hold as much money in anticipation of those purchases. Consequently, we observe that there is an inverse relationship between the hypothetical PPM and the quantity of money people demand to hold.

Now that we have a handle on the basic characteristics of the money stock and demand curves, we can put them together to see how the PPM is determined. Initially, the determination of the PPM appears rather simple. The price determination process for money is basically the same as that for all other goods. It should surprise no one then to find that the free market PPM is determined in the money market. If one wants to discuss the determination of the price of shoes they need to look to the shoe market. If we wish to examine the formation of the PPM, we should investigate the money market.

In doing so, we want to dispel some potential confusion regarding the term *money market*. Increasingly that term has been used for the short-term credit market. There is today an active market for loans by corporations who want to borrow money for only three months or less. Because such loans are usually taken out because the borrowers wish to quickly increase their access to present money, this market has been tagged with the title *money market*. However, do not forget that the market for any good is the network of voluntary exchanges for the good in question. The short-term credit market is just that, a credit market. It is actually part of the time market, the market in which present money is traded for future money. However, the money market is much broader; it is the market in which money is exchanged for goods. Brief contemplation should reveal that the money market is everywhere. As mentioned earlier, money is traded in all markets so the money market really encompasses the entire economy. When we use the term money market, we are not only speaking of the short-term credit market, but everywhere money is traded.

Having correctly identified the money market, we are free to pursue further knowledge regarding the determination of the PPM. When we say that the PPM is determined by the money market, we merely mean that the free market PPM will be that price of money where the quantity of money supplied equals the quantity of money demanded. This can be seen in the money market graph illustrated in figure 12.3.

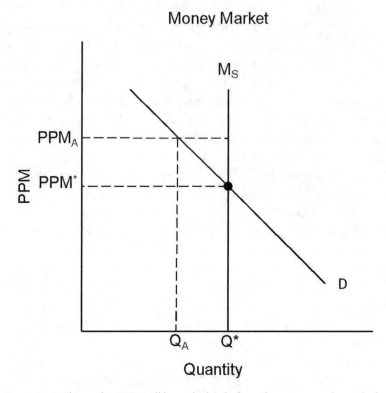

Figure 12.3. The market PPM will be at the level where the money stock equals the quantity of money demanded.

You can see in the money market graph that the PPM will move to that level at which the money stock and demand curves intersect. At this purchasing power of money, *PPM**, the money market is cleared because the quantity of money that people want to hold is equal to the amount that they are actually holding. In the money market, just as in other markets, the price of a good tends toward that one price where everyone who wants to buy can buy, and everyone who wants to sell can sell. There is one PPM that the market tends toward, the market PPM.

Suppose, for example, that the PPM of money is actually higher than the market clearing PPM. Suppose it is at PPM_A. At PPM_A the quantity of money that people are actually holding—the stock of money in existence—is greater than the quantity that people demand to hold. In other words, there is an excess supply of money. People are holding more money than they want to. What do people do if they are in this situation?

They spend it. Consumers increase their spending on consumer goods. Producers increase their spending on producer goods. Consequently, the demand for these goods increases. As the demand for goods increases, what happens to the prices for these goods? They increase.

At the same time, if producers already think they have cash balances larger than they want, they are going to be less likely to rush their current inventory to sale in order to stimulate cash flow. They will be willing to sell fewer goods at every hypothetical price. This will result in a decrease in present supply of goods already produced. Economic theory tells us that such a decrease in supply of goods will result in an increased price for these goods.

An excess supply of money stimulates people to act in such a way that sets in motion a process by which the prices of consumer and producer goods increase. As overall prices of goods increase, the PPM of money falls.

How long and far will the PPM fall? It will fall as long as people have an excess supply of money. As long as there is an excess supply, they will continue to increase spending and hold back current supply of goods. A decreasing PPM, however, will shrink the size of the excess supply. As the PPM falls, people demand to hold a greater quantity of money, because it takes more money to make the same quantity of anticipated purchases. Eventually, the PPM will fall to the level at which, because the prices of goods are higher, the quantity of money they are actually holding equals the quantity that they want to hold. People will then cease increasing spending and reserving inventories of saleable goods. They will have no further incentive to increase spending and withhold selling any further. The money market will be *at rest* at the market PPM. So if, for whatever reason, the PPM is above the market PPM, humans will act so that the PPM will fall until the quantity of money supplied is equal to the quantity of money people want to hold.

What happens, however, if the PPM is below the market clearing PPM? Suppose, for instance, that the PPM is, as seen in figure 12.4, at PPM_B.

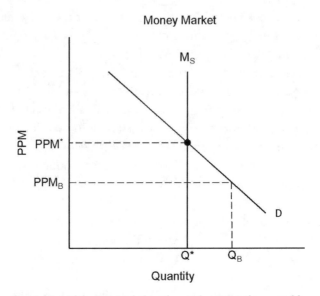

Figure 12.4. If the PPM is below the market PPM, there would be an excess demand for money and people would decrease their spending and be more willing to sell goods. The price of goods would fall and the PPM would increase until the money market is in equilibrium.

If this is the case, the money market is in a condition of excess demand. At PPM_B people demand to hold more money than they are actually holding. If people are in that position what can they do to alleviate their excess demand? They will tend to engage in what is popularly called *belt tightening*. They will spend less. They will demand fewer consumer goods. Businesses will try to cut costs by buying fewer producer goods. This decrease in the demand for goods will tend to lower the prices for those goods.

At the same time, some of those who are able will increase their supply of readily available goods. Some may take on second jobs. Some may hold rummage sales or begin selling things on Ebay. Some producers will be more eager to sell down their inventories in order to stimulate cash flow. All of these actions will tend to increase the supply of producer and consumer goods. Economic theory teaches us that an increased supply of goods will result in lower prices for those goods.

Consequently, when people experience an excess demand for money at a particular PPM such as PPM_B, they will act in such a way that sets in motion a process by which the prices of producer and consumer goods decrease. As the prices for these goods decline, the quantity of those goods that can be bought with a dollar increases. As overall prices fall, the PPM rises.

How long and far will the PPM rise? It will rise as long as people continue to reduce their spending and increase their current supply of saleable goods. This will occur as long as they want to hold more money than they are actually holding. They will experience this excess demand for money as long as the PPM is below the market clearing PPM. Once the PPM rises to that level at which the quantity of money demanded equals the quantity supplied, however, everyone is actually holding the quantity of money they want to hold. At this PPM, people have no further incentive to reduce their spending or increase their current supply of goods in order to get more money. Therefore, as soon as the PPM reaches the market clearing level, it will tend to stay there and the market will be at rest.

CHANGES IN THE PURCHASING POWER OF MONEY

Because the purchasing power of money is determined by the total demand to hold money and the total stock of money in existence, changes in the PPM will be the result of a change in either the stock of or demand for money. Suppose for example, that there is an increase in the demand to hold money. This is graphically illustrated in figure 12.5.

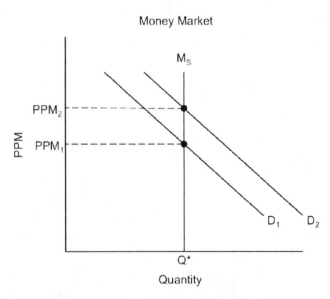

Figure 12.5. An increase in the demand to hold money will result in a higher PPM.

If there is an increase in the demand for money from D_1 to D_2, there would be an excess demand for money at the original purchasing power of money, PPM_1. As we have seen, when there is an excess demand for money, people will reduce their spending on goods, decreasing their demand for goods. At the same time, some will increase their supply of goods. Both of these actions will result in lower prices for producer and consumer goods, so the PPM will rise. The PPM will continue to rise until the amount of money people are actually holding is equal to the quantity that they demand to hold. This occurs at the new equilibrium purchasing power of money, PPM_2.

   If there is a decrease in the demand to hold money, the opposite will occur. As illustrated in figure 12.6, if the demand for money falls from D_1 to D_2, there will be an excess supply of money at PPM_1.

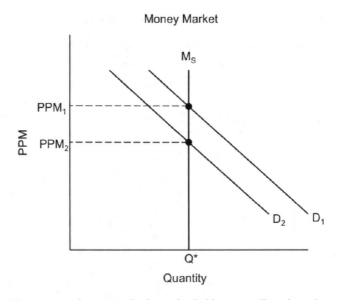

Figure 12.6. A decrease in the demand to hold money will result in a lower PPM.

As indicated earlier, when people hold more money than they want, they rid themselves of the excess supply by increasing their spending on goods and by withholding their supply of saleable goods. These actions result in an increase in overall prices for goods, and a reduction in the PPM. The PPM will continue to fall until people no longer perceive an excess supply. This occurs at PPM_2, at which the quantity of money demanded is, once again, equal to the stock of money in existence.

We have seen, then, how changes in the demand to hold money will affect the PPM. Like all other goods, however, the stock of a good in existence will have an impact on the price of that good. Changes in the stock of money will also affect the PPM.

Suppose that there is an increase in money production. We can see the effect of such an increase by looking at the money market graph in figure 12.7.

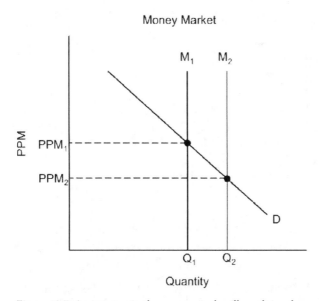

Figure 12.7. An increase in the money stock will result in a lower PPM.

An increase in the stock of money is shown by the rightward shift in the money stock curve from M_1 to M_2. Because of the increase in money production, we can see that there is an excess supply of money at the original purchasing power of money, PPM_1. People will spend more and supply fewer goods so that overall prices will rise and the PPM will fall. It will continue to fall until it reaches PPM_2, where the quantity of money demanded equals the quantity of the money stock.

A decrease in the production of money will have precisely the opposite effect. If the stock of money in existence decreases, shown in Figure 12.8 as a leftward shift in the money stock curve, the PPM will increase.

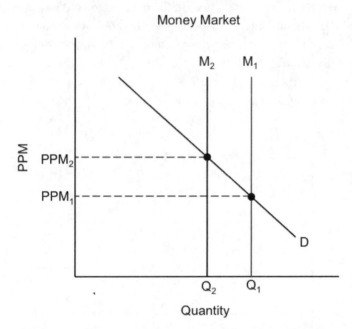

Figure 12.8. A decrease in the money stock will result in a higher PPM.

Why is this so? If the money stock decreases from M_1 to M_2, then at PPM_1 there would be an excess demand for money. People would spend less on consumer and producer goods. Some would also be more eager to sell those goods. Consequently, the market prices for those goods would decrease, increasing the PPM. The PPM would increase to PPM_2, the level at which the quantity of money people want to hold is equal to the quantity of money they are actually holding.

MARGINAL UTILITY
AND THE DEMAND TO HOLD MONEY

We have seen that the price of money—the PPM—is determined according to the same principles by which the prices of all other goods are determined. Likewise, all of the actors in the money market are just as motivated by their subjective value scales as suppliers and demanders of every other good. To better understand the true source of the PPM, we need to go behind the curves on a graph and investigate the foundations of money demand and the stock of money.

We begin with the demand to hold money. Although money is an economic good, and the value people attach to it is subjective just like all other goods, money is different from all other goods, because its value is derived only from the fact that it can be used in exchange. Money as money has no direct use value. People value money, not because they want to consume it or use it to produce some other good they can trade or consume. People value money precisely because they can trade it for other goods that help them satisfy their most preferred ends. Because people value and hold money only because they are able to exchange it for other goods that they want, the value that people impute to money is closely related to its purchasing power. This reality is what has for centuries made understanding the determination of the value of money a unique challenge for economic thinkers.

The subjective values of consumer and producer goods are determined by ideas that people have that these goods can be used as means to serve their ends. Thus it is people's perception that these goods can help satisfy their ends that results in value being imputed to them. The utility that people reckon toward consumer and producer goods depend on the use-value of the good and the subjective ranking of the ends those goods can serve. The subjective value imputed to consumer and producer goods is not related to whether these same goods have exchange value.

Money is different. The subjective value people impute to money is derived from its purchasing power, because money has utility precisely because it can be exchanged for other goods. Money, then, has no subjective value without exchange value. If money had no value in exchange—if it had no purchasing power—it would never be used and no one would want to hold it. The reason money is valued by someone is because he expects that it has a certain purchasing power. The reason you hold a few dollars in your pocket is because you think that you can readily buy whatever you may want with that money, as long as the price is right.

Because people only demand to hold money if they expect it to have purchasing power, we can conclude that the demand for money is a composite of the demand for producer and consumer goods and the marginal utility of money. This implication, however, leaves us with a bit of a puzzle, a circular argument really: it is illogical to explain the PPM of money as dependent on the marginal utility of money and then to explain the marginal utility of money as dependent on its purchasing power. This is like saying the PPM is determined by the marginal utility of money which is determined by the PPM which is determined by the marginal utility of money which is determined by... well, you get the picture. This reasoning is like a hamster that runs and runs and runs in the little wheel in its cage without getting anywhere.

It was the Austrian economist Ludwig von Mises who in his 1912 *Theory of Money and Credit* explained how to get us out of this circle. Mises developed his regression theorem to solve our logical puzzle. As Mises explained, on any given day, the money price of each good is determined by the interactions of supply and demand schedules for money and by the buyers and sellers of the goods in question on that day. The money prices for the goods at the end of the day are determined by the marginal utility of the goods and of money that existed at the beginning of the day during which the exchanges were made.

The marginal utility for money, however, is based on the previously existing PPM from the immediate past. Therefore the economic analysis of the value of money is not circular. We are *not* saying that the PPM is determined by the marginal utility of money that is determined by the PPM that is determined by the marginal utility of money. Our explanation says that Friday's PPM is determined in part by Friday's marginal utility of money, but Friday's marginal utility of money is *not* determined by Friday's PPM, but instead by Thursday's PPM. We do not have a logical circle, but a regression back in time.

It appears, however, that we may not be out of the intellectual woods yet. Pushing the source of today's marginal utility of money back a day does not fully explain everything, because we have to ask from where Thursday's PPM comes? Thursday's PPM is determined in part by Thursday's marginal utility of money that is determined by Wednesday's PPM. Someone could be reasonably concerned that we are merely pushing the problem back into the infinite past without ever really providing an answer.

Here again, however, Mises delivers the solution. The regression outlined above is not infinite. We can theoretically trace the PPM back in time to the day before whatever commodity is being used as a medium of exchange was used as a medium of exchange. For example, on the day before gold was used as money, the price of gold—its exchange value—was determined by the supply of and demand for gold for its use value. The utility that gold provided to people as a consumer or producer good on that day was not dependent on any previous purchasing power. So the regression does not go back in time indefinitely, but stops on the day before the money commodity was actually used as the general medium of exchange. Money, therefore, is an economic good that can be integrated into the general theory of value. Like all goods, the exchange value of money is determined by the money market.

DEMAND FOR MONEY

Because the demand to hold money is one of the determinants of the PPM, it is clear that changes in the demand for money will result in a change in the PPM. What factors affect the marginal utility of money, and hence, how much money people demand to hold at each hypothetical PPM?

The primary factor that has had a long-term positive impact on the demand for money is the supply of goods and services. People obtain money through exchange. In order to increase our cash holdings we must sell goods. Therefore, an increase in the supply of goods is synonymous with an increase in the demand for money. Historically, the long-term trend has been for the supply of goods to increase in progressing economies, as markets developed and as the structure of production has lengthened. As people supply more goods, they demand more money in exchange. We can observe how an increase in the supply of goods affects the PPM by looking again at Figure 12.5. As the supply of goods increases and the demand for money increases, overall prices of goods are bid down by the more eager sellers and the PPM increases.

In times of a regressing economy, when markets fall into decline or are eliminated altogether, the supply of goods falls, decreasing the demand for money. We can see the effects of such a change in Figure 12.6. If the demand for money falls as the supply of goods decreases, this can be illustrated by a leftward shift in the demand for money curve. Because the supply of goods has fallen, there would be excess demand and frustrated buyers at the original prevailing prices. Consequently, the more eager buyers would bid up the overall prices of goods for sale so that the PPM will fall.

The most important factor that has helped to offset the increased demand for money resulting from a progressing economy is the development of clearing system technology. The clearing system is a method by which accounts are cleared on paper rather than cash.

Suppose Peter raises horses, Paul produces wool cloth and Mary raises sheep. In order to increase their level of satisfaction, the following two trades are made. Paul trades a few bolts of cloth to Mary for a sheep. Paul then trades the sheep to Peter in exchange for a horse. In the end Peter ends up with a sheep, Paul has a horse, and Mary has the cloth. A bit of thinking reveals that this same arrangement could have been handled by all the participants merely passing their goods around once. Paul could have simply given his cloth to Mary who could have transferred her sheep to Peter who could have advanced his horse to Paul. The same result would have occurred with one fewer exchanges.

The development of money and credit makes such account clearing even easier. Suppose that the town butcher has a weekly tab amounting to $10 at the Bakery. At the same time, the baker's tab is $50 at the butcher shop. By offsetting the accounts on paper, the butcher and baker can undertake their transactions with less money. The baker's bill of $50 at the butcher shop is offset by the bill of $10 that the butcher owes the baker. Consequently at the end of the week, the only amount that is really owed by anyone is $40 owed by the baker to the butcher. Because of the clearing system, only $40 in cash is necessary to accommodate $60 worth of exchanges.

Without the clearing of accounts, at least $50 would have to be on hand to cover the transactions. At the end of the week, the baker would have to take $50 to the butcher who would have to turn right around and give the baker $10 back. By clearing the accounts on paper, the amount of cash necessary to undertake the exchanges is minimized. The clearing system, consequently, is an institution that allows for economizing the use of money.

Because the clearing system allows for more transactions to be undertaken with less money, the development of the clearing system decreases the demand for money. Therefore, as the clearing system develops and the demand for money decreases, people increase their spending. As spending increases, the demand for goods increases. The prices of goods will then tend to rise, decreasing the PPM.

Another important but more intangible factor that affects the demand for money is the confidence people have in money. By confidence, we mean how certain are people that their money is what it says it is, and how well it will maintain its purchasing power. People's confidence in money is positively related to their demand to hold money. If members of society are worried that the money that they are holding is not very sound (will not be readily accepted in trade), they will not want to hold it. Instead, they will want to get rid of it as soon as possible while it still has purchasing power. Because people are less willing to hold the money, the demand for money falls and people increase their spending. This results in higher prices for goods and a decreased PPM.

Closely related to the confidence people have in their money is their expectations regarding future prices and hence PPM. If people expect that overall prices in the future will decline, they are at the same time expecting the PPM to rise. They will be more likely to put off purchases now and hold on to their money until the prices of goods decrease so that they can buy more for their dollar. Because they wish to hold more money, the demand for money increases and spending on producer and consumer goods would decrease. Because people would be spending less on these goods, the demand for these goods would fall along with their prices. The fall in overall prices results in an increase in the PPM.

On the other hand, if people expect that future overall prices will be higher than they are in the present—meaning that the PPM of money will be lower in the future—they will not want to hold onto their money while its purchasing power constantly ebbs away. They will decrease their demand to hold money and increase their spending. The demand for consumer and producer goods will increase. The more capable buyers will bid up the prices of those goods and the PPM will fall.

From this analysis, we can glean two implications. First, in general there is an inverse relationship between expected future overall prices and the demand for money. Second, to the extent their judgments are correct, their expectations about overall future prices tend to be self-fulfilling. If people expect that future prices are higher, they will begin acting in such a way to more quickly bring about the higher prices.

How much expectations about future overall prices will affect the demand for money depends on four factors. One factor is the number of people who hold the expectation. Obviously the more people who hold certain expectations and act on these, the larger will be the impact felt in the money market. If only one person expects that overall prices will be higher in the future, this will have no noticeable effect on the market demand for money. Also the strength of the expectation will affect the strength of the affect on money demand. If people are relatively certain about their forecasts, their expectations will have a stronger impact on the demand for money. But if their expectations are merely mental suggestions of what might happen, such expectations will not have as great of impact on actual money demand.

Additionally, the distance into the future when prices are expected to change will affect the impact on the demand for money. The farther into the future people expect prices to change, the less effect such expectations will have on the demand for money. Expected future overall price changes will be discounted by time preferences. The further into the future such price changes are expected to occur, the more such price differences will be discounted, so the less of an impact such changes will have on the present demand for money.

Finally, the level of people's time preferences will also affect the impact expectations of future prices have on the demand for money. The lower are people's time preferences, the greater will be the effect of expected future overall price changes and the present demand for money. The higher social time preferences are, the less effect because any future price change will be discounted at a higher rate.

THE STOCK OF MONEY

As we discovered in our discussion about the money market, it is not only the demand for money that determines its purchasing power. It is the demand for money in relation to the quantity in existence. Consequently, we now take up a more in-depth investigation into the stock of money.

Plainly speaking, at any given moment the money stock is equal to the total quantity of money units in an economy. Under the gold standard, when gold was used as the sole medium of exchange, the money stock equaled the amount of gold in society used as money.

Although many societies voluntarily chose gold and silver as money throughout history, people did not always find it convenient to carry gold on their persons all the time. Neither did they always feel comfortable having their entire gold holdings in their homes. This presented the opportunity for certain entrepreneurs to undertake the service of deposit banking. Deposit banks were essentially money warehouses where people who owned gold could deposit their gold in return for a warehouse receipt called a bank note guaranteeing a right to the holder to redeem the note at the bank for his gold at any time on demand. These banks would reap income by charging a storage fee, like any warehouse storing producer or consumer goods. In this case, the gold was not looked upon as an asset of the bank, but still owned by the depositor, who only stored it at the bank.

Three facts resulted in these bank notes themselves being used as money. For one thing, paper notes were easier to physically carry and exchange than gold. Also, the notes themselves were payable to the bearer. They were redeemable for gold not only by the original depositor, but by whoever turned in the bank note for redemption. Consequently, people could be confident that if they received a bank note in exchange they could redeem it for gold. Finally, gold is a homogenous good so that one ounce is just as good as any other. Therefore, it did not matter to people which ounce of gold they received upon redemption. When people went to redeem their notes, they did not require the exact same coins or bullion they deposited. As long as they received the amount that they were entitled they were satisfied. Consequently, the bank notes representing the gold began to circulate as if it were the gold itself. People accepted them as payment just as if they were gold because at any time a holder of a note could redeem it for gold.

Once bank notes representing gold began circulating as money, the composition of the money stock changed. No longer was the money stock merely the quantity of monetary gold in an economy, but it was gold being used as the medium of exchange plus the number of bank notes. This change of composition of the money stock, however, did not alter its size.

For example, imagine three people in a community, each holding a certain sum of money. Their society uses gold and defines the dollar as one twentieth of an ounce of gold. Ernie Bishop owns $100,000 in gold, Bert Bond owns $50,000 in gold, and Violet Bick has another $50,000. If these are the only three members of society, the money stock for this society equals $200,000, the total quantity of monetary gold in the economy.

Now suppose each of the three decides to deposit half of his gold in the Acme bank presided over by Henry F. Potter. Bishop deposits $50,000 in gold and Potter gives him $50,000 worth of bank notes redeemable for gold on demand. Similarly both Bond and Bick deposit $25,000 of their gold holdings in return for $25,000 in bank notes each.

As a result of these deposits, the stock of money in existence has merely changed forms, not size. Bishop's cash holding is still $100,000 only now it is made up of $50,000 in gold plus $50,000 in bank notes. Likewise, the cash holdings of both Bond and Bick equal $50,000 each made up of $25,000 in gold and another $25,000 in bank notes. The total money stock for the community equals $100,000 in gold plus $100,000 in bank notes or $200,000, the same quantity as before any bank deposits were made.

One might ask why the money stock has not increased by 50%. The gold deposited in Acme Bank does not, after all, cease to exist. To answer this question, let us examine Acme Bank's balance sheet below:

Acme Bank's Balance Sheet			
Assets		Equity & Liabilities	
Gold	$100,000	Bank Notes	$100,000
Total	$100,000	Total	$100,000

A balance sheet is a table showing the market value of a firm's assets and the claims held against those assets. In our example, Acme bank holds $100,000 in gold while they have issued $100,000 on bank notes as claims against that gold. Acme must hold that gold in order to redeem the bank notes whenever they are returned. Because the gold is being held in reserve to redeem the bank notes and the notes are what actually are being used in exchange, economically we do not count both the bank notes and the gold as part of the money stock. The bank notes replace gold as a medium of exchange while the gold is necessarily held on reserve to redeem the bank notes as people demand. Counting both the gold reserves and bank notes held by the public would result in an artificially large money stock.

The gold that Acme holds in order to redeem outstanding bank notes is called *reserves*. Therefore, when a bank holds enough gold to cover every outstanding bank note it has issued, the bank is said to be engaging in 100% reserve banking. A bank's reserve ratio is equal to its reserves divided by its outstanding money warehouse receipts. Mathematically, it can be calculated with the following formula: $RR = \frac{R}{MWR}$, where RR = the reserve ratio, R = reserves on hand at the bank, and MWR = the amount of outstanding money warehouse receipts. In this case Acme's reserves are $100,000 in gold and their liabilities are $100,000 in outstanding bank notes that it is required to redeem on demand. Reserves of $100,000 divided by $100,000 in outstanding money warehouse receipts equals one or 100%.

Another form of money warehouse receipt is the demand deposit, often referred to as a checking account. Instead of depositing gold in a bank in exchange for a bank note the depositor could receive an open account equivalent to the quantity of gold he deposited. He could draw upon this account anytime and anywhere checks are accepted in payment. Suppose our three depositors each deposit half of their remaining gold holdings in Acme bank and open checking accounts. Bishop deposits $25,000 in gold, opening a checking account for the same amount. Bond and Bick both deposit $12,500 and both open a checking account for the same amount.

The money stock still has not changed. Again, only its composition has. Bishop still has $100,000: $25,000 in gold, $50,000 in bank notes, and $25,000 in demand deposits. Bond and Bick both still own $50,000. Each now has $12,500 in gold, $25,000 in bank notes, and $12,500 in demand deposits. The total money stock is equal to the sum of gold not in reserves, outstanding bank notes, and demand deposits. In our example, the total money stock is still $200,000. This is the sum of $50,000 in gold + $100,000 in bank notes + $50,000 in demand deposits.

Just as the money stock has not changed, the percentage of money warehouse receipts covered by gold on reserve has not changed. As a result of the checking accounts opened as described, Acme Bank's balance sheet now appears as follows:

Acme Bank			
Assets		Equity & Liabilities	
Gold	$150,000	Bank Notes	$100,000
		Demand Deposits	$50,000
Total	$150,000	Total	$150,000

Acme now has $150,000 in gold on reserve and there is a total of $150,000 of warehouse receipts that Acme must redeem on demand. Therefore, Acme is still running a 100% reserve bank. Of the $150,000 in warehouse receipts, $100,000 is in bank notes and $50,000 is in demand deposits.

So far we can deduce two implications from our investigation into deposit banking. First, money warehouse receipts serve economically as the medium of exchange and should be considered as such. Second, depositing money in a 100% reserve bank does not alter the quantity of the money stock, but merely its composition.

We are left with the general conclusion that in a free market gold standard, the only time the money stock would increase would be if there was an increase in the quantity of monetary gold in existence. The money stock can increase only if there is more gold minted into coin. Such gold could come from increased industrial mining or from melting jewelry down and turning the gold into coins. When will this occur? Consider the case of industrial mining. It is helpful to remember that mining gold is an act of production. Consequently, there are costs associated with mining and minting just like there are costs that must be incurred to produce any other good.

Suppose that Fred C. Dodds is contemplating whether to engage in mining the treasure of the Sierra Madre. What will determine his choice? Dodds will compare the costs of mining with its benefits. How much will he have to invest and what will be his return? He will choose to mine only if the value of the gold mined is greater than the cost of mining. If there is an increase in the demand for money, and the PPM increases enough so that the value of a gold ounce is greater than the cost of mining a gold ounce, more entrepreneurs will undertake more mining and the production of money will increase.

On the other hand, if the PPM falls to such a level that that value of an ounce of gold is less than the costs of mining an ounce of gold, entrepreneurs will cease their mining operations and the production of gold will not increase. There are similar considerations related to the conversion of jewelry into money. Because gold is so durable (that was after all one of the reasons gold became money), gold rarely wears out. Consequently, there is a long term trend for the quantity of gold to increase.

It is also possible for the money stock to increase via fraud. Fraud, of course, is not permitted in a free market because it is a type of theft. Nevertheless, we can use economic analysis to examine the effects of such activity. One important type of monetary fraud is counterfeiting. Counterfeiting is passing off an inferior object as real money. In the past kings who controlled the monetary mints would occasionally see to it that coins that were supposed to contain 100% gold would instead only be, say 50% gold and the rest copper or bronze or some other cheaper metal. This of course, allows for more coins to be produced with the same quantity of gold. As these additional coins enter the economy, the money stock increases, as illustrated in Figure 12.7. Increasing the money stock will result in an excess supply of money at the prevailing PPM. People will be holding more money than they want to hold and will increase their spending, driving up the demand for consumer and producer goods. Consequently, the price of those goods will rise, decreasing the PPM.

Another more subtle form of monetary fraud occurs in the form of fractional reserve banking. Fractional reserve banking is made possible by the fact that people readily use and accept bank notes and demand deposits as the medium of exchange. As bankers began to recognize that very rarely would they be called upon to redeem their gold, they could increase their incomes by making loans in the form of bank notes and demand deposits not backed by gold reserves.

Fractional reserve banking was made possible in great part because over time customers' deposits became treated as assets owned by the bank rather than a bailment—money still owned by the depositors while merely being stored at the money warehouse. Such a view of things became legal as the result of a series of rulings in English banking law. In a series of cases in the early to mid-1800s, judges ruled that cash deposited in banks became assets of the banks and bank notes and demand deposits were bank liabilities. Banking law in the United States tended to follow English law although there was still some confusion between debt and bailment. After all, if deposits are to be treated as debt and investment, why are the money warehouse receipts considered payable on demand? Nevertheless, the law is the law and banks were allowed to treat their clients' deposits as their own assets.

This opened the door to fractional reserve banking as banks began issuing *fiduciary media*. To get a handle on what we mean by fiduciary media, it is useful here to consider the linguistic root of our term. The etymological root of fiduciary is the Latin word *Fide*, meaning faith. One of the slogans popular during the Reformation was *sola fide*, which means *faith alone*. You may have seen the U. S. Marine Corps logo along with the phrase *semper fidelis*, or *semper fi* for short, meaning *always faithful*. Fiduciary money has everything to do with faith. Banks are able to issue fiduciary bank notes because people believe that they can be redeemed for money on demand. They are acceptable only because people believe they are actually fully redeemable for money but in fact they are not.

Consider our example of Acme Bank. They have $150,000 in deposits as assets. Their liabilities total the same amount in the form of $100,000 in outstanding bank notes and $50,000 in demand deposits. Suppose that Potter at Acme Bank recognizes that, because rarely do depositors decide to redeem all of their bank notes and demand deposits at the same time, the bank can loan out even more warehouse receipts in order to reap larger profits. Suppose that Acme Bank loans out $75,000 to Gussie

Finknottle who is thinking of building a large newt aquarium. If the bank loans the money to Finknottle by opening a checking account for him to draw from, Acme Bank's balance sheet would appear as illustrated below:

Acme Bank			
Assets		Equity & Liabilities	
Gold	$150,000	Bank Notes	$100,000
Loans	$75,000	Demand Deposits	$125,000
Total	$225,000	Total	$225,000

Note that Acme's total assets have increased because they now include the $75,000 loan to Gussie Finknottle that is like an accounts receivable. At the same time Acme's liabilities have increased by the same amount, because their outstanding demand deposits are now $125,000, bringing the total outstanding bank notes and demand deposits to $225,000. For the first time, Acme Bank does not have enough gold on reserve to cover all of the outstanding notes and deposits that they have pledged to redeem on demand. For the first time, Acme's reserve ratio is below 100%. It still only has $150,000 in gold on reserve, but know has $225,000 in outstanding money warehouse receipts in the form of bank notes and demand deposits. Acme's reserve ratio is now $\frac{\$150,000}{\$225,000}$ or 67%.

What has happened to the money stock? Remember that at any given moment the money stock is equal to the quantity of gold used as a medium of exchange plus outstanding bank notes and demand deposits. As a result of Acme's lending fiduciary money, the money stock is now $50,000 in gold + $100,000 in bank notes + $125,000 in demand deposits. A sum of $275,000 total. Before the issuance of the fiduciary media the money stock equaled $200,000. The money stock increased by $75,000.

Issuing fiduciary money substitutes increases the money stock, which, as we have seen earlier, results in a lower PPM. Fiduciary media is called money because when it is issued it is accepted by people as a medium of exchange. There is nothing stamped on fiduciary bank notes that indicates that they are fractional reserve notes. There is no asterisk on checkbooks for fiduciary demand deposits directing the holder to a footnote stating *Acme Bank does not have enough reserves to fully redeem this account.* On the contrary, the bank promises that they will redeem such notes and deposits, which is precisely why people use them as money.

Therefore, an increase in fiduciary money media is an increase in the money stock.

Because people accept fiduciary media as money, an increase in its stock results in an excess supply. Spending on consumer and producer goods increases. The demand for these goods increases, resulting in increased prices for these goods. Consequently, as the prices of overall goods increase, each dollar can buy less and the PPM falls. By investigating the nature of fiduciary money substitutes in the context of fraud, we are not using the term loosely. Fraud is a violation of the right to private property and as such a form of violent intervention in the free market. We will examine the nature and consequences of monetary debasement and inflation in more detail in the following chapter.

THE PPM AND THE OPTIMAL SUPPLY OF MONEY

The above analysis makes two things clear. The first is that in a free society the money market, just like any other market, can regulate itself. If there is an increase in the demand to hold money, for instance, people will act in such a way that sets in motion a process by which prices will fall and the PPM will rise. Prices of goods will not fall forever. They will cease falling when the quantity of money that people desire to hold is equal to the quantity that they are actually holding. The PPM will adjust to its equilibrium level in response to any changes in money demand. Those who are concerned that the state must step in to ensure there is enough money to grease the wheels of exchange need not be. Human action will always restore equilibrium. If the market PPM increases above the cost of producing the money commodity, more money will be mined and minted. Just like we do not hear of people anxiously wringing their hands worrying about whether the supply of Coca-Cola will be at just the right level to meet demand, we do not need to worry about whether the market will supply us with enough money.

A corollary implication is that the ability of money to effectively serve as a medium of exchange does not hinge on there being a specific optimal quantity. Any quantity of money is sufficient for it to fulfill its function as a medium of exchange. Money is different from other goods in that it is neither a producer nor consumer good. People do not eat money, they do not wear it, nor do they drive it or listen to it on their iPods or MP3 players. This is because money is only used as a medium of exchange. People do not value money for itself, but because it can be used to buy things they can use more directly to satisfy their ends. We have seen that if there is an increase in the money stock, there will be an excess supply of money at the prevailing PPM leading to an increase in spending on consumer and producer goods. The prices of these goods will increase resulting in a decreasing PPM. While people will have more money in their cash balances, each dollar will be worth less. There will be no net increase in societal wealth.

An increase in food, clothing, shelter, and musical entertainment devices would be an increase in wealth because more people would have their ends satisfied. An increase in the stock of money, however, does not result in more consumer and producer goods being at the disposal of society. Consumer goods do not spontaneously generate whenever the quantity of money increases. Having more dollars does not mean there are more consumer goods in existence. It just means there is more money to buy those consumer goods. Likewise, increasing the money stock does not cause the available stock of land, labor, or capital goods to increase either. Without an increase in factors of production, there will be no increases in the production of consumer goods.

While an increased money stock does not result in an increase in the supply of consumer and producer goods, it does result in an increase in prices. Consequently, an increase in the stock of money will lead to people paying higher prices for the same quantity of goods. People have more money, but the purchasing power of each dollar is diminished. Society generally does not benefit. A larger stock of money does not allow society to consume and produce more.

Because an increase in the money stock does not result in an increase in the quantity of or ability to produce consumer goods, some might be tempted to conclude that an increase in the quantity of money has no effect on production. One might assume that money is neutral (like the country of Switzerland regarding foreign conflicts). The neutrality of money refers to the claim that any change in the money stock has no effect on economic reality, because such change in the quantity of money will have no real affect on either the production structure or the distribution of consumer goods.

In fact, changes in the money stock have a very real impact on the production and distribution of producer and consumer goods. If the money stock increases or decreases, changes regarding the marginal utility of money will affect the specific demands for goods at different prices by different people in varying proportions. Suppose that we are considering two friends, Abbot and Costello, who have different value scales, but both have a cash balance of $20. Suppose that the price of apples is $1 per pound and the price of oranges is $2 per pound. Abbot's value scale is such that if Abbot has $20 he will buy two pounds of apples and nine pounds of oranges. On the other hand, Costello's value scale is such that if Costello also has $20, he will buy six pounds of apples and seven pounds of oranges. The ratio of apples to oranges purchased would be eight pounds of apples to sixteen pounds of oranges, or one to two.

Suppose that now Abbot and Costello each receive another $20 so their cash balance doubles. It is not unlikely that Abbot and Costello will spend the additional money differently than they spent their first $20. At least there is no reason to think they will not. Suppose that Abbot does buy another two pounds of apples and nine more pounds of oranges. Costello, however, with his additional $20 buys four more pounds of apples and eight more pounds of oranges. This changes the ratio of apples to oranges purchased to fourteen pounds of apples to thirty-three pounds of oranges, or a ratio of less than one to two. In other words, demand for oranges increased not just absolutely, but also relative to apples. The increased stock of money resulted in a larger increase in demand for oranges than the increase in the demand for apples.

Because the demand for apples and oranges increased by different degrees due to the increase in the stock of money, the price of both goods will also increase by different degrees. Therefore, an increase in the money stock will not affect all prices the same. Changes in the money stock are not neutral but will have an impact on real economic conditions. In our example, citrus growers will receive a larger increase in income relative to apple producers.

In reality of course, when the money stock is increased it is not dropped uniformly out of a helicopter hovering over the nation. The new money always enters the economy in the hands of particular people. In the case of fractional reserve banking, the new money is first in the hands of those to whom the banks loan the fiduciary money. The demand for the goods they purchase increases first. Then the producers of those goods spend their new income on factors of production. Then the owners of those factors gain income that they can spend on what they want to purchase. As the new money ripples throughout the economy so do the increases in demand for goods and the prices for those goods. Prices do not increase all at the same time all at once but do so gradually. Some prices might not increase at all or may even decrease depending on the value scales of those who receive the new money early in the process. In any event, it is easy to see that an increase in the money stock does not change prices of all consumer and producer goods at the same time to the same degree. So money is never neutral. As the money stock increases, different prices for different goods increase at different rates.

These changes in relative prices result in very real changes in the distribution of wealth. Those who get the new money first will be able to make purchases before any large-scale increases in prices, while those who get some portion of the new money later or not at all will pay higher prices for goods once the full effects of the increased money stock have worked their way through the economy. Those who receive the new money first will have their wealth increase while those who receive the new money late or not at all will suffer a decrease in wealth.

MONEY IS AN ECONOMIC GOOD

In this chapter we have considered the implications of money being an economic good. We have discovered that, as with all economic goods, the value of money depends on the subjective marginal value each person ascribes to money. Consequently, the PPM, the price of money, is determined similarly to the prices of consumer and producer goods in that it depends upon the stock of money in existence and people's demand to hold money. We have examined a number of factors that affect how much money people demand to hold as well as those factors affecting the size of the money stock. Particularly, we have seen that in a free market gold standard, the only thing that can result in a legitimate increase in the money stock is mining more gold and minting it into coin. However, the money stock can also increase due to fraud in the form of either counterfeiting or fractional reserve banking.

Because the money market, like any other market, can regulate itself we do not need any central monetary planner, such as the Federal Reserve, to regulate the quantity of money in order to attempt to provide an optimal supply. In fact, when monetary authorities do so the economic effect is never neutral, and very real negative consequences follow. In the next chapter, we will further investigate the nature and consequences of monetary inflation.

SUGGESTED READING

Mises, *The Theory of Money and Credit*, 41–193. In these first nine chapters of the seminal book bringing money into the main body of economic theory, Mises explains the nature of money and the determination of its purchasing power.

Rothbard, What Has Government Done to Our Money?, 11–54. Rothbard's brilliant introduction to the nature of money in a free society.

———. *The Mystery of Banking*, 1–110. An excellent explanation of the principles of money and banking in a free society. Rothbard provides an accessible exposition of the nature of money, the determination of the purchasing power of money, and the nature of both free-market and fractional reserve banking.

13

Inflation and Recession

FROM THE BEGINNING OF this text until this page the fundamental principles of economics have been explained by considering the nature and implications of human action. Fundamental economic laws have been exposited and applied by examining the case of the unhampered market. With the present chapter, we consider the nature and consequences of government intervention in the economy for the first time. We will examine the case of monetary inflation, by investigating its nature and consequences. First, however, it is necessary to discuss a few principles that apply to the evaluation of state intervention in general.

ECONOMICS, ETHICS, AND ECONOMIC POLICY

Economics is a social science that helps us to understand the created order. It is the systematic relationship of the laws which God has established as they relate to human action and exchange. It is value free in the sense that it discovers and describes universally true laws that are actually implied by human action. These laws are true no matter whether we like them or not. They are true no matter a person's ethics. Economics, as such, describes what is, not what *ought to be*. It describes how people act and not how they *should* act.

Economists do make value judgments, however, as soon as they advocate a particular economic policy. Economic policy is action that the state takes in the economy. The moment that an economist recommends lowering taxes, increasing the money supply, decreasing business regulation, increasing the minimum wage, and so forth, he is making a statement about how things *should be*. The economist is saying that by following the recommended policy society will be better. Therefore the policy is a *good* policy. Issues of good and bad and better and worse are matters of ethics.

A complete analysis of economic policy, therefore, must take into account not only economics but also ethics. Such analysis requires answering two questions. First, is the policy economically sound? By economically sound we mean is the policy the proper means to achieve its goals? Does the policy accomplish what it is supposed to do? To evaluate economic policy, we will use the economic principles explained thus far to answer these questions.

Second, is the policy ethically sound? By ethically sound we mean does the policy seek to accomplish a morally lawful end? Does it seek to accomplish this end in a morally lawful way? Only a policy that can answer all of these questions in the affirmative can be embraced as a truly good economic policy. Any ethical evaluation of economic policy must take into account the nature of economic transactions. The market economy is a vast society brought together by the exchange of property. Therefore, any promotion of any public policy from a Christian perspective must be consistent with the Biblical ethic as it regards property.

Chapters 13 through 18 examine various forms of government intervention from both economic and ethical perspectives. In general it will be shown that government intervention in the economy is bad economic policy because it neither achieves its ends nor does so in an ethical way. Not only is interventionism unhelpful in providing a net social benefit, it is an unethical means of doing so.

INFLATION DEFINED

In the previous chapter, we examined the fundamental principles of monetary economics and the value of money. We examined the nature and determinants of the purchasing power of money, and we saw how changes in the stock of and demand for money resulted in changes in the PPM. We learned that in a free market gold standard, the stock of money would only increase as entrepreneurs willing to undertake production mined more gold and minted it into money.

Throughout history, however, rulers have not been content to allow the stock of money to be that stock that the market provided. Kings, prime ministers, and congresses have not allowed the PPM to be determined by market stock and demand alone. They have repeatedly sought to intervene in the production of money to accomplish their ends. Almost always, official monetary policy involves some degree of inflation.

When the word *inflation* is used in contemporary parlance, it most often carries with it the meaning of increases in overall prices. A general rise in prices for consumer and producer goods has usually been the direct result of an increase in the stock of money. Increases in the money stock can occur in the free market as the result of increased production of the commodity used as money; however, such changes tend to be gradual and generally do not result in large-scale economic change. Sudden and rapid increases in overall prices have been the result of increases in *fiduciary media*. We have seen in the previous chapter that higher overall prices are due to a change in the relationship between the demand for and stock of money. Generally, as economies expand and develop, the demand for money increases, which works to lower overall prices and increase the PPM. Higher overall prices, therefore, are the consequence of an increase in fiduciary media—money certificates and demand deposits not fully covered by the money commodity held on reserve by the issuing banks. Consequently, from an economic perspective *inflation* is best defined as the process of issuing bank notes and demand deposits beyond any increase in the money commodity.

DEBASEMENT

There are three primary ways in which inflation can be and has been undertaken. The first is called *debasement*. Debasement is the act of reducing the quantity of money commodity in the same monetary unit. There are three main ways this is undertaken. One way is called clipping coins. As coins are traded back and forth, certain coin issuers could clip tiny pieces of each coin that comes through its establishment. Of course, the missing fragment would have to be small enough not to be obviously noticeable or the perpetrator immediately would be found out as a thief. After clipping enough coins, the minter would have enough metal to make a whole additional coin. This practice results in an increase in the quantity of coins in the economy. It also results in many coins that say they contain a certain weight of gold actually containing a fraction less.

Another form of debasement—diluting the quantity of money commodity in the coin with other base metals—allowed for keeping the coins the same size. As mentioned briefly in the previous chapter, those issuing coins could mix a bit of bronze or copper in with the gold. As long as the populace could not tell the debased coins from the pure coins, minters could pass them off as the real thing. Again the quantity of coins would increase with the same amount of gold, thus the quantity of gold actually in the coin would be less than what the coin said.

A third form of debasement is to merely redefine the size of the coin. This practice was made much easier once the minting of coins became monopolized by the state. The names for the various monetary units originally merely indicated a certain weight of precious metal. When different governments monopolized the minting of coins in their regions they kept the names commonly used for their coins. Originally a government's stamp was simply a seal guaranteeing its weight and fineness.

Unfortunately, as governments took over the exclusive right to mint coins, the name for the monetary unit began to be associated more with the specific government minting the coin and less with the weight of gold or silver that was supposed to make up the coin. The government separated the name of the currency from the weight of gold of which the coin was made.

This separation opened the door for monetary debasement by redefining the monetary unit as a lesser weight of gold. Those familiar with the popular 1950s satirical novel, *The Mouse That Roared* (subsequently made into a movie starring Peter Sellers), may remember that all of its action was set in motion because of a financial crisis faced by Gloriana the Twelfth, Grand Duchess of Grand Fenwick. The solution that she settled upon was to declare war on the United States and then surrendering in defeat so that the United States will lavish it with a large *Marshall Plan*. Before striking upon that plan, however, one of the suggestions mentioned by the Grand Duchess was a devaluation of the currency.

Such a devaluation could be undertaken by debasement. Suppose for example, that Gloriana the Twelfth inherits from her mother, Gloriana the Eleventh, a monetary system in which the *florin* was defined as 1/20th of an ounce of gold. If Gloriana wishes to solve the Duchy's financial crisis, she could raise more taxes from the people. Taxes, however, have never been overly popular.

She reasons instead that she could make out better by devaluing the *florin*. Gloriana calls in all of the existing *florins* in Grand Fenwick, all of them dirty from wear and still bearing the image of her great grandmother, Grand Duchess Gloriana the Ninth. A citizen of Grand Fenwick who had, say fifty *florins* could turn them in and be presented with fifty bright shiny new *florins*. However, the new *florins* are no longer 1/20th of an ounce of gold, but only 1/40th of an ounce of gold. The new coins given back to the citizens are only half as weighty as the ones they turned in, with the Grand Duchess keeping the excess and minting for herself *florins* that are also lighter than the previous coins. Grand Duchess Gloriana then spends this newly minted money on estate construction expenses and the new coins enter the economy. While the quantity of gold in Grand Fenwick remains the same, the money stock doubles because the monetary unit is now associated with the name *florin* instead of the weight $1/20^{th}$ of an ounce of gold.

This increased money stock will produce an excess supply of money at the original PPM, resulting in increased spending on consumer and producer goods. Overall prices will increase and the purchasing power of *florins* will decline. While the PPM will decline, Gloriana makes out like a bandit because she increases her wealth by the quantity of new coins and spends them before overall prices change. This increased wealth of the state resulting from inflation is called *seigniorage*.

Such devaluations are not restricted to successful Hollywood movies. History is replete with such examples of monetary debasement. In the year 1200 for example, the French *livre tournois* contained ninety-eight grams of silver. In 1600 it contained only eleven grams. In 1350 the Spanish *dinar* was made of sixty grains of gold. By the early 1400s it was debased to contain only fourteen grains of gold. Soon the *dinar* became too small a coin to circulate. The Spanish rulers debased it right out of existence.

PAPER MONEY

Before we proceed further in our analysis of inflation, it will be helpful to sketch the process by which we moved from a gold standard to what we have now, which is paper fiat currency. How did we get here from there? Remember that because money is a commodity voluntarily chosen as the most marketable good, no one—including the state—can force anyone to use a certain commodity as money. This principle applies to government fiat currency as well. We do not use paper money because one day the state destroyed all gold against our will and then forced us to use paper. The process by which we moved from a gold standard to a fiat paper standard and from a 100% reserve banking system to a legal fractional reserve system was gradual. We now take up the study of this process.

Our economy no longer operates with a gold standard. We have fiat paper currency. The Federal Reserve bank notes that we now use in the United States are not redeemable for anything. Not any old piece of paper, however, can be used as money. Only those printed by the U. S. Treasury (or are made by counterfeiters to look identical to those printed by the U.S. Treasury) are accepted as the medium of exchange. The paper bills are made money by government fiat, which is an authoritative decree. *Fiat* is a Latin word meaning *let it be done*. It is as if the monetary authority is like Yul Brenner's portrayal of Pharaoh in the Cecil B. DeMille classic, *The Ten Commandments*. From his office in the halls of power he declares paper is money: "So let it be written. So let it be done." For us it means that our money is simply whatever our government calls a dollar. While we will see that our system is certainly different than a free market gold standard system, the same economic principles we discovered in the previous chapter will allow us to analyze our current monetary system and government intervention that results in monetary inflation.

Once governments adopted a fiat paper money system, another obvious method of inflation is merely printing more fiduciary money (see the previous chapter). The monetary authority simply can print more bank notes that say they are redeemable for gold on demand.

The development of government paper money greatly facilitated the ability of governments to inflate. We have already seen how paper money was first issued by private banks as bank notes redeemable in gold or silver on demand. The public would not have initially accepted the paper without its promised redeemability.

Once paper becomes acceptable as the medium of exchange, inflation becomes much easier for rulers than if they had to resort to the rather cumbersome method of debasement. If the government wants to finance increased spending, it can simply print more money. If it wants to buy more tanks and planes for its military, it can pay the tank and plane manufacturers with this newly printed money. The makers of these military vehicles buy more factors of production and consumer goods for themselves. Their workers receive higher wages and spend more. Increased demand for consumer and producer goods ripples throughout the economy, increasing prices and decreasing the PPM.

Paper money was first used in the country in which printing was invented: China. In the years 811 and 812, Chinese rulers printed paper notes called *flying money* so that it would be easier to collect taxes in these notes rather than by collecting and transporting gold.

The first paper money issued in North America occurred in 1690 in the colony of Massachusetts. The state paid its soldiers in paper certificates, promising that they could redeem them for gold or silver in the future and that no more notes would be issued. Both promises were broken, resulting in an increase in the Massachusetts money stock.

Four times during the history of the United States the government invoked irredeemable fiat paper money. The first was during the American War for Independence. The Continental Congress issued fiat paper notes called *Continentals* to pay for war time expenses. It issued so many so fast that they quickly became worthless. Fiat paper currency was again printed by the U.S. Government to help finance the War of 1812. Gold redeemability was resumed two years after the war's conclusion. The third period of fiat currency was during the Civil War. Both the Union and the Confederate governments printed fiat paper money, again to finance war spending. From 1860–1863 the money stock increased over 92%. This resulted in an increase of approximately 110% in overall prices.

We are currently living in the fourth era of fiat paper currency. This era began for U.S. citizens in 1933 when the administration of Franklin D. Roosevelt suspended gold payments to U.S. citizens and made it illegal for U. S. Citizens to use gold as money so that domestically the dollar was now fiat paper currency. This allowed for much inflation because the government did not have to worry that U.S. citizens would take their notes to a bank and demand gold. The money stock increased and the dollar was officially devalued from 1/20th of an ounce of gold per dollar to 1/35th of an ounce.

The United States remained on a gold standard internationally, redeeming any Federal Reserve note presented by a foreigner for gold. This link to gold, however, was cut by President Nixon in 1971 when he suspended gold payments to foreigners as well. This suspension seems to have turned into an expulsion.

We should not be surprised that the money stock has increased approximately 825% since 1959. We should also not be surprised that, because of large increases in overall prices since then, overall prices in 1998 were approximately five and a half times what they were in 1959.

FRACTIONAL RESERVE BANKING REVISITED

Inflation via issuance of paper currency not covered by metal money was facilitated primarily by fractional reserve banking. As seen in the previous chapter, fractional reserve banking occurs when a bank issues more money warehouse receipts—either in the form of bank notes or demand deposits—than they have money commodity on reserve to redeem them.

We also saw that this type of banking is inherently inflationary. In a free market, issuing fiduciary money would be considered a form of fraud and hence illegal. Once the monetary deposits by clients of the bank were ruled assets owned by the bank, fractional reserve banking became legal.

One might think, consequently, that banks in such an environment are allowed to inflate *hog wild*, throwing sobriety to the wind. While bankers may be tempted to do so, and some indeed succumb, there are still market constraints that limit the amount of inflation that fractional reserve banks can undertake.

Do not forget that in a free market no one has to engage in an exchange if they do not want. No one is forced to deposit money in any bank if they do not want to. Consequently, before people will accept any bank's bank notes or accept their demand deposits, they must have built up a critical mass of trust at the outset. If not, no one would accept their bank notes or checks and such a bank could not successfully issue any fiduciary money. In a free market, therefore, it is not in the bank's interest to be recklessly profligate in their inflation. They can afford only to be *carefully* profligate.

If bankers are carelessly profligate, there is the threat of the bank run looming over the fractional reserve bank's head. Depositors may lose confidence in a bank meaning that they feel that the bank will not be able to redeem the bank notes and demand deposits of everyone who wants to redeem them. If they do lose confidence in the bank's ability to redeem their money warehouse receipts, depositors will rush to the bank and try to redeem them as soon as possible because the situation will become one of first come first served. They do not want to be the ones holding merely paper and a checkbook. If enough depositors do this, the bank will not be able to meet its obligations, and so will go bankrupt.

The threat of a bank run is a very effective deterrent to reckless fractional reserve banking for two reasons. In the first place, once a bank run starts it cannot easily be stopped. Bankers know that once depositors make a run on the bank they cannot be easily swayed to keep their money in the bank. A bank run is not something the bank can allow to start and then turn off like an electric light or water faucet. Therefore, the bank has an incentive to avoid playing with financial fire. In fact, bankers have every reason to not even get close to the edge of a bank run.

Another reason that the bank run is an effective deterrent is that the run itself calls attention to the unsoundness of fractional reserve banking. Fractional reserve banks are always in danger of violating the *financial golden rule*. The financial golden rule states that the date on which a bank's obligations fall due must not precede the date on which its corresponding claims can be realized. Suppose that in order to raise money that it can then lend out, Acme Bank issues a $100,000 certificate of deposit to a saver for one month on June 1. Suppose Acme then lends that $100,000 to a homeowner who agrees to pay it back plus interest over the course of the next thirty years. If these two transactions are the only business that Acme has, then Acme is setting itself up for rough financial sailing. On July 1, Acme is obligated to give back $100,000 plus interest to the holder of the CD, but that $100,000 has not been fully paid back by the homeowner. It will not be fully paid back for thirty years. Consequently, Acme does not have enough money to make good on its obligation to the holder of the CD and is insolvent.

Obviously bankers try very hard to follow the golden rule by not overextending themselves. They do not want their payments to come due before they receive an adequate inflow of cash from savers to be able to pay their obligations. By assuming the risk associated in making time transactions for present and future money, they provide a productive service that makes it profitable for successful banks to be financial intermediaries. By taking upon these time transactions in the face of an uncertain future the banks reap either profits or losses.

Fractional reserve banks are always in a precarious position because they are always in danger of violating the golden rule. A fractional reserve bank's assets are always longer term than their liability. Bank notes and demand deposits are payable on demand. They do not call them demand deposits for nothing. It takes time, however, to sell their assets in order to raise cash to make good on their liabilities. Therefore, a bank run highlights the fact that fractional reserve banks never have enough money in their reserves to redeem all of the bank notes and demand deposits they have promised to redeem on demand.

While the bank run is quite effective, it is not the most important daily restraint on fractional reserve banking in an unregulated market. A bank run usually occurs only after a long period of inflation, when depositors are getting worried over the potential insolvency of their banks. The day-to-day constraint on inflation via fractional reserve banking is the size of the clientele of each bank. Suppose that Acme Bank loans out $100,000 to Ernie Bishop in the form of demand deposits that the bank created out of thin air. Because this money is not covered by gold on reserve, Acme is engaging in fractional reserve banking and the money stock increases by the quantity of the loan.

Now, Bishop is not borrowing $100,000 just for the fun he can have looking in a checkbook with six figures in it. He borrows the money to buy a house from his friend Bert Bond. If Bond is also a client of Acme bank, then everything is okay for now at Acme. Bishop writes a check for one hundred grand to Bond, who deposits the check in his account at Acme. Acme is not called to redeem the demand deposit, because neither Bishop nor Bond wants to take money out of Acme.

On the other hand, suppose that Bond has an account at 1st National. When Bond deposits the $100,000 check that Bishop wrote him in his account at 1st National Bank, 1st National sends the check to Acme and Acme is required to subtract $100,000 out of the checking account of Bishop and send $100,000 in money to 1st National Bank. This requirement to redeem the check presented by 1st National Bank can put Acme in danger of insolvency if it does not have enough reserves to cover the checks it is called to redeem.

The broader a bank's clientele the less the bank has to worry about redemption of its bank notes and demand deposits. This is true because the broader a bank's clientele, the more likely that new loans will be issued to borrowers who will buy goods from other people who also are clients of the same bank. Therefore, they are freer to engage in loaning out fiduciary money. The broader a bank's clientele, the less restraint there will be on inflation.

The narrower a bank's clientele, however, the greater the restraint will be on inflationary fractional reserve banking. If a bank's customer set is relatively narrow, it is more likely that other banks will end up redeeming that bank's notes and checks for cash, because it is more likely that their borrowers will write checks to people who are not customers of that bank.

Finally, even if there is only one bank in business, and all depositors who want to bank must bank there, another constraint on fractional reserve inflation still exists. There is gold redemption by foreign central banks with which to contend.

Suppose, for example, that Acme Bank is the only bank in which depositors can deposit their money in the United States. This also means that Acme bank notes are the only bank notes in circulation and the only demand deposits are checking accounts written on accounts at Acme. If Acme inflates the money stock by loaning out more fiduciary money, there will be an excess stock of money at the prevailing PPM. To disgorge themselves of their excess supply, people will increase spending on consumer and producer goods, some of which will be produced overseas. Upon receiving bank notes or checks from Acme bank, foreign merchants will deposit them in their local banks. Their local banks will deposit them in their country's central bank, so Acme bank notes and checks from the United States will pile up in foreign central banks. Because the Acme bank notes and checks do them no good, the foreign central banks will seek to redeem them by sending them to Acme in return for the promised gold. Acme then finds that its reserves decrease, so it has to slow or cease inflating before it completely runs out of gold to cover the notes and demand deposits being redeemed.

This scenario is precisely what happened to the United States in the late 1960s and early 1970s until President Richard Nixon took the nation off the gold standard for international transactions in 1971. In a free market in which private property rights are protected, the inflationary banks are held liable for their money warehouse receipts. If a bank cannot make good on the redemption of all notes and checks presented, it would be forced to declare bankruptcy and liquidate all of its assets necessary to raise the gold sufficient to redeem all of its bank notes and checks returned to them for redemption.

When Nixon took us off of our last link to the gold standard, he did so by what was termed *suspension of specie payments*. Specie is the commodity metal that serves as money and for which all bank notes and demand deposits can be redeemed. Whenever leaders have taken their citizens off their gold or silver standard, it has usually been characterized as a suspension of specie payments. In other words, these actions were all supposed to be temporary. When redemption of bank notes and demand deposits for gold is suspended, the rulers are in effect mandating that banks do not have to fulfill their contracts. They are allowing banks to violate their depositors' property rights. They are allowing fraudulent activity to turn into outright theft.

BANK CARTELIZATION

One of the primary ways inflation occurs is via bank cartelization. Remember that a cartel is a group of separate sellers who decide to act together in order to jointly maximize their profits. When makers of consumer or producer goods form a cartel, they usually attempt to establish a uniform cartel price and assign production quotas for each member.

The prices that banks compare to each other in order to make their money are prices of time. Banks generate profits through the positive difference between the interest rate they charge their borrowers and the interest rate they pay for their time depositors. It is profitable for a bank to loan out fiduciary money because they reap the interest from the borrower but pay no interest because they are loaning out money that they have created *ex nihilo*, out of nothing, as if they were God.

Recall that the primary day-to-day constraint on inflation of fiduciary money is the limited clientele of each bank. If a bank inflates too much compared to other banks, it will face a relatively high rate of bank note and demand deposit redemption, and its reserves will dwindle so that it will run the risk of insolvency.

If, however, different commercial banks are able to cartelize, they can increase their profits via loans of fiduciary money with a much lower risk of running out of reserves. This is because if banks act together by inflating at the same rate, the probability that they will experience a net drain on their reserves is greatly reduced.

Suppose that Groucho is a customer of Acme, Harpo is a customer at 1st National, and Chico is a customer at Farmers, and all three of them receive loans in fiduciary money. Suppose also that each bank has entered into a cartel agreement and they each agree to inflate by $1,000. Groucho, Harpo, and Chico each receive loans of $1,000 from their respective banks. Suppose further that Groucho writes a check for $1,000 to Harpo for a case of cigars. Harpo writes a check to Chico for $1,000 for a new Harp, and Chico writes a check to Groucho for $1,000 for a used piano. Each of the three deposit the checks they received in their bank and their bank in turn sends them to the bank of origin for redemption. Because each of the banks inflated by the same amount, each bank is both redeeming a check for $1,000 and is required to redeem a check for $1,000. In other words all of the transactions clear, so that none of the banks are in any riskier position than before the inflation. If banks can cartelize and inflate at the same rate, an important constraint against inflation is removed.

In chapter 10 we discovered that, in a free market, cartels tend to be inherently unstable due to competition from both within and without the cartel. More efficient cartel members have an incentive to increase profits by expanding output and selling at a price just below the cartel price. Entrepreneurs who are not a part of the cartel have an incentive to enter any industry in which profits are being earned. Because they are not bound by any cartel agreement they tend to charge lower prices in order to gain sales.

Banks are no different. If there are banks that are more efficient or more financially sound than other cartel members, they might get tired of carrying the weight of the other members. Additionally, new bankers might enter the industry and not inflate at the cartel rate. If a new bank enters the market and wishes to draw customers by promoting its soundness, it may not want to issue as much fiduciary money as the cartel issues. This puts the cartel in a dangerous position which would tend toward the breakup of the cartel. In a free market it is not necessarily easy for bank cartels to perpetuate themselves.

A number of forms of state intervention into the monetary system have arisen, however, that help keep bank cartel systems in place. Such interventions greatly remove the free market constraints on inflation. The preeminent intervention is that of central banking. The central bank takes advantage of the key government privilege in the monetary system: the monopoly right to issue bank notes. Often the central bank is owned by the government, but even if not it is at least chartered by the government. In either case, the central bank has the monopoly right to issue bank notes. In the United States the central bank is the Federal Reserve System, often referred to simply as *the Fed*.

The central bank is often referred to as the *bankers' bank* because commercial banks must get bank notes from the central bank. In a central bank setting, regular commercial banks, such as Acme, 1st National, and Farmers in our example, cannot legally issue their own bank notes. Like everyone else, they must use central bank notes. Commercial banks can obtain central bank notes by selling some sort of asset such as a bond or gold to the central bank. They can also withdraw the bank notes from their reserve account they have at the central bank. All commercial banks have their reserves deposited at the central bank. In the United States, all commercial banks must deposit their reserves at the Federal Reserve.

Central banking removes many of the limits on inflation found in a free market. It is true that if the public demands to hold only gold instead of central bank notes, inflation by the banking system is very limited. However, legal tender laws greatly encourage the use of government paper money. Legal tender laws are regulations mandating that certain things must be accepted in exchange for goods and services. A glance at our current fiat paper currency—the Federal Reserve Note—reveals a statement reflecting legal tender laws. Every piece of paper money issued by the Federal Reserve bears the statement "this note is legal tender for all debts, public and private." Because the law mandates that Federal Reserve notes are good for paying debts to the government and by law for all private exchanges, people have an incentive to use such paper.

Additionally, central banking tends to reduce the threat of a bank run because the central bank is seen as a *lender of last resort*. This means that if the going gets financially tough for a bank that issues too many fiduciary demand deposits, it can call on the central bank to bail it out.

The reduction in the threat of bank runs is further helped along by the intervention of deposit insurance. This insurance protects individual deposits in the event of a bank run. Before such insurance was mandated, there were still bank runs even under central banking. After deposit insurance was put into place, the number of runs greatly diminished to the point that they were thought to be virtually extinct. However, they do still occur. In July 2008 depositors at IndyMac Bank in California became worried that their bank was insolvent. They rushed to withdraw their money, precipitating a bank run. The bank subsequently failed and was taken over by government banking regulators.

Deposit insurance provides that every depositor's account is insured up to a ceiling amount. If there happens to occur a bank run, and the bank cannot pay all its depositors, they would still get their money back out of the deposit insurance fund. In the United States, the Federal Deposit Insurance Corporation (FDIC) insures every depositor's account up to $250,000.

Deposit insurance is government mandated but funded by participating banks. But, as we saw in the aftermath of the savings and loans scandal of the 1980s, if the insurance fund is insufficient to cover all rightful claims, the government will step in and provide enough subsidies to cover all depositors. FDIC deposit insurance creates a serious moral hazard problem. If banks know that they will not be held fully liable for their outstanding demand deposits if they become insolvent, they have an incentive to inflate more liberally than otherwise. Because of the central bank standing ready as the lender of last resort and deposit insurance giving depositors a sense of security, the threat of the bank run is greatly diminished.

Furthermore, in a central banking environment supported by legal tender laws, inflation via expanding credit in the form of fiduciary money becomes much more uniform, relieving the commercial banks from the pressures of competition. Because each commercial bank is prohibited from issuing its own bank notes and all demand deposits are redeemable in the same central bank notes, citizens are no longer merely customers of different banks. Economically, all become customers of the central bank. There no longer exists competition between Acme bank notes and 1st National bank notes. All commercial banks use the same notes.

Therefore, the central bank can preside over a much more uniform inflation of the money stock. The central bank issues increased reserves to the commercial banks. The commercial banks are then free to issue even more fiduciary money inflating the money stock.

The only legal restraint on inflation in a central bank regime is the required reserve ratio. The required reserve ratio is the legal minimum reserve ratio. The Federal Reserve likes to portray itself as *the* institution, ever vigilant, in charge of keeping inflation in check. Commercial banks find it in their interest to loan out as much fiduciary money as possible in order to maximize profits. The quantity of fiduciary money that they can issue is constrained by the legal reserve requirement.

Suppose that the duchy of Grand Fenwick now went the way of the United States and is operating on the fiat paper currency system and that everyone must use the paper *florins* issued by the central bank of Grand Fenwick. If there have been one million *florins* deposited in commercial banks throughout the duchy we can represent the financial condition of the commercial banks by the following t-account:

Commercial Banks			
Assets		Equity & Liabilities	
Florins	$1,000,000	Demand Deposits	$1,000,000
Total	$1,000,000	Total	$1,000,000

We see that as a result of the demand deposits, commercial banks have assets of one million *florins* deposited in their reserve account at the Central Bank of Grand Fenwick. Commercial banks also have liabilities that total one million *florins* in demand deposits.

Commercial banks have an economic incentive to increase the quantity of fiduciary *florins* via loans. Such credit expansion will be limited only by the legal reserve requirement. If the required reserve ratio is 10%, this means that commercial banks must maintain reserves enough to cover 10% of their outstanding demand deposits. If commercial banks hold reserves totaling one millions *florins*, they will be able to legally increase their outstanding demand deposits to ten million *florins* and be within the legal limit. In which case, the aggregate t-account would appear as follows:

Commercial Banks			
Assets		Equity & Liabilities	
Florins	$1,000,000	Demand Deposits	$10,000,000
Loans	$9,000,000		
Total	$10,000,000	Total	$10,000,000

Commercial banks can loan out nine million *florins* in demand deposits and still be within the law. Commercial banks will have assets of one million *florins* in reserve and nine million *florins* in loans. They will have liabilities totaling ten million *florins* in outstanding demand deposits. The reserve ratio, then, is one million to ten million, or 10 %.

Notice that if commercial banks maintain only their required reserves, then their outstanding demand deposits will equal their quantity of reserves multiplied by the inverse of the reserve requirement, *RR*.

$$\text{DEMAND DEPOSITS} = \text{RESERVES} \times \frac{1}{RR}$$

As we see above, reserves are equal to one million *florins* and the reserve requirement is equal to 10%. Therefore, for every *florin* that Grand Fenwick commercial banks have in their reserves, they can have ten *florins* in outstanding demand deposits. Every increase in the bank reserves by one *florin* will result in an increase in the money stock of ten *florins*, or the inverse of the required reserve ratio. Consequently, this inverse reserve requirement is called the *money multiplier*. If the required reserve ratio is 10% then the money multiplier is ten. If the required reserve ratio is 25% or ¼ then the money multiplier would be four.

Consequently, the two key determinants of the money stock in a central banking regime are the legal reserve requirement and the quantity of total reserves. Likewise, if the central bank wishes to manipulate the money stock, it must do something that affects either of these two determinants.

In the United States, Congress allows the Federal Reserve to set the legal reserve requirement within a broadly defined range. If the Fed increases the reserve requirement, the money multiplier decreases because commercial banks have to keep a larger percentage of their outstanding demand deposits covered by reserves. Therefore, if the reserve requirement increases, commercial banks can have a smaller quantity of outstanding demand deposits with the same amount of reserves. The money stock would decrease as banks issue fewer loans in fiduciary money.

On the other hand, if the Fed lowers the reserve requirement, the money multiplier increases because commercial banks are able to issue more demand deposits with the same quantity of reserves. Banks would make more loans in fiduciary money, increasing the money stock.

Changing the reserve requirement is not something central banks like to do every day; it is a rather blunt instrument for daily altering the money stock. It is sort of like trying to hit a fly every day with a bazooka. The primary tool that the central bank uses to manipulate the money stock on a day-to-day basis is open market operations.

Open market operations refer to the central bank's buying and selling government bonds in the bond market. The central bank can engage in buying bonds, called open market purchases or selling bonds, called open market sales.

One way for commercial banks to build up their reserve account at the central bank is to sell bonds to the central bank in exchange for bank notes held in their reserve account. If commercial banks are selling bonds, this means that the central bank is buying them. The central bank is making an open market purchase. The central bank is able to buy as many bonds as they want, because they take a lesson from Don Corleone and make bond holders an offer they cannot refuse. They simply offer to pay a price above the current market price.

Because the central bank buys these bonds with new money, reserves at commercial banks increase. If the bonds they buy are sold to them directly by the commercial banks, commercial bank reserves obviously rise. On the other hand, if the bonds bought by the central bank are sold to them by private bond holders outside the banking system, they will deposit the new money they receive from the central bank in their accounts at their commercial banks thereby increasing the reserves held at their commercial banks. Regardless, open market purchases by the central bank result in increased commercial bank reserves.

If commercial bank reserves increase, these banks are free to make more loans by issuing more fiduciary demand deposits. The precise quantity of the increase will be determined by the money multiplier. If the central bank of Grand Fenwick buys one million *florins'* worth of bonds, commercial bank reserves will increase by the same amount. If the reserve requirement is 10% banks can issue a total of ten million *florins* in demand deposits and still be within the legal limit. Open market purchases, therefore, result in an increase in the money stock.

On the other hand, open market sales on the part of the central bank result in a decrease in the money stock. The central bank can also sell as many bonds as they want by agreeing to accept a price below the market price. As the central bank sells bonds, they are paid for by commercial banks spending their reserves or by private citizens outside the bank that write checks against their demand deposits, which ultimately will have to be redeemed by commercial banks. In either case, commercial bank reserves will fall resulting in a drop in the quantity of outstanding demand deposits they can maintain and still meet the reserve requirement. As banks reduce the quantity of outstanding demand deposits, the money stock decreases. Therefore, open market sales by the central bank result in a decrease in the money stock.

What can we conclude? We have seen that in a free market economic system in which the right to private property is protected, there would not be any inflation. The only instances in which the money stock would increase would be if there is an actual increase in the production of the money commodity. Even in an environment that allows for fractional reserve banking, competition between banks, their limited clientele, and the threat of the bank run provide constraints on monetary inflation.

The real source of inflation is government intervention in the monetary system. The development of central banking assisted by legal tender laws and deposit insurance has given the state many opportunities to inflate the money stock for whatever reason they see fit. The central bank can, as already demonstrated, increase the money stock by lowering the legal reserve requirement. However, the primary tool that they use to inflate is open market purchases. This injects more money into commercial bank reserves allowing those banks to increase the quantity of demand deposits they issue in loans by an amount equal to the quantity of new reserves times the money multiplier.

THE CONSEQUENCES OF INFLATION

Sometimes rulers seek to inflate the money stock to increase their own material wealth. Other times they inflate the money stock because they think it will grow the economy, making them more popular and hence more electable. Our job, however, is to analyze the nature and consequences of inflation from an economic point of view.

We can begin to get a handle on the results of inflation by first considering what it does not do. By now you should understand that inflation is *not* an increase in prices. In our contemporary setting, inflation is an increase in the money stock resulting from issuing fiduciary money.

It is true that whenever the government central bank increases the money stock, prices for consumer and producer goods will be higher than otherwise. We have seen in figure 12.7 in the previous chapter why this is so. If the state increases the money stock, there will be an excess supply of money at the prevailing PPM. People will be holding more money than they want to at that PPM. In order to alleviate the excess supply, people will increase their spending on consumer and producer goods. They will also be less willing to sell their inventories, reducing the supply of readily saleable goods. Both the increase in demand and decrease in supply will cause overall prices of goods to increase. Higher overall prices decrease the PPM because as prices rise each unit of money can buy less.

This brings us to what inflation does not do. Inflation does not result in an increase in general wealth for society. Increasing the money stock does not spontaneously generate more land, labor, or capital goods. Because inflation does not yield an increase in the stock of factors of production, neither does it result in an increase in consumer goods. As a result of inflation, people do get more money; but this larger quantity of money is being exchanged for the same quantity of goods. Therefore, the prices for consumer and producer goods increase so that society is no better off generally as a result of inflation.

The fact that there is no general social benefit does not mean, however, that no one benefits. Someone must think it is in their interest or they would not inflate. Who does benefit from inflation? To answer this question, we must look more deeply into the process by which inflation occurs.

We have already discovered that money is not neutral. While overall prices increase as a result of inflation, they do not all increase at the same time or by the same amount. When the government decides to inflate, everyone's checking account does not magically increase in size all at once. Money does not drop uniformly throughout the country from a helicopter.

Prices increase following an increase in the money stock as people seek to remove their excess supply. Precisely who are those with an excess supply of money following inflation? We can answer this by thinking about the way the central bank inflates the money stock. Remember that the central bank inflates mainly by engaging in open market purchases of bonds, injecting more reserves into commercial banks. The commercial banks then increase the money stock by making particular loans in the form of fiduciary demand deposits to particular people. Recipients of the new money are those who first increase their spending.

Suppose the Central Bank of Grand Fenwick buys a 10,000 *florin* Grand Fenwick Treasury Bond from the Mountjoy Bank, a commercial bank respected throughout the duchy. The commercial bank is able to expand its lending by 100,000 *florins* and still be within the legal reserve requirement of 10%. Whomever Mountjoy Bank lends the 100,000 *florins* to first will be the first person to see his cash balance increase. Suppose that Mountjoy loans the new money to Tully who spends it on a new forest adjacent to his house. The first good that experiences an increase in demand is forests. The owner of the forest receives the new money second and uses part of the money to pay his rent, buy a new car, buy more food, and clothing, and increase his CD collection. Then the demand for all of these goods increases.

But notice that the demand for all goods does not increase all at once. Therefore, prices for all goods do not increase all at once. Prices for consumer and producer goods increase as their respective demands increase, and they increase only as the new money works its way through the economy.

Those who benefit from inflation are the people who receive the new money first. They are able to spend the new money before the prices of any goods have actually increased. Therefore, their real wealth increases. However, as the new money works its way through the economy, more and more prices increase so that at some point the recipients of the new money are able to spend it on goods for which the prices have already increased. The welfare of these people does not change at all. Finally, there are some, living on fixed incomes, like retired pensioners, who do not see any of the new money but still have to pay higher prices for the goods they buy. These people find themselves in a worse economic condition. They do not have any more money but the purchasing power of the money they do have decreases.

While inflation does not provide a general social benefit, it does benefit some and harm others. Inflation benefits those who receive the money first at the expense of those who receive it late or not at all. Inflation does not result in an increase in societal wealth, but it does redistribute it to the early recipients of money and away from the late recipients of money.

Inflation also redistributes wealth between those people who participate in the time market by borrowing and lending money. In general, inflation hurts the lender and benefits the borrower. This is because as inflation occurs and the PPM falls borrowers are paying back money that is worth less than the money that was lent to them.

Suppose that Mountjoy Bank loans that 100,000 *florins* to Tully and Tully buys that forest. He agrees to pay the 100,000 *florins* plus 6,000 *florins* in interest back to Mountjoy in one year's time. Mountjoy plans to take the money at that time and buy a crazy fast sports car that also costs 100,000 *florins*. Now, if the Central Bank of Grand Fenwick increases the money stock such that overall prices increase and the price of crazy fast sports cars double, Mountjoy comes out way behind on the loan. Tully pays Mountjoy 106,000 *florins* at the appointed time, but Mountjoy cannot buy his crazy fast sports car like he anticipated; its price increased from 100,000 *florins* to 200,000 over the course of a year. Tully gets his forest but Mountjoy does not get his crazy fast sports car. The money that Mountjoy lent Tully was worth twice as much as the money Tully paid back to Mountjoy at the end of one year. We see that inflation does not merely redistribute wealth to those who get the new money first from those who get it late or not at all. Inflation also redistributes wealth from lenders to borrowers.

Another potential result of inflation has devastating consequences. If monetary inflation is carried out for a long enough time, it can result in the complete meltdown of the monetary system. Long-term, persistent inflation can usher in what is called the *crack-up boom*.

Inflation can have such disastrous results because of the role expectations play in determining the demand for money. When people expect that prices in the future will be higher than they are now, this is the same as expecting that the future PPM will be less than it is currently. Such expectations result in a decreased demand to hold money. People tend not to desire to hold money in their cash balances as its purchasing power is being eaten away like a wheat field before a plague of locusts.

If people see prices increasing and the PPM decreasing because of inflation, they might not change their demand for money at first. However, as it becomes apparent that the inflation will continue for some time, people's psychology will tend to change. At some point they will think that prices will never come back down and will keep going up. Once people develop inflationary expectations, their demand to hold money decreases. As seen in figure 13.1, a decrease in money demand will result in an even lower PPM.

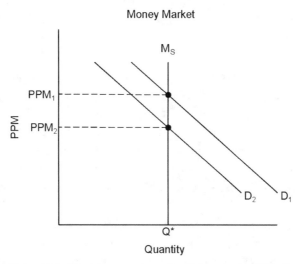

Figure 13.1. If people expect a continual decrease in PPM, the demand to hold money will fall, further decreasing PPM.

The PPM will be even lower because the prices of consumer and producer goods will increase faster as people spend money faster in order to avoid paying higher prices in the future. The decreased demand for money results in increased spending on goods. The increased spending results in increased prices for goods and a decreased PPM.

As people begin to face a severe shortage of goods because the demand for goods begins to increase faster than the money stock is increasing, they begin to complain that prices are rising too fast for their paycheck and call for more inflation of the money stock. If the central bank complies with these wishes, the stage is set for the final crack-up: *hyper-inflation.*

If the money stock increases, prices rise and the PPM falls even further. People develop almost panic-level expectations. They begin to think, "I must get rid of my money right away." People do not demand to hold money at all for any length of time and spend it as soon as they receive it. Prices rise many times in the same day and the money becomes worthless. The monetary system completely implodes as the value of money withers away to nothing.

This scenario is not merely a hypothetical, theoretical outcome of mad monetary policy. Hyper-inflation destroyed the German currency in the early 1920s and was experienced as recently as the 2000s in Zimbabwe. In June 2008 a loaf of bread, in Zimbabwe dollars, cost $2 billion. A pint of milk cost $3 billion. Car batteries were being sold for $2.4 trillion. In less than a month the price of a newspaper in Zimbabwe increased from $200,000 to $25 billion.

That hyper-inflation is a relevant possibility is demonstrated by the fact that the popular Nobel laureate economist and *New York Times* columnist, Paul Krugman, actually promoted massive inflation to help lift Japan out of its 1990s depression. He advised the Bank of Japan to publicly commit to increasing the money stock until the Japanese developed inflationary expectations. He reasoned that they would then stop hoarding their money and begin to spend it, thereby pulling themselves out of their depression. He failed to recognize that once inflationary expectations set in, Japan would be on the tipping point of hyper-inflation. In light of sound monetary economics, such recommendations are dangerously reckless. Hyper-inflation is not something that can be easily turned on and off like an electric light. Once hyper-inflation starts, the only thing that can be done to try to stop it is to cease expanding the money stock. If the demand for money has already fallen enough, however, such a move could be too late. That Krugman would suggest such a policy as a way to improve an economy is shockingly irresponsible.

RECESSION AND DEPRESSION

Even if the government monetary authority does not completely destroy the entire monetary system via hyper-inflation, its inflationary activity will be destructive to some extent. This is because not only does central bank inflation result in increased overall prices, and a redistribution of wealth, but it also ushers in the business cycle. It is monetary inflation that is at the root of the other great macroeconomic problem: the *recession*.

Recessions are part of what has been termed the business cycle. As the term implies, a business cycle refers to an up and down pattern in the economy during which the economy seems to progress and then regress. People generally like the *up* part, so when using the term business cycle, they are usually referring to the negative, down side—the decrease in prosperity that is associated with economic recession. However, to gain a good understanding of business cycles it is important to remember that every cycle has both a *boom* and a *bust*. We have already seen that in a free market, money prices allow for entrepreneurs to calculate expected profit and loss and thereby allow for the coordinating of all market activity. The big question is what causes recessions.

Before we answer this question, it is important to define the term *recession* exactly. A recession is a general downturn in the economy. It is not merely a decline in one particular industry. As low-carbohydrate diets such as the Atkins and South Beach diets became popular in the early 2000s, there was a decrease in profitability in the market for high-carbohydrate foods such as potatoes, doughnuts, and cookies. This did not illustrate an economy-wide recession, however. This decrease in profitability from selling high-carb foods was offset by an increase in demand for steak, chicken, and other low-carb foods. When entrepreneurs reap losses in one market or industry because of market changes, this does not mean that a recession is underway. Some particular entrepreneurs forecasted the future market conditions worse than their competitors.

A recession, therefore, is not the result merely of a few entrepreneurial errors. A general downturn in the economy must be the result of a large number of entrepreneurs making mistakes in a large number of industries throughout the economy. A recession has been characterized as a cluster of entrepreneurial errors. Our investigation into the seeds of the business cycle must begin by seeking out the source of this cluster of large-scale entrepreneurial error. Entrepreneurs who make a habit of making mistakes do not remain entrepreneurs for very long. Such entrepreneurs will be quickly ushered into the ranks of laborers. If a large mass of entrepreneurs make errors throughout the economy, there must be a reason. Such a cluster of entrepreneurial errors occurring at the same time must be the result of something that impacts all of their markets. The only good that is traded in all markets is money. Therefore, it is reasonable to conclude that the source of recession must be monetary in nature.

Additionally, the element of time plays an important part of the business cycle. Time is important because entrepreneurial error manifests itself as production that has been undertaken but that should not have been from the viewpoint of the realized *future* market conditions. Recall that the price of time is the interest rate. The interest rate is what coordinates and brings into equilibrium the supply of and demand for present money in exchange for future money.

In our discussion of the production structure and the social economy we discovered that net saving and investment results in lower interest rates and a progressing economy. Net saving is a product of lower time preferences. If time preferences decrease people are willing to save more. On the other side of the economic coin, if the rate of saving and investment increases, the rate of consumption expenditures falls. This decrease in consumption spending decreases the demand for and subsequently the price of consumer goods.

At the same time, the production structure lengthens. Capitalist-Entrepreneurs increase investment in production of producer goods higher up the structure of production. The demand increases for the original factors of production used in higher stages. The prices for these factors follow suit. The prices of the highest order capital goods rise the most while those at the lower stages rise less.

As this adjustment process plays out, the price differentials between final products and factors of production diminish throughout the entire production structure. The pure interest rate decreases, and, as we have seen in our previous discussion of interest, the interest rate in the loanable funds market follows the pure rate and also decreases.

The lowering of time preferences causes more saving and investing and results in a larger stock of capital, increased productivity, and an increase in the standard of living throughout society. Sometimes, however, government officials wish to see the positive effects of investment without requiring their citizens to do the necessary saving that allows for such investment. Their impatience leads them to attempt to expand the economy by increasing the money stock through credit expansion.

CREDIT EXPANSION AND THE BUSINESS CYCLE

Credit expansion on the part of commercial banks lowers the interest rate that banks charge for loanable funds. This rate is called the money interest rate. The market loanable funds rate is determined by the supply of and demand for loans *before* there are any new issues of fiduciary money. The structure of the loanable funds market is ultimately determined by time preference in the broader time market. At the market money interest rate the quantity of loanable funds demanded is equal to that supplied by banks. Therefore, the only way to expand credit by inducing more people—especially entrepreneurs—to borrow more money is to lower the money interest rate on loans. Therefore as banks expand credit, businesses are able to acquire new present money at lower interest rates. The banks find this credit expansion profitable because it does not cost them anything to create more money and loan it out, with negligible loan processing costs.

Because businesses have access to more money, entrepreneurial ambitions expand, ushering in the boom of the boom-bust business cycle. New businesses are started with the expectation that the capital necessary to complete production projects can be obtained by borrowing. In any given economic situation, the opportunities for production that can actually be undertaken and carried out are limited by the supply of capital goods, the produced means of production. The prices of these capital goods, compared to the prices of the final products, determine the interest rate. Monetary inflation via credit expansion makes new investment projects *appear* profitable without actually making them profitable. Entrepreneurs are able to borrow new money at a rate of interest below the natural rate established in the time market. At the same time, there are no more capital goods than there were before the inflation.

Suppose that Tully no longer wants to be the head forester of Grand Fenwick but instead begins a new vocation as entrepreneur-at-large. In order to choose an investment project, he looks at his investment opportunities. Suppose that he is considering investing in the manufacturing of small machine parts that are used in the production of motors for automated grape pickers. These marvelous machines pick grapes that are then stomped and used to make the Duchy's chief exportable commodity, Pinot Grand Fenwick. Tully considers the prices of the factors of production he needs to obtain in order to make the machine parts and compares that to what he anticipates will be the price of the finished parts. He then calculates his expected rate of return on his investment and finds that it is 6%.

Suppose that the going market interest rate is 5%. Because the expected rate of return on producing machine parts is greater than the return he could make generally in the time market, Tully invests his money in the purchase of factors used to produce machine parts. If he has forecasted the future market correctly he reaps a profit. If he reaps a profit and others see this as well, Tully and other entrepreneurs will increase investment in the machine part industry, which sets in motion a process by which the rate of return will fall to the pure rate of 5%.

Remember from our discussion of the interest rate that the pure rate of interest is determined by time preference. The same time preference of individual people that determines the interest rate is the time preference that also determines how much they save and hence how much they will invest, and hence how large the stock of capital goods gets. As such, it is time preference that determines how much production is able to be undertaken.

Now suppose that the Central Bank of Grand Fenwick, with the blessing of the Grand Duchess, wishes to grow the economy and engages in open market purchases. Reserves increase at commercial banks that take advantage of this increase and issue more fiduciary demand deposits by expanding credit. In order to make it worthwhile for businesses to borrow, the banks reduce the money interest rate. Now the money rate no longer tracks the pure interest rate, but instead dips below. Suppose that the rate charged in the loanable funds market falls to 4%, while the pure interest rate does not decrease. Credit has expanded but time preferences have not changed.

Tully can borrow money at a rate of 4% and use it to buy factors of production to make machine parts that he can sell at a 5% rate of return on his money. Things look good. As long as this situation holds up, Tully is guaranteed a 1% profit night and day. Additionally, previously unprofitable investments look profitable. Suppose that the rate of return on producing steel from which the machine parts are made was 4.5%. If time preferences of individuals are such that the going interest rate is 5%, entrepreneurs will want to invest less in that industry. Those who did so would reap a half per cent loss. However, if the state increases the money stock by facilitating credit expansion so that the money rate of interest decreases to 4%, investments in steel production that previously looked like losers now appear quite profitable. Therefore, when inflation takes place via credit expansion, entrepreneurial ambitions expand and production activity increases.

More entrepreneurs will therefore have more money in their hands with which to spend on factors of production. More money, however, does not mean more factors of production. Businesses that get the new money must use this money to bid away factors from other uses. Because the stock of these factors at the time of the inflation is fixed, the prices of these factors will be bid up by the most eager entrepreneurs—the ones with the new money. Because such business expansion tends to take place in the higher orders of the production structure, the first prices that tend to rise are those of raw materials, partly finished goods, other higher order goods, and wage rates. The structure of production lengthens. The prices of the highest order goods increase the most. Resources are shifted away from the production of lower order goods to the production of higher order goods. This increase in money incomes and economic activity is rightly characterized as a boom.

If this was the end of the story that economic analysis tells us, then inflation certainly would appear as a beneficent policy. Unfortunately, for prosperity the buck literally does *not* stop here. The main economic problem with inflation as far as sustained prosperity is concerned is that while the money stock has increased and credit has expanded, people's time preferences do not really change.

When entrepreneurs receive the new money in loans they use it to buy the services of factors of production. The owners of these factors receive increased monetary income. Because an increased money stock does nothing to lower time preferences, the recipients of this new money allocate their incomes at the same old ratio of consumption to savings.

Given the existing money stock, suppose that total income earned in the Grand Fenwick economy is one hundred ten billion *florins*. Of this income, citizens hold ten billion *florins* in their cash balances. Suppose that the time preferences of the citizenry are such that Fenwickians spend sixty billion *florins* of their income on consumption and they save and invest forty billion. Their consumption/savings ratio is 60/40 or 3/2.

Now suppose that the money stock is increased. Twenty billion *florins* of new money are created in the form of loans to businesses that spend it on additional investment. Because of credit expansion, investment spending increases to sixty billion *florins* while consumption remains at the same level, also sixty billion. We see, consequently, that as a result of the increase in lending fiduciary money, the consumption/savings and investment ratio initially becomes 50/50 or ½. However, once this new money is paid by the recipient entrepreneurs to the owners of the factors of production they are buying or renting, the owners of the factors spend the new money at the previous ratio. Income is again allocated so that 60% of the money not held in cash balances is spent on consumer goods and 40% is saved and spent on investment. While the money spent on both consumer and producer goods has increased, the quantity of factors of production and hence consumer goods have not. Prices of both sets of goods have increased. The important thing, however, is that the *ratio* of the price differential that is the interest rate moves back to its original level. Spending shifts back toward lower order goods, increasing the price differentials at every stage, which results in an increase in the money rate of interest so that once again the loanable funds rate follows the pure interest rate.

Due to inflation via credit expansion, production no longer reflects voluntary time preferences. Businesses are led to invest in higher stages of production as if more real savings are available when in fact, they are not. The money interest rate gives every impression that there are more real savings available, but there is only more money available. Why is more money not more real savings? Think of it this way. In order for investment in the production of any good to be profitable, the demand for the final product has to be great enough so that the good's market price will be more than enough to cover the sum of the prices of the factors of production necessary to produce that good plus the capitalist's foregone interest. If increases in the production of higher order goods are facilitated by actual savings, the money available for investment is matched by the availability of factors of production for these projects. Because people would be spending less on consumer goods, non-specific factors such as computers, labor, and common tools would be freed up for use at the higher stages of production. Because there would be an increase in the supply of factors of production for use in the higher stages, these higher order projects would be economically feasible.

On the other hand, when there is more money for investment in higher order goods due merely to inflation via credit expansion, people are not really engaging in additional savings. Entrepreneurs have more money to invest in lengthening the structure of production, but people want to consume the same quantity of consumer goods. Therefore, there are not enough factors of production freed up from consumer goods production. So the supply of factors available for production at the highest stages of production is not large enough for these production projects ultimately to be profitable.

Because of the artificially low loanable funds rate, businesses have tended to over-invest in higher stages of production and under-invest in lower stages of production. This analysis allows us to correct a common misunderstanding. Too often those who rightly understand that inflation can wreak havoc in an economy have the mistaken view that the havoc is due to over-investment creating a boom that must eventually turn into a recession. In fact, what we have observed is that the total number of producer goods does not increase as a result of inflation. Consequently, the quantity of real investment that occurs does not increase. What inflation via credit expansion generates is *malinvestment,* not over-investment. It is not that the economy is investing too much, but rather entrepreneurs are directing investment to the wrong stages of production for them to be profitable over time.

The malinvestment ushered in during the inflationary boom has been likened to a carpenter beginning to build a house, but being unable to finish because he committed too many materials to a too large foundation. Before he completes the upper floors, he runs out of materials. He had enough materials to complete a house, but only if he would have directed fewer bricks, mortar, lumber, and nails to the foundation and lower floor and more to the upper floors and roof.

The production structure is being stretched in two directions. Entrepreneurs who receive the new money in loans are trying to pull non-specific factors of production to work in the highest orders. On the other hand, those factor owners who receive the new money from the entrepreneurs allocate their income at the old consumption/savings rate, so resources are also directed back toward the production of consumer goods. As the Johnny Mercer song says, "Something's gotta give, Something's gotta give, / Something's gotta give."

What gives are projects that appeared profitable only as a result of the monetary inflation, but in fact are not. Once the money interest rate moves back up toward the pure interest rate, the nature of the malinvestments begins to show. Entrepreneurs who have already begun projects that are now appearing to be unprofitable increase their demand for even more loanable funds in the hope of saving their enterprises. But desires do not necessarily produce new money. If the central bank does not continue to inflate, there will be no new money for the commercial banks to lend. The loanable funds rate will spike up as the demand for loanable funds increases, while the supply of loanable funds remains the same.

At the same time, in an effort to increase their cash flow, those firms that have finished products in their inventory will try to boost sales by slashing prices. They will begin to offer bargain basement sales in order to maintain a positive cash flow and stay afloat financially. This is one reason why prices of goods tend to fall during a recession.

New investments at higher stages of production will have to be either liquidated or abandoned. Many new buildings will remain uncompleted. Other factories already completed will be shut down. Some will still continue to operate because after writing off their losses, they will still pay *some* income, which is better than none. Labor that has been working in these liquidated ventures will be laid-off. Others will be down-sized as companies seek to cut whatever costs they can. The result of all this is the bust—an economic recession.

The truly destructive part of the inflationary boom and resultant bust is that not all capital goods are easily convertible to some other line of production. Labor is generally a nonspecific factor. People can apply their labor in many different lines of production. Likewise land is also often relatively nonspecific. Many capital goods are also relatively nonspecific. A Hewlett-Packard computer for example, can be used in a myriad of businesses from colleges to retail stores to the office of the CEO of an iron ore mining operation.

Other capital goods, however, are not easily transferable to other lines of production. Suppose that during the inflationary boom some capital was invested in the production of expensive machine tool presses that are only suited to be used by entrepreneurs like Tully to make parts for the motors of grape picking machines. If Tully's operation is one of those forced to liquidate, he would lay off his labor force and sell off his equipment and perhaps use some of the proceeds in his next venture. However, because the tool presses are suitable for doing only one task—making specific parts for grape picking machines—and since there is not enough demand for those parts, the capital invested in those tool presses is completely and utterly wasted. These sunk costs cannot be retrieved. The inflationary boom, therefore, does not generate merely malinvestment and a certain amount of unemployment, but also consumes capital, resulting in a regressing economy and a lowering of the standard of living until the capital stock can be built up again by sufficient savings. The process of saving and investment and accumulating capital can begin again as the losses associated with the business cycle are washed out of the system during the bust and the entrepreneurs are again able to allocate resources to their most valued use, guided by free market prices.

This analysis provides the basic explanation of the nature and cause of the business cycle. The business cycle is a boom/bust phenomenon sparked by monetary inflation via credit expansion to businesses. The boom occurs as malinvestment bids up wages and boosting monetary incomes. When it becomes evident that some of the projects begun in the boom are unprofitable, they are liquidated and workers are laid off. Entrepreneurs begin the process of redirecting resources into other areas of the economy. During the bust, factors of production are reallocated back to their most urgent uses as determined by consumers. The recession is the necessary correction allowing the market to return to a surer foundation. In general, the longer the inflation, the worse the corrective bust.

SECONDARY DEVELOPMENTS

The bust following the boom is not always a tidy event. Often there are secondary developments that occur during the recession. One such development is a decrease in overall prices. Lower overall prices often accompany a severe recession for two reasons.

Once it becomes apparent that a large number of businesses are failing, banks begin to reduce the credit they issue as their business borrowers face financial difficulties and bankruptcy. As banks shrink their credit and make fewer loans, they do not readily distinguish between those projects that would be malinvestments and those that would be relatively sound.

Indeed, there may be no easy way of identifying either before the fact. Lending in general will fall. As banks reduce loans in the form of demand deposits, the money supply will decrease.

This contraction of credit actually hastens the adjustment process back to a more sound economy by increasing the interest rate. The rise in interest rates encourages saving, which is necessary for the accumulation of new capital.

Also working toward a decrease in overall prices is the fact that in a recession, businesses tend to increase their demand for money. Borrowers that get themselves into a bind need cash to pay off their debts and stay financially afloat. Potential investors become more cautious due to the rush of bankruptcies as a result of the recession.

As expected, such an increase in the demand for money will result in a decrease in overall prices. Prices could eventually decline even further if the demand for money increases because people begin to expect that the prices for consumer and producer goods will fall even lower.

So far we have been describing the boom/bust cycle that occurs from a one-shot inflationary issuing of fiduciary money loaned out to businesses. Politicians and government officials are not usually eager, however, to reap the recessionary consequences of their inflationary booms. They will attempt to keep the boom going, or at the very least provide for a so-called soft landing. The only way for the government to attempt to keep an inflationary boom going is to engage in even more inflation. If it becomes apparent that the economy is tipping into recession due to prior inflation, monetary authorities will try another injection of inflation. This sets in motion another process of malinvestment that eventually turns into a bust once the new money works itself through the economy. At some point the state must cease inflating the money stock. If it does not, it runs the risk of hyper-inflation and the crack-up boom. We can conclude that there are only two possible consequences of inflating the money stock via credit expansion—the boom/bust cycle or hyper-inflation, neither of which provides a happy ending to an inflationary day.

SOLUTIONS

Given what we have learned about the nature and causes of the business cycle, what can be done to solve the problem of recurring booms and busts? What is the solution for economic recessions and depressions? A time-worn cliché actually proves relevant in this case. An ounce of prevention may indeed be worth a pound of cure.

To avoid the pain associated with economic downturns, the best thing is to not initiate the business cycle to begin with. The business cycle is not a phenomenon inherent in the free market. On the contrary, in a free market entrepreneurs have the ability and incentive to use market prices to coordinate all economic activity so that markets tend to equilibrate. Those who serve consumers better than anyone else are rewarded with profits and those who waste capital are punished with losses.

The business cycle is a product of monetary inflation via credit expansion. The recessionary bust is the necessary consequence of an inflationary boom set in motion by central bank inflation. To prevent recessions, we should prevent unsustainable booms that precede recessions. In other words, we should stop inflating. It is as simple as that. If central banks stop creating bank reserves out of nothing and cease encouraging and enabling commercial banks to issue fiduciary money in the form of loans to entrepreneurs, then these same entrepreneurs would not be encouraged to unwisely invest in projects that have only the appearance of profitability. There would not be the tug-of-war with the production structure that is the inevitable result of the malinvestment caused by credit expansion. There would be no unsustainable boom followed by the inevitable bust. Any economic expansion that would take place would be sustainable, because it would be set in motion by real savings.

But what if we find that our leaders do not live according to sound economic principles? What if they have inflated and the economy has lived through the boom and is now experiencing the recessionary bust? What should be done then? If our goal is to return to an economy with a sound economic foundation, economic analysis tells us that we should simply allow the market to adjust.

The recession should be allowed to run its course. This sounds like a bitter pill to swallow, but sometimes taking nasty-tasting medicine is what the doctor orders. Remember what causes recessions. Entrepreneurs are led into starting wasteful, unprofitable projects by the siren song of artificially low money interest rates. Once such investments are made, they cannot be unmade. The best thing that can be done, once the true nature of the unprofitable projects is made apparent, is to allow those unwise entrepreneurs to reap their losses. Only this will give them an incentive to cut such losses so that they no longer waste any more capital in these ventures.

Some have argued that since the boom occurred with an increase in the money stock, we can alleviate economic recessions by inflating the money stock a little more. However, as should be clear, inflation of the money stock does not encourage markets to coordinate. Increasing the money stock to solve a recession would merely serve to usher in another boom/bust cycle resulting in another recession. This has been the history of any economy that has existed under the authority of a central bank charged with the task of providing full employment with stable prices. As the central bank inflates more rapidly or more slowly as the case allegedly demands, the economy moves in fits and starts through the business cycle. Inflating the money stock will neither stop nor ameliorate the business cycle; it causes the business cycle. At best inflation merely puts off the inevitable bust.

Can anything else be done—besides allowing the market to freely adjust—so that resources are once again directed toward their most highly valued uses? Economic theory suggests an affirmative answer. The primary reason that inflationary booms are not painlessly corrected by a seamless adjustment process is that capital is wasted during the boom phase of the cycle. Remember that some of the capital invested is spent on capital goods that are not convertible to other uses. Therefore, it is consumed and cannot be recovered.

An economic recovery then, requires rebuilding the capital stock. Savings is necessary in order for capital to be accumulated. Therefore, people need to save more so that entrepreneurs can get back to the original level of investment and can rehire the labor thrown out of work during the bust phase.

In addition to staying out of the market adjustment process, if the government really wants to help an economy in recession, it should *decrease* government spending and taxation. We will examine the consequences of government taxing and spending later, but suffice for now to know that government spending is akin to consumption because governments do not produce; they merely direct resources. Government spending directs resources away from those uses that are most highly valued by consumers. If this were not the case, the government would not have to undertake the spending; consumers would allocate money there anyway. Taxation works to reduce a nation's capital stock because it leaves people with less income to save and discourages productive activity because profits are smaller than they would be otherwise, reducing the incentive to invest. Reducing both government spending and taxation would do much to assist in the accumulation of capital necessary for a quick recovery.

Finally, the state should also allow labor markets to freely adjust so that human resources can also be directed toward their most highly valued uses. This brings us to the popular topic of the relationship between recession and unemployment.

ECONOMIC RECESSIONS AND UNEMPLOYMENT

Because unemployment almost always accompanies economic recessions, people usually think of unemployment as a macroeconomic problem. Workers are let go from their positions in the wasteful projects during the bust until they can be worked back into the structure of production at lower stages. Because the recession is a macroeconomic problem, some people think the unemployment that goes with it is also macroeconomic. The notion of unemployment as a macroeconomic problem that must be addressed by the government is also advanced by the calculation of the national unemployment rate by the U. S. Bureau of Labor Statistics. After all, if there is something called the *national* unemployment rate, unemployment must be a problem for the entire nation and the national economy.

However, when we recall that the wage rate is merely the price of labor and that this price is determined by suppliers of and demanders for labor, we see that the same principles that we have discovered in earlier chapters apply to the labor market just like they do to the market for every other good. Unemployment is in fact, a problem concerning the labor market, making unemployment a *microeconomic* issue and certainly not warranting any special economic policy on the part of the state.

Certainly there will be fewer people employed after the economic boom turns into the bust. As unprofitable firms are liquidated they lay off their employees. Because the stock of capital is partially consumed during the boom, the marginal product of the labor that could use that capital decreases lowering the MRP and DMRP of labor. Consequently, the demand for labor services will also decline.

In a free market, however, the labor market will tend to equilibrate like any other market. As illustrated in figure 13.2, a decrease in the demand for labor results in an excess supply of labor at the prevailing wage rate.

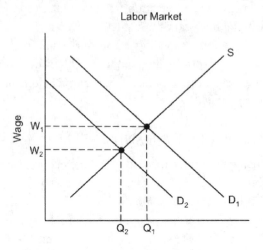

Figure 13.2. A decrease in the demand for labor will not cause unemployment in a free market. The market wage will fall until everyone who wants to work at that wage can work.

As in other markets, excess supply will be removed by human action. More eager workers will bid down the wage for which they are willing to work, and they will do this until a market wage in reached at which everyone who wants to work can work.

In a free market, we see that there is no need to worry about persistent involuntary unemployment. To the extent that people remain unemployed, they do so voluntarily in the sense that they are not willing to work at the market wage. Those who are willing to work will be employed. Those who refuse will not be. However, this is a voluntary decision on the part of those not willing to accept the market wage. At the new market wage rate as labor markets adjust to the new equilibrium, labor becomes cheaper encouraging entrepreneurs to hire.

Involuntary unemployment can persist, however, if something hinders wages from adjusting downward following a decrease in demand. If the government fixes wages at the original level, then the excess supply of labor turns into a perpetual surplus which will exist as long as the wage rate is fixed above the market rate. A surplus of labor is what we call unemployment. The quantity of labor workers want to supply is greater than the quantity entrepreneurs demand at the fixed rate. This will tend to prolong the adjustment process and prolong unemployment as long as wages are kept artificially high. Unemployment will persist as long as there are obstacles to wage flexibility.

A TALE OF TWO RECESSIONS

These principles are vividly portrayed in the economic history of the 1920s and early 1930s. Almost everyone has heard of the Great Depression and is somewhat familiar with the trials and tribulations large numbers of people had to live through at that time. The Great Depression was a recession that began in 1929 and turned into the worst economic period in modern history.

Not as many people are aware of the fact, however, that our economy also experienced a sharp recession as part of the 1920–22 business cycle. As the result of an inflationary boom in the late 1910s and early 1920, the inevitable bust took hold; from July 1920 through July 1921, industrial production fell sharply by 27.5%. As expected, this was accompanied by a noticeable decrease in the demand for labor and a 13% fall in wages. Nevertheless, the economy experienced full recovery by 1923 because markets were allowed to adjust. Wage rates were allowed to fall so the cost of production decreased allowing for more real investment in production as resources were directed back toward their most valued uses.

As you most likely know, the story turned out very different during the Great Depression. Many people believe that it was the stock market crash in October of 1929 that brought on the Great Depression. In fact, the economic downturn began in the middle of the year, before the Great Crash. From June of 1929 through July of 1930 industrial production fell by 21.3%. Although this was a sharp decrease, it was not as great as the drop in production in 1920–21. Be that as it may, this recession turned into the Great Depression, resulting in almost one third of the nation's labor force being unemployed for a time during the 1930s.

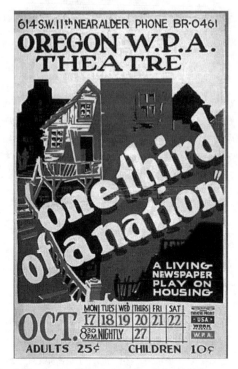

A Depression Era poster advertising a play produced by a government theater project designed
to provide jobs to out-of-work actors.

What explains the sharp difference between the 1920-22 recession
and the Great Depression? Why did the economy have a full recovery in
1923 but did not recover from the recession in 1929–30 until after World
War II? The answer is found in obstacles put in the way of the market
throughout the 1930s that hindered the necessary adjustment process.

Although demand for labor fell sharply during late 1929 and early
1930, wages were kept high. Certain businessmen, economists, and politicians thought that keeping wages high would get the economy going
again more quickly. They believed that if higher wages resulted in more
purchasing power in the hands of the masses, the aggregate demand for
goods would increase so that there would be a recovery led by increased
spending. President Hoover jawboned businessmen and convinced several of them like Henry Ford to keep wages high. With wages kept high,
the excess supply of labor continued.

Other disastrous measures were also enacted. The United States
passed the Smoot-Hawley tariff, the largest tariff increase since before

the Civil War. This greatly reduced the number of exports purchased by foreigners. The Hoover Administration also lobbied for and received a huge increase in taxation drastically reducing the ability and incentive to save and invest.

Unofficial wage floors were made official after the election of Franklin Roosevelt and the beginning of his New Deal. The National Recovery Administration (NRA) Law authorized the blue eagle NRA sign to be posted in a business only if it agreed to pay a minimum wage. Other New Deal legislation established legal preferences for labor unions making it easier for labor unions to lobby and obtain higher than market wages. Social Security and unemployment insurance were added to wage laws and the minimum wage was finally mandated for all employers in the United States. Minimum wage laws and union privileges kept wages from adjusting downward as they would have. Social Security and unemployment insurance reduced the incentive to accept work by subsidizing unemployment. The end result was a large increase in the official unemployment rate.

While trying to help right the economy, state intervention only made the situation worse. Instead of allowing the market to adjust in order to regain its solid footing, two presidential administrations, unintentionally acted in such a way as to turn the recession of 1929–1930 into the Great Depression.

What conclusions can we draw from our economic analysis of this period of history? The early 1920s recession and recovery instructs us that unemployment during the bust of the business cycle will not persist if wages are free to adjust to equilibrium. If wages are not allowed to adjust, and the state increases intervention in the economy, it only exacerbates the problem it is trying to solve.

THE ETHICS OF INFLATION

A complete evaluation of economic policy from a Christian perspective must take into account both economics *and* ethics. The ethics of inflation begins with the relationship between monetary inflation and property rights. When the state monopolizes the money production process and engages in monetary inflation it violates the Christian ethic of private property and hence, inflation cannot be accepted as moral economic policy.

It is perhaps easiest to begin understanding the ethical problems of inflation by considering the debasement of a metallic currency. When the state debases metallic money either by clipping coins or by diluting coins with base metals, it represents a good (the debased coin) as something it is not (a pure metallic coin of defined weight). Recall that metallic money developed in the free market and the different monetary units were different amounts of precious metals. For example, the money mentioned in the bible—the skekel and the talent—were different units of weight. Debasement is an effort to use unjust scales by issuing money that says it is a certain weight of gold or silver, when in fact it contains less. In chapter 4 it was explained that using false weights and measures is an act of fraud akin to theft. God requires just balances and finds false scales an abomination (Lev. 19:35–36; Prov. 11:1). Furthermore, when prophesying against the unfaithfulness of Jerusalem, Isaiah specifically cites monetary debasement as one of their sins when he says, "Your silver has become dross" (Isa. 1:22a).

Fractional reserve banking also is an example of fraud in that the fiduciary media created by commercial banks with the full support of the central bank are advertised as being payable on demand. However, because banks only hold a fraction of the money reserves necessary to cover all of their outstanding money warehouse receipts, there is no way for them to make good on all of their promises.

Additionally, inflation results in a redistribution of wealth. It increases the wealth of those who receive new money sooner and takes it away from those who receive the new money later or not at all. Such redistribution of wealth is not the result of voluntary exchange of private property, but solely due to the state coercively monopolizing money production and using that monopoly to increase the money supply. Government inflation violates the right of private property on many levels and, therefore, cannot be considered an ethical economic policy.

MONETARY INFLATION:
AN ECONOMIC AND ETHICAL EVIL

Monetary inflation is unsound for both economic and ethical reasons. Increasing the money supply can never provide a general benefit to the whole of society. Inflation merely results in higher prices paid for the same quantity of goods. At most it benefits those who receive the money sooner at the expense of those who receive it later or not at all. What is worse, inflation via credit expansion creates the business cycle. Inflationary credit expansion sets in motion a boom that necessarily resolves itself in a bust whose twin children are capital consumption and unemployment.

State monopolization of the money production process and government inflation violates the Christian ethic of private property. It violates the ethical principles of voluntary exchange and is inherently fraudulent. Inflation is both economically dangerous and ethically unjust.

SUGGESTED READING

Heurta de Soto, *Money, Bank Credit, and Economic Cycles*, 347–508. From his recent monumental study of money, banking, and business cycles, Heurta de Soto provides the most up-to-date exposition on the consequences of monetary inflation via credit expansion.

Mises, "Monetary Stabilization and Cyclical Policy," 97–128. An exposition of Mises' seminal theory of how recessions are caused by inflationary credit expansion.

North, *Honest Money*. This is North's excellent treatise on the economics and ethics of money.

Rothbard, *America's Great Depression*. Rothbard's application of business cycle theory in explaining the origins of the Great Depression.

——, *Man, Economy, and State*. 989–1025. Rothbard's masterful treatment of the nature and consequences of inflationary credit expansion through fractional reserve banking.

——, *The Mystery of Banking*, 127–77. A clear explanation of how central banking removes constraints on monetary inflation.

Schlossberg, *Idols for Destruction*, 88–102. From Schlossberg's monumental Christian critique of American culture comes this exposition of the ethical problems of inflation.

Vedder and Gallaway, *Out of Work*, 61–149. From their outstanding study of unemployment in the twentieth century, Vedder and Gallaway describe the pattern of unemployment during both the recession of 1920–1922 and the Great Depression.

14

Macroeconomic Policy

E CONOMIC RECESSIONS ARE NEVER pleasant experiences. Entrepreneurs do not like reaping losses or going bankrupt. They take no joy in laying off workers and the workers do not like losing their jobs. For those who lose their jobs it can be painful to be forced to cut back on purchases, watch their families do with less than they could previously afford. Worse still, some people lose their cars or homes if they are unable to pay their loans.

Such unpleasantness behooves us to have some understanding of how to avoid a recession and how to get out of one once it begins. In modern society it has become all too common for people to look to the state for answers. Citizens want elected officials and the bureaucrats they appoint and hire to do something—anything—to fix economic problems. What the state should do regarding the economy is a matter of economic policy.

Policy is a word that means the actions or plan of the state regarding some area of life. Compulsory attendance laws and state certification of teachers, for example, are part of government education policy. Government subsidization of artist, art museums, and symphony orchestras are facets of a state arts policy. Questions about what the government should do to avoid recession or lead to recovery are matters of macroeconomic policy.

ECONOMIC THEORY AND MACROECONOMIC POLICY

When attempting to formulate economic policy that best fosters sustained economic progress, prevents recession, or encourages recovery in the middle of a recession, policy makers must keep in mind that economic law is indeed economic *law*. It is not merely economic opinion or economic suggestion. Sound economic principles are sound because they are derived from the reality of human action. Just as Congress cannot repeal the law of gravity, it cannot repeal the laws of economics either. For a policy to effectively achieve its goal, it must be in full accord with sound economic analysis. This means that good macroeconomic policy must not ignore the causes of the business cycle.

In the previous chapter it was shown that the business cycle is the result of monetary inflation via credit expansion. Recessions are the necessary consequence of an inflationary boom set in motion by central bank inflation. Preventing recessions requires preventing the unsustainable booms that precede them. The solution to preventing the business cycle is to stop inflating. Without artificial credit expansion entrepreneurs would not be encouraged to unwisely invest in projects that have only the appearance of profitability. There would not be general malinvestment of capital resulting in an unsustainable boom followed by the inevitable recessionary bust.

We also noted that the way for an economy in recession to return to a solid economic foundation is to allow market adjustments. A recession should be allowed to run its course. Once unwise capital investment has been made, it cannot be unmade. The best course is to allow entrepreneurs who made unproductive investments to reap their losses. Such a policy provides an incentive for losing entrepreneurs to cut their losses by liquidating bad investments, so that they no longer waste any more capital in these ventures. Such action will hasten the allocation of factors toward more profitable and truly productive investments. Because of the loss of nonconvertible capital during the recession, economic recovery necessitates rebuilding the capital stock, which requires additional saving. The government can assist the saving process by decreasing government spending and taxation, which would increase both the ability and incentive to save and invest in productive capital accumulation. The state could also help the situation by allowing labor markets to freely adjust so that human resources can also be directed toward their most highly valued uses. In short, the state should stay out of the way and allow entrepreneurs in the market to direct factors of production without hindrance and intervention.

Sound economic analysis teaches that the boom/bust cycle is best prevented by the state abstaining from monetary inflation via credit expansion. Recovery from a recession already in progress is best achieved if the state refrains from intervention in the economy. Not all economists accept sound analysis however.

A sizable number of economists and economic journalists view recessions as caused by insufficient aggregate demand. The adjective *aggregate* is a fancy name for total. Aggregate demand means the total demand for consumer and producer goods in an economy. Some argue that if aggregate demand is allowed to fall below an optimal amount there will be overproduction in the economy. There will be an excess supply of goods which will lead to a drop in prices for goods. Entrepreneurs will, therefore, reap losses. Firms will decrease production and output and having less demand for workers, layoffs will occur. A recession results as national income falls and unemployment increases.

Advocates of this view think that the solution to preventing and curing recessions is to properly manage aggregate demand so as to ensure an expanding economy with stable prices. Effective economic management in their view requires interventionist monetary or fiscal policy or some combination of both. As the name implies, monetary policy has to do with government authorities manipulating the money stock to achieve their ends. In the United States, the Federal Reserve System—the official monetary authority—is charged by Congress to promote economic expansion, stable prices, and low unemployment. Fiscal policy is policy relating to the government budget. Fiscal policy is the result of the state increasing or decreasing taxes or government spending in order to achieve its ends.

Politicians, journalists, and economists who look to interventionist monetary and fiscal policy in order to manage the economy traditionally seek to adopt what they think are the appropriate monetary and budgetary measures to promote aggregate demand. The intellectual source of such interventionist macroeconomic policy comes primarily from two theoretical traditions in economics: Monetarism and Keynesianism.

MONETARISM

Monetarists are so named because of their insistence on the importance of money for macroeconomic phenomena, especially inflation. They place great importance on the size of the money stock and think that the economy will run more smoothly and prosperously if the state sees to it that the money stock is at a particular optimal level.

Monetarists arrive at this conclusion by following the *Quantity Theory of Money*. This theory was originally developed by classical economists David Hume (1711–1776), David Ricardo (1772–1823), and John Stuart

Mill (1806-73). The classic modern treatment is Irving Fisher's *Purchasing Power of Money*, published in 1911. Fisher lived from 1867 to 1947.

In his book, Fisher stated that he wanted to investigate the "causes determining the purchasing power of money." Fisher began his treatise by rightly defining the PPM as the different quantities of goods that a unit of money can buy. As the price of goods decreases, the quantity of goods bought with a given quantity of money—the PPM—increases. Consequently, Fisher saw the study of the PPM as synonymous with the study of price levels.

The quantity theory of money uses what is called the *equation of exchange* to argue that the price level is determined by three aggregative factors. The equation of exchange states that MV = PT, where M = the quantity of money in circulation, V = the velocity of money, P = the price level, and T = the total quantity of goods bought with money.

When quantity theorists refer to the quantity of money in circulation, they mean the same as what we have already referred to as the money stock or the total stock of money in existence. This is so because, at any given moment, all money is held by someone. There is no such thing as a division between money that is in circulation and money that is hoarded in someone's cash balance. All money is always in someone's cash balance. Every dollar in existence is held by someone. There is no moment of time when money is in circulation as opposed to being in someone's cash balance. Even when a good is being purchased with money, the money is always either the property of the buyer immediately before the trade, or the property of the seller immediately after the trade; it never just floats in circulation.

The term *velocity of money* refers to the average number of times that a unit of money is exchanged for a good over a period of time. Suppose that a student gets paid $20 for mowing a very large lawn at the beginning of June. On June 14 that student took his sweetie out for dinner for Flag Day, spending the $20 on a good (and economical) dinner of Syrian fried kibbe, shunkleesh salad, hummus, and baklava. The owner of the Syrian restaurant then used the $20 to pay part of his wait staff's wages. The employee, on the last day of the month, spent the $20 on the hard-to-get imported Henry Mancini CD featuring music from soundtracks to the films *Arabesque* and *Charade*. What is the velocity of that $20? The answer lies in how many times the money changed hands. A quick summary of the transactions reveal that the money was part of four transactions in

June. Consequently, the velocity of that $20 during the month of June equals four.

When defining P, notice that the quantity theory does not define it conceptually as the level of overall prices. Instead, it defines P as a single number measuring *the* price level. In order for a mathematical equation to be operational, a single number measuring the level of prices is necessary. The total volume of goods bought with money, T, is simply the total quantity of all goods purchased with money during a given time period.

According to the quantity theory of money the price level, P, is determined by the three other variables: M, V, and T. Using their definitions, quantity theorists arrive at the equation of exchange by correctly defining total expenditure on a good as total expenditure on a good. Suppose that Bud Abbot buys ten pounds of apples from Lou Costello at a price of $1.00 per pound. Abbot gives up $10 to Costello and Costello gives ten pounds of apples to Abbot. From the above transaction we see that the total expenditure on apples is equal to the price of apples multiplied times the quantity of apples bought. Mathematically, quantity theorists express the expenditures with the equation $E = pQ$ where E is the total expenditure on apples, p is the price of apples and Q is the quantity of apples bought. In this case, $10 = $1 x 10 pounds of apples.

The equation $E = pQ$ implies a number of things. The money side of the equation equals the goods side. The total money paid equals the value of goods bought. Using this equation, Fisher and his followers derive an equation for the price level. If $E = pQ$, this also means that $p = E \div Q$. In this case, $p = \$10.00 \div 10$ lbs. $= \$1.00$.

The quantity theory of money seeks to explain the determinants of the price level by aggregating individual expenditures. $E = p_1 Q_1 + p_2 Q_2 + p_3 Q_3 + \ldots$ etc. Consequently, total expenditures depend on the money stock and how often each unit of money is spent. $E = MV$, so $MV = PT$. Rearranging the equation of exchange in terms of P results in $P = MV \div T$, so the price level, and hence the PPM, is determined by the stock of money, the velocity of money, and the total quantity of goods purchased with money.

There is a pair of primary conclusions that monetarists take from the quantity theory of money. The main theoretical conclusion is that a change in the money stock will provide a direct, proportionate change in the price level—*ceteris paribus*. Suppose the money stock increases by 10% and according to our *ceteris paribus* assumption, the velocity of money and the volume of goods bought with money stays the same. If MV = PT, a 10% increase in the money production will result in a 10% increase in the price level. Because the equation requires it, the price level will not merely increase, but must increase at the same rate that the money stock is increasing.

The quantity theory implies that a change in the money stock leads to a proportional change in the price level, and this has led many to conclude that money is neutral. If the money stock increases or decreases and prices will change by the same proportion, then nothing changes regarding people's wealth or the level of satisfaction they are able to attain with their current disbursement of income. Because the pattern of spending does not shift among goods, their prices do not shift relative to each other. Therefore, the profitability of the existing pattern of production does not change and producers' real incomes stay the same.

Consider our example of the economy containing apples, gasoline, shoes, and computers from chapter 12. If a consumer named Jenny had $905 in her cash balance, she would be able to buy one pound of apples for $1, one gallon of gas for $4, one pair of shoes for $100, and one computer for $800. If the money stock doubled so that Jenny now possessed $1810, and if—as the quantity theory of money suggests—that the prices of each good also doubled, she could still only purchase one pound of apples, one gallon of gas, one pair of shoes, and one computer. On the other hand, if the money stock decreased by half along with the prices of each of those goods, Jenny could still only buy one of each. Because of the alleged neutrality of money, changes in the money stock have no real impact on economic realities. This is the main theoretical conclusion of the quantity theory of money.

The quantity theory equation of exchange also points monetarists to their main policy conclusion. Because MV = PT, if we increase the money stock at a rate equal to that of the increase in goods, we can have prosperity with a stable price level and hence a stable PPM. Generally, the quantity of goods traded in an economy—variable T—does not remain constant over time. In a progressing economy, the quantity of goods produced for sale tends to increase. Therefore, if T is increasing by 5% a year, M can also increase by 5% without increasing P.

In light of our analysis of the PPM, what are we to make of the quantity theory of money? The quantity theory does contain a kernel of truth. Other things equal, there is a direct relationship between the money stock and overall prices. If the stock of money increases, the demand for money remaining the same, overall prices will increase and the PPM will decrease. If the stock of money decreases and the demand to hold money does not change, overall prices will in decrease and the PPM will increase.

However, we have already identified the relationship between the quantity of money and prices of goods using the money market analysis above. No reliance on an inferior mathematical approach is necessary. Using the equation of exchange to derive meaningful monetary principles is unnecessary at best and deceiving at worst.

In the first place, there is no equality of values in economic exchange. Fisher's derivation of the equation of exchange is built on his claim that, in our example, ten pounds of apples equals $10.00. If Abbot buys the apples for $10.00, he demonstrates that he values the apples more than he values the $10.00. Likewise, Costello values the $10.00 more than he values the apples.

At most the equation of exchange is a truism and conveys no economic information. The equation of exchange basically suggests that $10.00 equals $10.00 and provides the startling revelation that the total money received in an exchange is equal to the total money spent in an exchange. If the equation of exchange is attempting to tell us the determinants of the price level, it tells us that $10.00 = $1.00 x 10 pounds of apples, so that the price of apples is $10.00 ÷ 10 pounds of apples; it tells us that the price of apples is determined by $10.00 and ten pounds of apples, and that the price of apples is determined by things—inanimate objects. We know, however, that things do not determine prices, because things do not act. Human beings in the act of buying and not buying and selling and not selling determine the prices of all goods. At best the equation of exchange is a mere truism; at worst it is wrong and misleading.

Even if the equation of exchange is true when constrained to describing a single transaction, it is less than useful for describing the role of money in the economy. We cannot merely add up the equations from all of the individual exchanges in an economy and represent them all with one equation of exchange. Go back and think about our four-good economy from which we derived the concept of the PPM. The price of apples was $1.00 per pound. The price of gasoline was $4.00 per gallon. The price of shoes was $100 a pair, and the price of a computer was $800. Suppose that all four of these goods were purchased. The equation of exchange requires that all four of those transactions be added together. How are we supposed to add a pound of apples to a gallon of gas to a pair of shoes to a computer? There is no common unit that we can sum. It makes sense to add the twenty miles it takes to get from Glenwood, Iowa to Council Bluffs, Iowa to the twenty miles it takes to get from Council Bluffs to Missouri Valley, Iowa, because we are adding units that are the same. We are adding miles to miles. However if we are attempting to sum all of the exchange equations for each purchase in the economy, we are attempting to quite literally add apples to oranges, with gas, shoes, and computers also thrown into the mix. Without a common unit, the quantity theory's T is meaningless.

If T is meaningless, then P is as well. Arranging the equation of exchange in terms of the price level results in $P = MV \div T$. However, if the denominator on the right hand side of the equation is meaningless, then the entire equation is meaningless.

In fact, prices of different goods can never be averaged. As alluded to earlier, in order to average something, the different quantities of that something have to be in the same units. We can average the heights of three different people together. Suppose Huey is 5'6" tall, Duey is 6'0" tall, and Luey is 6'6" tall. We can add up the three heights, divide by three and find that the average height of this group is 6'0". This is possible, because we are adding up feet and inches for everyone. We cannot, however, sum up different goods such as apples, gas, and shoes. Therefore, we cannot get a true average.

Some economists think we can solve this problem by using a weighted average to derive an index number that purports to measure the price level. The most well-known price index is computed by the United States Bureau of Labor Statistics (BLS) and is called the Consumer Price Index (CPI). The CPI attempts to measure the price level of a representative basket of goods and services that it is reasonable to think consumers purchase.

Any weighted average, which is what a price index is, involves assigning a particular weight to each good sold. Suppose that the BLS thinks that the typical consumer buys four apples and three oranges. The CPI will be computed as if *every* consumer purchases that many of each fruit for every time period. A weighted average, however, does not solve our problem because the goods represented in the denominator of the average—the apples and oranges—are still different units of different goods. A price index such as the CPI can be calculated and reported to the press; however, the precise weight assigned to each good is an arbitrary judgment. Therefore, because the weighting system used is arbitrary, there is no price index that is scientifically objective. A price index is more art than science.

If there is no such thing as a scientific price index, there is no way to scientifically measure the price level. There is no single number that we can calculate that is an objective measurement of all prices. Therefore, it turns out that we cannot really say that if monetary expenditures (MV) double while T remains the same, then P must double, because there is no single number that objectively measures P. All we are entitled to say is that when MV doubles, PT must double. In other words when expenditures double, expenditures double. Consequently, it makes no economic sense to refer to any numerical weighted average or index number as the price level.

As we have seen in chapter 12, it does make sense to consider the level of overall prices because at any moment a price structure exists. The price structure includes every price for the vast array of goods bought and sold in the economy. Entering into every price is supply and demand for goods and the stock of and demand for money. Both of these rest upon human value scales as people compare the marginal utility of money to the marginal utility of goods they could purchase with their money.

Because there is no way to scientifically measure the price level by a single number, it should be clear that we cannot stabilize it by increasing the M at the same rate as T is increasing. This is because money is not in fact neutral regarding economic realities.

Consequently, the monetarist policy of attempting to hold the mythical price level stable by increasing the money stock at the same rate that T increases is doomed to failure. As the money stock increases, different prices for different goods will increase at different rates. We do not need to increase the money stock in order to accommodate the increased demand for money corresponding to an increased stock of goods for sale. The voluntary actions of people in the market will do this themselves. As long as voluntary exchange is allowed, the markets for all consumer and producer goods will clear, lowering the PPM until the quantity of money people want to hold is equal to the quantity of money they actually hold. The monetarist policy of increasing the money stock at a rate equal to the rate of increase in T merely serves to put in motion perpetual inflation. We have already discovered the nature and consequences of inflation in chapter 13.

KEYNES AND KEYNESIANISM

Another group of economists who reject non-interventionist macroeconomic policy are the followers of that most famous rejecter of "economic orthodoxy," John Maynard Keynes. Keynes is surely the most famous economist of the twentieth century and his book, *The General Theory of Employment, Interest, and Money* is surely the most famous economics text of that century. His book and the policies of his followers are so popular because Keynes provided a theory of recession and what is necessary for recovery that is very palatable for politicians and government officials. Keynesian economics has been popular with them because Keynes provided intellectual justification for a whole host of economic interventions they already favored.

Keynes denied that recessions are the inevitable consequence of an inflationary boom. He claimed instead that economic downturns are the result of insufficient aggregate demand. People simply do not demand enough consumer and producer goods. Because spending is not enough to buy the goods produced, a surplus of goods in the economy results. Production falters, employment falls, incomes fall, and as a result people spend even less so the cycle becomes self-enforcing. Recession ensues.

In the world according to Keynes, aggregate demand is the product of private spending on consumption plus private spending on business investment plus spending by the government. Keynes began with only consumption and investment, but his followers—noting Keynes' prescription for an interventionist government in smoothing out the business cycle—added government spending to the equation. Keynes represented national income and aggregate demand with the following equation: $Y = C + I$ where Y = national income, C = consumption spending, and I = business investment. It is because of Keynes that so many people are fixated on Gross Domestic Product as an indicator of the health of the national economy. As we have seen, the very calculation of the GDP is built within the Keynesian framework summing consumption, investment, and government spending. Not surprisingly, consumption spending is defined as income spent by consumers on consumer goods. In Keynes' opinion, the only good kind of investment is direct investment on factors of production; in other words, capitalists fund the purchase of services of land, labor, and capital goods to produce economic goods. Government spending is merely the money spent by the government on various purchases.

Keynes thought that consumption spending was a stable positive function of income. He thought that the consumption function was positive because as people earned higher incomes, they would spend more money on consumption goods. Additionally he thought it a stable function because he thought people spent a relatively constant fraction of each dollar of income on consumption. This fraction of each dollar spent on consumer goods he called the marginal propensity to consume (MPC). Keynes took a dim view of savings and considered savings merely the residual left over after we have spent all that we want on consumption.

Because consumption is considered a stable function of income, investment becomes in effect the sole determining factor of present and future income and consumption. In a huge break with the economists who came before him, Keynes separated the theoretical link between savings and investment. Here is the crux of the matter for Keynes and Keynesians. Capital accumulation, it is argued, can take place without saving and saving can take place without investment courtesy of what is called *the spending multiplier*. The multiplier is a product of the mathematics inherent in the Keynesian system. Suppose that value scales of Grand Fenwickians are such that their aggregate consumption function is as follows: $C = 5 + 0.6Y$. The 0.6 is the fraction of every *florin* of income spent on consumption, or the MPC. Because consumption is a stable function of income, the consumption function generates a multiplier that allows investment to increase without an increase in savings.

Suppose that investment in Grand Fenwick increases once by five billion *florins*. Because $Y = C + I$, national income will also increase by five billion. You might think that an increase in savings and investment of five billion would result in a total increase in income of five billion *florins*, but you would not be living in a Keynesian world. Remember that as the money is spent by the receivers of the investment money—the owners of factors of production—they spend 0.6 of every *florin* of increased income on consumption. Because $C = 5 + 0.6Y$, consumption spending now increases by 0.6×5 billion $= 3$ billion. Because $Y = C + I$, a three billion *florin* increase in consumption will boost income by the same amount. You might think that that the total increase in income is equal to the initial five billion plus three billion. But you would be wrong. Remember that $C = 5 + 0.6Y$. The additional consumer spending is someone else's income.

Therefore consumption spending will increase again this time by 0.6 × 3 billion = 1.8 billion.

By now you may have noticed a pattern. Every time consumption spending increases it generates income, a fraction of which is spent again on consumption. Each iteration results in an increase in income but a smaller increase than the iteration before. This process continues to play out until the incremental changes in consumption are negligible. The cumulative effect of the process can be seen by looking at the table in table 14.1.

Table 14.1

Keynesian Spending Multiplier	
Change in Investment	$5 billion
First change in Consumption	MPCx $5 billion
Second change in Consumption	MPCx (MPCx $5 bil.) = MPC_2 x $5 bil.
Third change in Consumption	MPCx (MPC_2 x $5 bil.) = MPC_3 x $5 bil.
•	•
•	•
Total Change in Demand	$(1 + MPC + MPC_2 + MPC_3 + \ldots)$ x $5 bil.

The total effect of an increase in investment is not merely the initial increase in income but the increase *times* the spending multiplier. As illustrated in table 14.1, the Keynesian multiplier = $(1 + MPC + MPC^2 + MPC^3 + \ldots)$. A principle of mathematics allows us to express the multiplier as $1 \div (1 - MPC)$. In our case, the MPC = 0.6, so the multiplier is equal to $1 \div (1 - 0.6) = 2.5$. Therefore, given Grand Fenwick's MPC, an increase of five billion *florins* in investment spending will actually increase national income by five billion times 2.5, or 12.5 billion *florins*. A one shot increase in investment will increase income by more than amount of the additional investment.

Because aggregate demand drives the economy in the Keynesian framework, the economy appears to grow as a result of the multiplier, and not because of increased savings. Savings is seen as a pool of income that leaks out of the consumption spending stream. Investment comes from a completely different decision process.

Keynes departs from economics derived from the principles of human action by shrinking the choice set for income earners. As you read in chapter 7, when a person earns income he decides to allocate it in either of three ways. He can spend income on consumption, save and spend it on investment, or not spend it and add to his cash balance. Therefore, total income will always equal consumption spending plus investment plus the change in cash balances.

In the Keynesian framework, however, it is thought that people first decide how much of their income to spend on consumption and how much to save. Then they decide how much of their savings they want to spend on investment and how much they want to hold in their cash balance—a practice tagged by Keynes with the pejorative *hoarding*.

From Keynes' perspective, saving is practically evil because of what he called the *paradox of thrift*. If people try to increase their savings, and thereby reduce their consumption, they spend a smaller fraction of their income on consumer goods than previously. Consequently, the MPC falls, and if the MPC falls then so does the multiplier. Therefore, while individuals are attempting to increase their savings, saving actually is reduced in the aggregate because the margin between consumption and income shrinks as income decreases. Keynes himself proposed a radical cure for the paradox of thrift. He suggested that the government should allow no choice between spending income on consumption and investing it in physical capital goods on the one hand, or hoarding money or investing it in stocks and bonds on the other. Ultimately, Keynesians opted for lesser measures such as monetary inflation and government spending.

While consumption is a fairly stable function of income, investment for Keynes was a very different matter. Investment, in Keynes' opinion, is inherently unstable due to the changing psychology and whims of investors. He thought that business investment is determined by average expectations of stock market investors and in turn the average investor bases his expectation on arbitrary convention that assumes current market prices reflect current information. Keynes characterized this convention as determined by the state of confidence possessed by the mass of investors. Keynes is famous for describing such an arbitrary convention as being driven by animal spirits. At any moment, expectation and convention could be such that investors abandon productive investment in factors of production and instead spend their capital funds buying merely stocks and bonds.

If people either invest their savings in bonds or stocks or increase their cash balance, aggregate demand falls. If direct investment in production declines because the animal spirits turn sour, aggregate demand will fall and a surplus of goods will exist. Keynes draws this conclusion because he makes a radical distinction between the rate of return earned in the loanable funds market (which he calls the interest rate) and the rate of return earned throughout the production structure (which he calls the marginal efficiency of capital). If the interest rate is greater than the marginal efficiency of capital, capitalists will direct their savings away from productive investment and into financial investments. Less money is spent on land, labor, and capital goods. Production decreases, which results in recession and unemployment. Keynes' solution to this problem is inflation via credit expansion.

Keynes' recommendation of inflation to cure recessions follows from his separation of the marginal efficiency of capital and the money interest rate; therefore, it profits us to examine more closely this separation. Keynes defined the marginal efficiency of capital as the prospective rate of return on an increment of monetary investment in the current production of capital goods. This rate is the primary factor by which expectations of the future influence present action according to Keynes. The marginal efficiency of capital is a good and proper return to investment in physical production, therefore this kind of investment is a productive activity. Interest, however, is seen as a bribe paid to nonproductive savers in order to induce them to part with their money.

Keynes' view of interest provides the key to understanding his liquidity preference interest theory. Liquidity refers to the ease with which an asset can be sold for money. The most liquid good of all is money itself. Liquidity preference is merely Keynes' term for the demand for money.

The liquidity preference theory of interest artificially divides the decision making process into fixed dichotomies. It views people as first deciding what they want to spend on consumption and what they want to reserve in some form. Only then do the liquidity preferences of these same people decide how much of the reserved money they want to hold in their cash balance and how much they want to spend on investment in physical production. Interest, therefore, is not seen as a return for waiting, because if a person reduces his consumption spending and adds to his cash balance, he is not earning any interest. Instead, interest is viewed as the price that must be paid to get people to part with their money, or the price of money. As such, the interest rate is alleged to be determined by the demand for and stock of money, not the supply of and demand for present goods in exchange for future goods. Keynesians think the interest rate is a purely monetary phenomenon.

Keynes theorized that people had three different motives for holding money. The first is referred to as *transactions demand*. Transactions demand for money refers to the demand people have for money in order to make purchases. People demand a certain quantity of money solely for purchasing goods and services. How much they want to spend on goods is unrelated to the interest rate, therefore, so is transactions demand.

The second motive for holding money is referred to as *precautionary demand*. This refers to the fact that people will want to hold money so that they can make unexpected yet necessary purchases in the future. Remember the account of a father driving his son and college friends back to campus following a long weekend break documented in chapter 7. Dad's car breaks down, but fortunately just as they pass a truck stop that has a repair garage. The problem is the water pump, so it is replaced. Dad has neither a credit card nor enough cash on hand to pay, so he has to borrow money from his passengers to pay for the repair. After he gets home, the father, deeply embarrassed, repays the money he borrowed from the less-than-flush college students. From then on the father carries a couple of large bills in the back of his wallet in the event something like this ever happens again. He holds this money as a precaution. Because precautionary demand is also related to purchases of goods and not of bonds or stocks, this motive is also unrelated to the interest rate. People will want to hold a certain quantity of money for unexpected but necessary purchases regardless of the interest rate.

The third motive, however, is related to the interest rate. This motive, called *speculation demand*, is said to derive from people viewing money as an asset that is compared to other assets when making decisions regarding how to allocate wealth. People can hold their wealth in many forms including money or financial instruments such as stocks or bonds. As the interest rate, which is the rate of return earned by holding stocks or bonds decreases, holding these financial instruments becomes less attractive so people demand to hold more money. As the interest rate rises, the cost of holding money—the foregone interest—increases so people demand to hold less and instead spend it on more stocks and bonds. Therefore, the claim is made that there is an inverse relationship between the quantity of money people demand for speculative purposes and the interest rate.

We see the implications of Keynes' money demand theory by examining the money demand curve in figure 14.1.

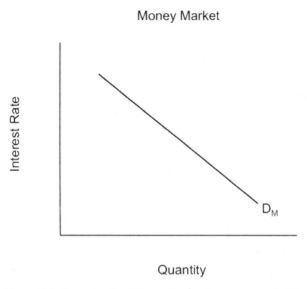

Figure 14.1. Keynesian liquidity preference theory asserts that the quantity of money people demand to hold is inversely related to the interest rate.

In Keynes' theory, the price of money is the interest rate, so the interest rate replaces the PPM on the vertical axis. The horizontal axis still tracks the quantity of money. Because both transaction demand and precautionary demand are unrelated to the interest rate, the demand curves that represent both of these motives for holding cash are vertical. However, speculation demand is inversely related to the interest rate, and the demand curve that represents this motive will be downward sloping to the right. By summing the demand curves for the three different motives, Keynesians derive the liquidity preference function represented by a market money demand curve that is downward sloping to the right. This is the demand curve shown above in figure 14.1.

The equilibrium interest rate will be that rate that equates the demand for money with the stock for money as can be seen in figure 14.2.

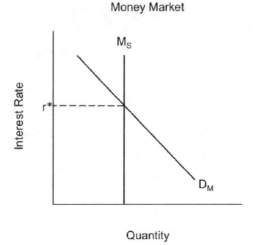

Figure 14.2. Keynesian liquidity preference theory asserts that the interest rate is determined by the supply of and demand for money.

As with our previous analysis, the money market will tend toward an equilibrium price. The big difference, however, is that this price is not the PPM but the interest rate. If the interest rate happens to be above equilibrium, it is argued that there will be an excess supply of money held; people would demand more bonds and the interest rate would decrease until it reached equilibrium. If the interest rate happens to be below equilibrium, there will be an excess demand for money; people would sell bonds in order to obtain cash. Interest rates will rise until equilibrium is reached.

Notice that changes in the demand for bonds results in an inverse effect on interest rates. Why is this so? Because of the nature of bonds. Recall that bonds are certificates of debt. A bond is in essence a document that indicates the bond holder has a legal right to receive fixed interest payments through the life of the bond—the term of the loan—at which point the bond holder is also entitled to receive the principle amount. For example, suppose that to raise cash, Henry F. Potter, President of Acme Bank, authorizes the issuing of $1,000,000 in bonds. Each bond is sold for $10,000 and entitles the bond holder to an interest payment of $500 at the end of the year for thirty years at which time Acme bank must pay the holder the $10,000 in principle as well.

If Ernie Bishop purchases one of these bonds, he is loaning money to Acme Bank; Acme is borrowing money from Bishop. For Bishop the bond is an investment. Because his $10,000 investment yields him an annual return of $500 interest, the price of present money in exchange for future money is 5%. Bishop, however, does not have to hold on to that bond for the entire thirty year term if he does not want to. He can sell it in the bond market if he so chooses, in which case the new owner is entitled to the interest payment.

Suppose that, because of perceived brighter future prospects for Acme Bank, Acme bonds become in heavier demand. Just like any other good, if there is an increase in the demand for Acme bonds, the price of those bonds will increase. If the demand for Acme bonds increases so that their market price increases from $10,000 to $11,000, a change in the rate of interest earned on Acme bonds will result. If Bishop sells his Acme bond to Bert for the new price of $11,000, the rate of return Bert earns will decrease. Remember that the rate of return is equal to the net return divided by the initial investment. The net return received by the holder of the bond will be the fixed payment, in Bert's case $500. Bert, however, did not pay what Bishop paid for the bond. Bert's initial investment was $11,000. Therefore, the effective interest rate Bert is paid is $500 ÷ $11,000 or 4.5%. Because an increase in the price of the bond increases the bond holder's initial investment and because the payments to bond holders are fixed, an increase in the price of the bond results in a decrease in its interest rate.

On the other hand, if the market price for Acme bonds fell, the effective interest rate a buyer of that bond reaps will increase. If the price of Acme bonds decreased to $9,000 and Bishop sells it to Bert, an interest rate of 5.5% will be reaped by Bert, which we calculate by dividing the fixed interest payment of $500, by Bert's lower initial investment of $9,000. As the price of the bond decreased, the interest rate on Acme bonds increased. In general there is an inverse relationship between the price of bonds and the interest rate reaped by their holders.

This explains why Keynes thought it possible to affect the interest rate and hence, real investment, by manipulating the money stock. If the economy is in recession because of depressed animal spirits, an obvious remedy to Keynes is to increase the money stock, the effects of which are illustrated in the money market graph in figure 14.3.

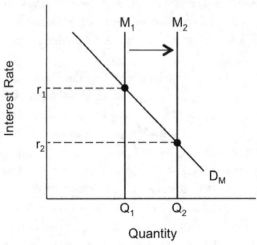

Figure 14.3. According to Keynesian theory, an increase in the money supply will result in a decrease in the interest rate.

As the money stock increases, there results an excess supply of money at the original market interest rate. People remove this excess supply by increasing spending on bonds. This drives the price of bonds up and decreases the interest rate. The interest rate will fall until the quantity of money people want to hold is equal to the new increased stock.

How is this lower interest rate supposed to get the economy out of recession? Remember that the interest rate that we are talking about in the Keynesian framework is not the price of time or the price of present money in exchange for future money, but the price of money determined only in the loanable funds market. It is the marginal efficiency of capital that is the rate of return on physical investment. If the money stock is increased enough, the money interest rate will fall below the marginal efficiency of capital. If this occurs, investments in bonds will be less appealing than investment in buying the services of factors of production. This results in productive investment according to Keynes, which in turn results in an increase in national income. Remember too that the total increase in national income is even greater than the increase in investment made possible through the magic of the multiplier. In Keynes' opinion, by engaging in inflationary credit expansion, we are able to take Satan up on his offer and perform the "miracle of turning a stone into bread."[1]

1. Keynes, *Paper of the British Experts* cited in Mises, "Stones into Bread, the Keynesian

The only limitation on inflation's stimulative effect, from the Keynesian perspective, is the Liquidity Trap. Keynes taught and contemporary Keynesians believe that if the interest rate falls, it is possible for it to reach a level at which liquidity preference might become absolute. In other words, there might be some relatively low interest rate at which people demand to hold only money and purchase no bonds. A liquidity trap could occur according to Keynes, because of people's expectations of further interest rate changes. Keynes argues that at a very low rate of interest everyone comes to expect the rate to rise so they hold money until bond prices fall again. Consequently, further increases in the money supply will induce no further decreases in the interest rate and hence no further increase in business investment. Instead, people will engage in the almost unpardonable Keynesian sin of hoarding money. At this interest rate, consequently, the demand to hold money becomes virtually infinite.

As seen in figure 14.4, in a liquidity trap the demand for money curve becomes horizontal at some interest rate.

Money Market

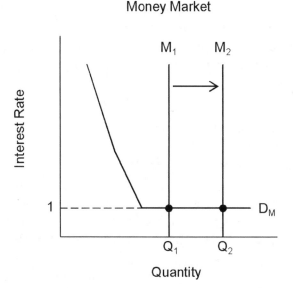

Figure 14.4. The Keynesian liquidity trap. It is argued that if the interest rate falls low enough (in this case 1%), the demand for money curve becomes horizontal and further increases in the money supply will not further decrease the market interest rate.

Suppose, that the money stock is at M_1 so that the interest rate is 1%. If the central bank is desperate to boost economic activity so as to relieve a recession, it might decide to inflate the money stock in the hopes of decreasing the interest rate. In a Keynesian framework, the lower interest rate on bonds makes investment in physical production more attractive, resulting in an increase in investment and national income.

However, if the money market experiences a liquidity trap, further increases in the money stock will do nothing to lower interest rates. If the money stock is increased from M_1 to M_2, the interest rate remains constant at 1%. In the Keynesian framework, if the economy is experiencing a liquidity trap, the effectiveness of inflationary monetary policy in battling a recession comes to naught. Increases in the money stock do not result in lower interest rates. Lower interest rates are not able to stimulate investment in factors of production. Because there is no increase in investment, there is no increase in national income to begin with and we cannot take advantage of the alleged multiplier.

How will an increase in demand for money affect the interest rate? According to Keynesians, if people want to hoard more money, (if their demand to hold money increases), people will spend less money on bonds resulting in a decrease in bond prices and an increase in interest rates. Higher interest rates discourage investment in productive investment which results in recession and unemployment. But what does economic theory tell us about all this?

THE TROUBLE WITH KEYNES

Although Keynes has often been complimented for giving due importance to the crucial role that expectations play in economic action, there are a number of problems with the Keynesian system. In analyzing Keynes' theory, we first examine his assertion that savings do not equal investment, and then evaluate his liquidity preference theory of interest. In the first place there can be no investment without saving. Investment spending due to credit expansion is malinvestment and is what leads into depression in the first place.

Additionally, cash holdings are not savings that are hoarded. As we have seen, people can do three things with monetary income: spend it on consumption, spend it on investment, or add to or subtract from their previous cash balance. The crucial determinant of the resources directed toward consumption or investment is time preference. The individual subjective time preferences of people determine the consumption/savings-investment proportions. The spending/cash balance ratio is determined by the demand for money.

An increase in the demand for money does not necessarily affect the consumption/savings ratio. With an increase in the demand for money, it is possible and indeed likely that if time preferences do not change, the resulting increase in cash balances comes from reduced consumption and investment in equal proportions. Suppose that before a change in the demand for money, the income allocation for citizens in the economy of Grand Fenwick is as provided in table 14.2.

Table 14.2

Monetary Income Allocation for the Economy of Grand Fenwick			
	Consumption	Saving/Investment	Cash Holding
Before Increase:	60 Billion	40 Billion	10 Billion
After Increase:	54 Billion	36 Billion	20 Billion

To begin with, people's value scales were such that, of the one hundred ten billion *florins* in income earned in the Grand Fenwickian economy, sixty billion were spent on consumption, forty billion on investment, and ten billion held in cash balances. The consumption/savings-investment ratio, then, is 60/40.

Now suppose that the demand to hold money increases such that the citizens of Grand Fenwick wish to increase their cash balances by ten billion *florins* to a total of twenty billion. If their time preferences do not change, and there is no reason to suspect that they would merely because of an increase in the demand to hold money, they will decrease *both* consumption and investment spending by the same 60/40 ratio. Consequently, 60% of the increase in cash holdings would come from a six billion *florins* drop in consumption spending. 40% of the increase in cash holdings would come in a four billion *florin* decrease in savings and investment spending. This change results in fifty-four billion *florins* spent on consumption and thirty-six billion saved and spent on investment. Notice that the consumption/saving-investment ratio remains the same 60/40.

There is no doubt that the overall prices of goods will drop due to the increased demand for money. However, because the decrease in spending is occurring for both consumer goods and producer goods, the price of both sets of goods will fall and they will fall so as to maintain roughly the same price differential ratios that were obtained before the increase in the demand for money.

An increase in the demand to hold money may possibly result in an increase in the interest rate if the increase in money demand is accompanied by a lowering of time preferences. Suppose that beginning from the same initial allocation, an increased demand for money occurs along with a decrease in social time preference. This case is illustrated in table 14.3.

Table 14.3

Monetary Income Allocation After a Decrease in Time Preferences			
	Consumption	Saving/Investment	Cash Holding
Before Increase:	60 Billion	40 Billion	10 Billion
After Increase:	50 Billion	40 Billion	20 Billion

Suppose that due to a lowering of time preference, the people of Grand Fenwick accommodate their increased demand for money by reducing only consumption spending. Spending decreases by ten billion *florins* but this time all of the reduction in spending is on consumption. Consumption spending decreases to fifty billion *florins* while investment spending remains the same. Now the consumption/savings-investment ratio changes. It is no longer 60/40, but 50/40. The demand for consumer goods falls not only in absolute terms but also relative to that for producer goods. Therefore, while the overall prices of all goods tend to decrease, the prices of lower order goods decrease at a greater rate than the prices of higher order goods. This alters the price differentials between factors of production and final products, shrinking them throughout the structure of production. Consequently, the interest rate falls. So it is possible for the interest rate to fall as the demand to hold money increases. Note that this decrease in the interest rate is not due to the increase in the demand for money, but because of a decrease in social time preferences that accompanied the increased demand for money.

Two conclusions follow from this analysis. First, there will be no change in the interest rate due to an increased demand for money. Because the consumption/spending ratio remains the same, the ratio between the sum of the prices of factors of production and the price of final products remain at the same rate. Therefore, an increase in the demand for money does not necessarily have any effect on the interest rate. The demand for money determines its purchasing power. The interest rate is determined by time preference.

Additionally, there is no reason to worry that an increase in the demand to hold money will result in large-scale losses resulting in recession and depression. Though the prices of final products will decrease throughout the production structure, so will the prices of the factors of production. While the prices of consumer goods will fall, so will the prices of the higher order goods necessary to produce consumer goods. Therefore, no firm is placed in financial difficulty merely because of an increase in the demand to hold money.

The concern over the liquidity trap is only as valid as the liquidity preference theory of interest. What we have already learned about the interest rate should be enough to make us question the soundness of this theory. We have seen that the interest rate is not merely a monetary phenomenon, but a *time* phenomenon. It is the price of *present* money in exchange for *future* money. Therefore, the interest rate is determined by people's subjective time preferences, not the stock of and demand for money.

Additionally, the marginal efficiency of capital is not a return radically distinct from the money interest rate. The marginal efficiency of capital is also a payment for time; it is the return capitalists receive for paying the owners of factors of production in the present in exchange for the services of those factors that represent future revenue. The marginal efficiency of capital is the interest rate earned throughout the structure of production. This interest rate—the rate of return throughout the production structure—is the important return. The loanable funds rate *follows* the pure interest rate earned in the production structure, not the other way around. Both rates tend to become equal as the economy moves toward the evenly rotating economy. Both are determined by social time preference. They are not radically different, but rather radically similar.

Because the interest rate is the price of time and not the price of money, the liquidity preference theory of interest is invalid. We have already seen that one can have falling interest rates that occur along with an increased demand for money if time preferences decrease. Further, if people increase their demand for money because they want to hold cash balances based on expectations of falling interest rates, this actually will help and not hinder the adjustment process that must take place after the boom. People always consume at some level, so they must produce at some level. Therefore, we do not have to be concerned about a perpetual downward economic spiral driven by an ever increasing demand to hold money.

Wages and the prices of other factors of production will fall along with the prices of consumer goods, so firms will not be placed under undo financial pressure by the increased demand for money. In fact, those who correctly anticipate an increase in the demand to hold money will act accordingly and reap profits once the increased money demand comes to pass.

Finally, the very notion that the economy is driven by aggregate demand, the largest fraction of which is demand for consumption, is erroneous. People cannot consume or use what is not produced. Production makes demand possible. Production makes consumption possible. One cannot provide for long-term prosperity by stimulating consumption. Economic progress is only possible through the accumulation of capital. Attempts to boost consumption spending in efforts to boost the economy necessarily result in reduced savings, consumption of capital, and a decrease in wealth and the standard of living over time.

Given Keynes' theoretical problems, what are we to make of Keynesian policy recommendations? Everything that we have learned up to this point suggests that it is extremely unwise to pursue a policy of inflation in order to alleviate a recession. Remember that inflationary credit expansion sets in motion the whole boom/bust business cycle to begin with. Consequently, inflating in response to a recession is merely asking for the whole scenario to occur over again. To engage in wild, persistent inflation in order to shake people out of their perceived hoarding is to encourage inflationary expectations that may lead to hyperinflation and the final crack-up boom. To recommend this policy path seems the height of irresponsibility.

We will examine the effects of government taxation and spending in chapter 16 devoted to the effects of government intervention in the economy through the budget process. For now it is enough to know that increased government spending must be paid by someone. If the money comes from increased taxes, this puts a drag on profits, savings, investment, capital accumulation, productivity, and ultimately the standard of living. If government spending is financed by monetary inflation, the results are everything we have explained in chapter 13. If spending is funded by borrowing, the capital funds loaned to the government are funds directed away from productive activity in the economy and toward state directed spending. Government borrowing contributes to the consumption of capital, which results in lower economic prosperity.

As indicated, the only way to respond to a recession so that the economy returns to a sound footing is to free the market. Only in a free market do the prices of goods and services reflect the subjective values of the members of society. In order for entrepreneurs to redirect resources toward their most productive use, they must be able to calculate profit and loss using market prices. Therefore, the markets for all goods should be allowed to adjust to their new equilibrium levels. The money stock should in no way be inflated. Taxation, government spending, and regulation on business should be reduced.

BEWARE BAD MACROECONOMIC POLICY

Economic analysis deduced from the reality that people act purposefully—an axiom we have drawn from the Christian doctrine of man being created in the image of God—teaches us that economic recessions and depressions are the result of government intervention in the economy. Specifically, monetary inflation through credit expansion spawns an inflationary boom that encourages malinvestment that must then be liquidated resulting in recession. Economic recovery is hampered or even prevented altogether by the state resisting the necessary reallocation process and by discouraging saving necessary to rebuild a depleted capital stock.

Monetarists and Keynesians argue that economic recessions can be alleviated by inflating the money stock. However, as should be clear from our analysis, inflation of the money stock does not encourage markets to coordinate. Increasing the money stock to solve a recession merely serves to usher in another boom/bust cycle, resulting in another recession. This has been the history of any economy that has existed under the authority of a central bank charged with the task of providing full employment with stable prices. As the central bank inflates more rapidly or more slowly as the case allegedly demands, the economy moves in fits and starts through the business cycle. Inflating the money stock does not stop nor ameliorate the business cycle, but *causes* the business cycle. At best inflation merely puts off the inevitable bust.

Putting our faith in fiscal stimulus such as increased government spending is likewise foolish. Increased government spending must be financed somehow. Clearly, paying for it by increasing the money supply does more harm than good. Paying for it by increasing taxes reduces the ability and incentive to save and invest because taxation reduces disposal income. Financing increased government spending by borrowing takes otherwise productive capital out of the private economy and redirects it into the hands of bureaucrats who consume it according to their statist ends. All of this tends to promote capital consumption and hampers the capital accumulation process necessary for the economy to get back on the path to prosperity.

SUGGESTED READING

Garrison, *Time and Money*, 123–243. These chapters provide Garrison's excellent critique of Keynesian and Monetarist macroeconomic theory and policy.

Huerta de Soto, *Money, Bank Credit and Economic Cycles*, 509–83. An outstanding recent evaluation and critique of Monetarist and Keynesian economic theory.

Mises, "Lord Keynes and Say's Law," 64–71. Mises' brief refutation of Keynes assertion that recessions are the result of insufficient aggregate demand.

——. "Monetary Stabilization and Cyclical Policy," 73–79. Mises explains that a completely stable money is impossible and calculation of a price index is ultimately arbitrary and unscientific.

——. "Stones Into Bread, the Keynesian Miracle," 50–63. An excellent and brief refutation of the notion that we can inflate our way to prosperity through artificial credit expansion.

Rothbard, *Man, Economy, and State*, 831–74. Rothbard's masterful critique of Monetarist and Keynesian theory and policy.

15

Price Controls

I N THE PREVIOUS TWO chapters we introduced detailed analysis of one facet of government intervention in the economy. We analyzed the nature and consequences of interventionist macroeconomic policy and discovered that neither monetary inflation nor government spending are efficient ways to expand an economy.

Macroeconomic policy, however, is only one category of state intervention in the economy. Another common form of intervention is price controls. Price controls are the result of laws regulating prices at which people can legally buy and sell. Rarely do governments force buyers and sellers to accept a single price to make an exchange. Instead governments prefer to set maximum and minimum prices.

PRICE CEILINGS

The form of price control governments often use in an attempt to thwart the negative consequences of monetary inflation is the *price ceiling*. As the name implies, a price ceiling is a maximum legal price. If you attempt to throw this textbook up into the air as far as it will go, what will stop it? What is the barrier above which it cannot fly? The ceiling. Just as the ceiling in an indoor room is the highest a thrown object can travel, a price ceiling is the highest price that buyers can legally pay and that sellers can legally accept.

There are two types of price ceilings: effective and ineffective. We will initially investigate the consequences of an effective price ceiling, but first need to understand what we mean by the words *effective* and *ineffective*. Typically, those words are taken to mean that something either works well or does not. In our case, however, the terms refer to whether the price ceiling has an effect on the actual price that buyers pay and sellers receive in an exchange. An effective price ceiling hampers voluntary exchange from negotiating a market price. An ineffective price ceiling is one that has no effect on the price that is charged and received in the market.

A price ceiling is effective if the maximum legal price is less than the market price. An effective price ceiling is illustrated in the graph in figure 15.1.

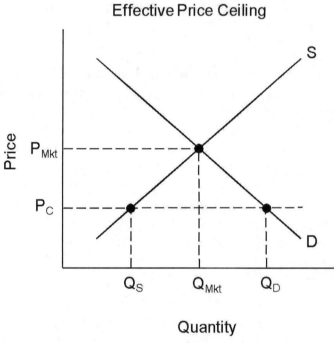

Figure 15.1. An effective price ceiling is a maximum legal price fixed below the market price. It always results in a shortage.

As can be seen, if the government enacts a price ceiling P_C that is below the market price P_{Mkt} the market will not clear. We know this by identifying the quantity demanded and supplied at the ceiling price. Remember that the demand curve tells us the maximum quantity buyers will buy at any given price. Consequently, if we trace a horizontal line from the price axis at P_C to the demand curve, we see that the quantity demanded is Q_D. Likewise, the supply curve tells us the maximum quantity sellers are willing to sell at any given price. If we follow the same horizontal price line to the supply curve, we see that the quantity supplied at the price ceiling is Q_S. We can determine the state of the market by comparing the quantity supplied with the quantity demanded.

Because the price ceiling is below the market clearing price, the quantity demanded is greater than the quantity supplied. Chapter 5, in dealing with price determination, illustrated that this is a case of excess demand. Excess demand results in frustrated buyers because there are some buyers who are willing to buy the good at that price but cannot because there is not enough supplied by sellers. In a free market, this excess demand withers away as the more eager buyers bid up the price until everyone who wants to buy can buy. This occurs at the market price, the price at which quantity supplied equals quantity demanded.

In the case of a price ceiling, however, this adjustment process is prohibited. It is illegal for more eager buyers to pay a higher price, just as it is illegal for sellers to accept a higher price. If either group attempts to do so they can be fined, sent to prison for failing to pay the fine, or shot escaping from prison—whichever they prefer. Because the price adjustment process is prevented, the excess demand does not abate but instead turns into a perpetual shortage that will last as long as the price ceiling is held below the market price.

The direct consequence of an effective price ceiling, therefore, is a shortage of the good that is covered by the price ceiling legislation. A simple, static shortage is not the only result, however. The shortage becomes worse the longer the effective price ceiling is in effect. As people become aware of the shortage resulting from the price ceiling, consumers rush to buy the products, increasing the quantity of the good demanded. At the same time, producers who cannot earn a profit at the artificially low price cease production. Consequently, the shortage is exacerbated.

The extent of the shortage is determined by how far below the market price the ceiling is mandated and the elasticity of demand for the good. Not surprisingly, the larger the positive gap between the market price and the price ceiling, the greater the resulting shortage will be. Additionally the more elastic is demand for the good, the greater will be the shortage. This can be seen in the graph in figure 15.2.

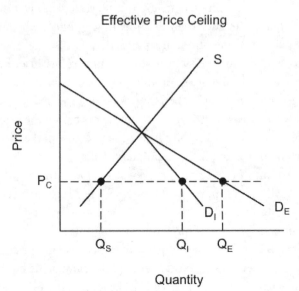

Figure 15.2. The more elastic is demand, the greater will be the shortage resulting from an effective price ceiling.

On this graph there are two differently sloped demand curves. The more elastic curve is flatter and marked D_E. The more inelastic curve is labeled D_I. To calculate the extent of the shortage resulting from the price ceiling, we must find the difference between the quantity supplied and quantity demanded. When demand is more elastic, the relevant demand curve is D_E so the quantity demanded at the price ceiling is Q_E. There is only one supply curve so the quantity supplied at the price ceiling is Q_S. The extent of the shortage in this case of more elastic demand equals $Q_E - Q_S$.

To find the shortage in the case of more inelastic demand, we do the same thing except that we use the relevant demand curve D_I. In the more inelastic case, the extent of the shortage is equal to $Q_I - Q_S$. We can see by the graph that $(Q_E - Q_S)$ is greater than $(Q_I - Q_S)$. This should not surprise us. The more elastic is demand, the more responsive buyers are to a change in the hypothetical price of a good. Consequently, the decrease in quantity demanded as a result of an artificially low price would be greater if demand is more elastic. The greater the decrease in quantity demanded, the larger the gap will be between quantity demanded and quantity supplied, hence the greater the shortage.

In the first half of the 1970s the United States government was called to do something about the increased prices that resulted from two decades of monetary inflation. The CPI was increasing at a rate from about three to five percent per year. That a three to five percent increase in consumer prices was something that alarmed people is almost quaint. Now we are assured that a three percent rate of overall price increases is not only something not to be worried about, but it is somehow supposed to be a positive good keeping disastrous deflation at bay. If Stanley Kubrick would have made a movie about the economic history of the 1970s through the 2000s, he could have very well titled it, *How I Stopped Worrying and Learned to Love Inflation.*

In any event, in response to public outcry due to higher prices for necessary goods like food, the government placed effective price ceilings on poultry and beef. The consequences were just as sound economic theory teaches. There developed an almost immediate shortage of dressed chicken fryers and broilers. The shortage became worse as chicken farmers ceased their chicken operations. The price farmers received for their chickens simply could not cover the expenses associated with housing and feeding their chicks while they grew to maturity ready for market. Consequently, farmers killed masses of baby chicks, which further reduced the supply of mature chickens into the next year.

The beef market experienced similar results. During this period, known as the *meat freeze* because the prices for meat were frozen at ceiling levels, consumers rushed to buy whatever beef was still available at meat markets. Television news reports documented the reaction to the beef shortage. One report showed a little old lady (perhaps from Pasadena) carefully looking over a package of steak deciding whether she wanted to make the purchase. Out of nowhere and with no warning whatsoever, a younger woman who had her shopping cart in overdrive careened into the older woman, grabbed the steak, threw it into her cart and tore off, leaving the older lady with no steak and a thrown out hip.

Elderly women's hips were not the only things thrown out during the meat freeze. My father, who worked at a beef packing plant while I was growing up, was thrown out of work for a few months when his factory ceased operations because it could not profitably produce beef while the ceiling was in effect. He had to take whatever work he could find in order to make ends meet. Leaving us for a few weeks to take a job putting up hay in Western Nebraska was necessary so that we had a place to live and food on the table. All of this was a result of the government's attempt to help us by keeping the price of meat affordable for American families. Beef suppliers ceased production, resulting in the layoff of hundreds of meat cutters. Thankfully for our family and others, the price ceilings on meat and poultry were lifted soon after it became apparent that they were not helping alleviate the inflation problem.

Effective price ceilings work completely against the stated goals of those putting in place the price controls. Government officials resorted to price controls because of complaints that people could not afford food due to inflation. The government claimed that the price ceilings would ensure that people could get the meat they wanted at a price they could afford; but the opposite occurred. Fewer people had access to meat after the price ceilings were enacted. Look again at the graph in figure 15.1. Without the price ceiling, the quantity actually sold would be Q_{Mkt} the market quantity. With the price ceiling in effect, the quantity actually bought and sold would be Q_S. Although buyers would like to buy more than that at the ceiling price, sellers are not willing to sell any more, so the quantity that is actually bought and sold is Q_S. This quantity is less than the market quantity. Instead of allowing everyone to buy as much meat as they want at a price they can afford, the effective price ceiling ensures that it is actually harder for people to buy what they want. The only thing ensured is that there will be frustrated buyers who cannot relieve their frustration by bidding up the price in order to ensure that they are the ones who will be able to buy the goods for which they are willing to pay.

An immediate and exacerbated shortage is a certain result of an effective price ceiling. However, as they say on those television infomercials, "But wait . . . there's more." Shortages are not the only consequences that result from effective price ceilings. There tend to be long lines of customers found wherever there are goods sold with ceiling prices on them.

These long lines form because the goods covered by the price ceiling legislation are scarce. The quantity of goods in existence is not enough to meet everyone's ends. Because goods are scarce not everyone can obtain a particular good they desire. Not everyone who would like to have a 1958 Cadillac, for example, can get one. Therefore economic goods must be rationed in some manner.

In a free market, economic goods like 1958 Cadillacs are rationed to those willing to pay the market price. This rationing process is what occurs as the more eager buyers bid up the price of a good, preventing excess demand. As the price of the good increases, the number of potential buyers falls and the less eager buyers turn from purchasing the good in question and look for a less expensive substitute.

In a regime of effective price ceilings, however, this price adjustment process is made illegal. The goods that have price ceilings placed upon them are no longer able to be rationed according to the price system. As we have seen, this does not automatically make them abundant. They are still scarce and still need to be rationed. They are merely rationed by another means. When the price system is not allowed to function, a common way for goods to be rationed is by waiting. People cue up and stand in line waiting for whatever is available to come on the shelves.

Also sellers can evade price controls by charging higher prices for ancillary activity, such as customer service. This is a less efficient method of rationing goods than free-market prices, but more efficient than queues. Such evasion is not illegal, but it pressures state officials to extend intervention to cover such ancillary activity as well.

Another form of rationing that sometimes takes place in lieu of the price system is an official quota system. During World War II, the Roosevelt Administration did not want to allow the prices of food and gasoline to freely adjust to wartime market conditions. Instead, the federal government enacted price controls and issued ration coupon books so that in order to buy certain goods people had to redeem a ration coupon. This limited the quantity that people could buy at the artificially lower price.

Besides shortages and rationing by waiting and quotas, an additional consequence of effective price ceilings is a decrease in quality. Because an effective price ceiling keeps the price at which sellers can sell artificially low, many sellers have difficulty breaking even on their investment. As we saw above, this very problem led some chicken farmers to kill their baby chicks after price controls were placed on poultry in the early 1970s. A less drastic response is for sellers to do what they can to reduce their costs.

One way to do this is to scrimp on quality. If an effective price ceiling is placed upon a manufactured good, then perhaps the seller will settle for second or third grade inputs instead of the highest grade like he would have used without the price ceiling. Businesses engaged in renting goods to the public may spend less money on maintenance. Customer service may decline.

Finally, if an effective price ceiling is left in place long enough for shortages to exacerbate, people often resort to what is called the underground economy. The underground economy is the network of voluntary exchange that is also illegal. Because such exchanges are illegal, this network is often referred to as the *black market*. In response to the shortage generated by effective price ceilings, some sellers take the risk inherent in breaking the law in order to supply the more eager buyers who are willing to pay a price above the legal ceiling. Such black market practices, although illegal, expand voluntary exchange. Effective price controls also often breed corruption as buyers or sellers may bribe price commissioners and law enforcers to look the other way as black market exchanges are being transacted.

RENT CONTROL

One of the most common forms of price ceilings is rent control. Rent control is a maximum legal price that landlords can charge for renting apartments covered by the price control regulation. The motivation for enacting rent control is always the same: to ensure affordable housing for those who cannot afford to rent or buy housing at market prices. Rent control is usually a policy embarked upon in urban areas where housing prices tend to be higher than in non-urban settings.

Wherever rent control has been mandated and enforced, the results have been as economic theory predicts. The quantity of apartments supplied decreases. Construction of new apartments slows or decreases. Apartments that cannot be rented at high enough prices to pay for maintenance will not be rented. Often landlords attempt to sell their apartment buildings as condominiums. Rent control obviously decreases their present capital value as apartment buildings; but rent control does not have the same constraining effect on a building's present value if it is sold as a condominium because rent control does not apply to condominiums. Therefore, there might be an increase in the supply of condominiums but a decrease in the supply of apartments for rent.

At the same time, the quantity of apartments demanded will be greater because the rent is held artificially low. There emerges inflexibility in buyers because people who are renting apartments with artificially low rent do not move out of their apartments. Tenants tend to hoard their apartments. Often apartments are mismatched with families. Small families hang on to their large apartments while large families crowd into small apartments because they are the only ones they can find.

An immediate shortage of apartments results. The shortage will be exacerbated as builders cease building apartments and convert existing apartments into condominiums, and as existing tenants hoard apartments. Additionally, when rent control is in force we also tend to see a decrease in the quality of apartments for rent. Landlords who attempt to supply the market quantity at the controlled rental ceiling can only do so by reducing costs, so they often cut back on building maintenance. In New York City, a heavily rent-controlled metropolis, one apartment building deteriorated so badly that one corner of the structure collapsed. Not surprisingly, soon afterward a *New York Times* editorial decried the fact that people were allowed to rent in such an unsafe building, criticizing both the landlord and city authorities for not enforcing safety regulations properly.

What the writer of the editorial failed to realize, however, is that one reason that landlords rent apartments in such rundown buildings and the reason why renters accept such dismal living conditions is that rent control makes it much harder to do otherwise. The reason that a rent controlled apartment is often not well maintained is not necessarily that the landlord is a greedy pig. He very well may be, but an improperly maintained apartment building does not prove that he is. It is difficult to see how a greedy, profit-hungry landlord could sustain profitability if his buildings fall down. He might merely be providing the highest quality housing he can afford to maintain while being forced to rent at the artificially low price. Likewise, renters may feel that although the apartment they are renting is in horrible condition, it is their best option because others are simply not available.

Landlords can attempt to evade the legal price maximum through evasion, which is not illegal. For example, suppose the market rent for an apartment is $2000 a month, but the city imposes a price ceiling of $1500 a month. The landlord could charge $1500 for the apartment and another $500 for a parking space, or charge $500 a month for furniture and offer no unfurnished apartments.

Black markets and corruption also often accompany effective rent controls. Because of the shortage of apartments that invariably arise due to rent control, there is usually a waiting list to get into most apartments. This is why the apartments tend to be hoarded by their tenants in the first place. As an apartment becomes open, a prospective renter may bribe the landlord or municipal authority to be pushed to the front of the waiting list. Some provide a little encouragement to the landlord to allow them to rent ahead of someone else, by paying a little something extra under the table so the actual amount of money is closer to the market rent. This, of course, is illegal.

In many cities, some apartments are not regulated by the rent control law. This tends to generate two housing markets: regulated and unregulated. Shortages will develop for regulated apartments, while demand increases for unregulated apartments as frustrated renters cannot find enough regulated apartments at the ceiling price. This increased demand raises the rent for unregulated apartments above what it would be if no apartments were regulated. Consequently, the average rent for all housing is raised close to a completely unregulated market price.

Once again, we find that the consequences of rent control are exactly the opposite of the stated goals of policy makers. Instead of providing affordable housing for all, a shortage of apartment housing is generated. Many people cannot rent housing at all. If this shortage is absorbed by an unregulated apartment sector, the prices of these apartments tend to rise so that the average rent is similar to the free market rate.

INEFFECTIVE PRICE CEILINGS

The other type of price ceiling is called *ineffective* because this type of price control does not have an effect on the price that is actually paid and received in the market. A price ceiling is ineffective if the maximum legal price is at or above the market price. Such an ineffective price ceiling is illustrated in the graph in figure 15.3.

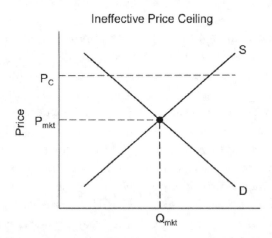

Figure 15.3. A price ceiling fixed at or above the market price has no effect on the market.

The reason a ceiling price above the market price is ineffective is that a price ceiling is only a maximum legal price. It is not the price that must be charged, but merely the highest price that can be legally paid. What would happen, for instance, if the price for a good that has a price ceiling placed upon it is at a level denoted by P_C? The quantity supplied is greater than the quantity demanded; the market is in a state of excess supply. Whenever this occurs, there are frustrated sellers who are very willing to sell their goods at this price but cannot because there is not enough quantity demanded.

In a free market, an excess supply is not permanent, but goes away as the more eager sellers bid down the price until everyone who wants to sell can sell. The price falls until the quantity supplied equals the quantity demanded. Because a price ceiling is a legal maximum, there is nothing preventing the price from falling lower than the ceiling. Therefore, if the price would ever be at P_C and the market price was lower than that, more eager sellers would be allowed to bid down the price so that the price that buyers actually pay and sellers actually receive is the market price. Therefore, if the mandated price ceiling is above the market price, nothing in the market changes. The price ceiling is, in this case, ineffective in changing the market outcome.

Consequently, all of the consequences of price ceilings described earlier are only the result of price ceilings *below* the market price. Even though so much of what politicians do is only symbolic, they do not like to leave the impression that what they are doing is in vain. They need to demonstrate to their constituents that they are working for them. Therefore, rarely do governments enact price ceilings they know will be ineffective. They have an incentive to only enact price ceilings that will have a noticeable impact in the market. The actual effects of their price ceilings are perverse. They can tell their constituents that they are working at keeping their prices down and blame wicked businesses for failing to provide constituents what they desire.

PRICE FLOORS

The second type of price control that governments impose on the market is a price floor. As the name implies, a price floor is a minimum legal price. Just as a ceiling constrains how high something can go, the floor constrains how low it can go. If you were to take this book and throw it down as hard and as far as it could go, the floor stops it from going any lower. A price floor is the lowest price that buyers can legally pay and that sellers can legally receive. Just as there are effective and ineffective price ceilings, there are effective and ineffective price floors.

An effective price floor affects the price actually paid and received on the market. As can be seen from the graph in figure 15.4, a price floor will be effective if it is greater than the market price.

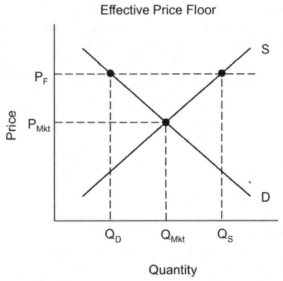

Figure 15.4. An effective price floor is a minimum legal price fixed above the market price. It always results in a surplus.

In this case, the price floor P_F is above the market price P_{Mkt}. The graph shows us that such a price floor will indeed have a noticeable effect.

The immediate effect of such a price floor is a surplus. The demand and supply curves tell us that at the price floor the quantity of the good demanded is Q_D and the quantity of the good supplied is Q_S. At the floor, the quantity supplied is greater than the quantity demanded. This results in excess supply. In a free market, there is no need to worry about such a state of affairs because it never lasts long. Excess supply results in frustrated sellers, to be sure, because some suppliers who are willing to sell at the price floor but cannot because there is not enough demand. Consequently, in an effort to ensure that they are the ones who actually do get to sell, the more eager sellers will bid down the price. As the price falls, the less eager sellers leave the market. At the same time the lower price increases the quantity of the good demanded. The more eager sellers will continue to bid down the price until everyone who wants to sell can sell. This occurs at the market price, the price at which quantity demanded equals quantity supplied.

Government intervention in the form of the price floor, however, does not allow this adjustment process to take place. The price floor is the minimum legal price that can be charged. If sellers tried to bid the price lower, they would be in violation of the law. If buyers tried to pay a price lower than this, they could be fined or incarcerated. Because the price floor is above the market price, sellers cannot bid down the price to the market clearing level. The excess supply does not wither away, but remains as a perpetual surplus as long as the price floor is above the market price.

The extent of the surplus is determined by how far above the market price the floor is mandated and the elasticity of demand for the good. Not surprisingly, the larger the positive gap between the market price and the price floor, the greater the resulting surplus will be. Additionally the more elastic is the demand for the good, the greater will be the surplus. This can be seen in the graph in figure 15.5:

Figure 15.5. The more elastic is demand, the greater the surplus that results from an effective price floor.

On this graph there are two differently sloped demand curves. The more elastic curve is flatter and marked D_E. The more inelastic curve is labeled D_I. To calculate the extent of the surplus resulting from the price floor, we must find the difference between the quantity supplied and quantity demanded. To find the quantity supplied and demanded, we must look at where the price line crosses the relevant supply and demand curves. When demand is more elastic, the relevant demand curve is D_E so the quantity demanded at the price floor is Q_E. There is only one supply curve, so the quantity supplied at the price floor is Q_S. The extent of the surplus in this case of more elastic demand, then, equals $Q_S - Q_E$.

To find the surplus in the case of more inelastic demand, we do the same thing, except that we use the relevant demand curve D_I. In the more inelastic case, the extent of the surplus is equal to $Q_S - Q_I$. We can see by the graph that $(Q_S - Q_E)$ is greater than $(Q_S - Q_I)$. This makes good economic sense. The more elastic is demand, the more responsive buyers are to a change in the hypothetical price of a good. Consequently, the decrease in quantity demanded as a result of an artificially high price would be greater if demand is more elastic. The greater the decrease in quantity demanded, the larger the gap between quantity demanded and quantity supplied, hence the greater the surplus will be.

The surplus resulting from an effective price floor becomes more pronounced the longer the price floor remains in place. As people become aware that the price of the good is being held artificially high, they react accordingly. Because the selling price is artificially high, investors are attracted to the market with the price floor. This increases the supply of the good further. At the same time, an artificially high price discourages buyers from buying as they seek out other, less expensive substitutes. Consequently, the longer an effective price floor remains in place, the more intense will be the results.

MINIMUM WAGES

One of the most common forms of an effective price floor is the minimum wage. The minimum wage is the lowest wage that firms can legally pay workers. It is often touted as an easy way to help the working poor. A common complaint heard from proponents of minimum wage increases is that a family cannot be supported by a person working full-time at the low wages firms want to pay. Therefore, it is argued, to raise such families out of poverty a minimum wage is necessary. The economic theory of effective price floors reveals that the effects of the minimum wage are largely the opposite of the stated desires of its supporters.

In order to get the analysis straight, we must keep in mind what market we are investigating. Not every worker earns the minimum wage because different workers have different marginal value products. Those workers who earn the lowest wages tend to provide the least productive labor because of a lack of capital goods or a lack of job skills or a combination of both. Most low paying jobs are entry-level and given to people who are entering the labor market for the first time. Consequently, the impact of a minimum wage will be felt the most in the market for lower skilled labor. This is labor that is primarily provided by teenagers and young adults without a high school education. Consequently, when examining the effects of a minimum wage on the market for labor, we must remember that we are discussing the market for lower skilled labor and not the market for specialized, highly-skilled labor.

The effects of a minimum wage on the market for lower skilled labor are what we should expect. An effective minimum wage results in a surplus of lower-skilled labor. This is because a minimum wage increases the cost of hiring a worker without increasing his productivity. Remember that the demand for labor curve is synonymous with the DMRP curve, which is determined by labor's MRP and the interest rate. An increase in the minimum wage does not affect the stock of capital goods or the quality of labor so labor's MRP remains the same. Increasing the minimum wage does not alter time preferences or interest rates either so a change in the minimum wage does not affect the DMRP for labor. Consequently, the demand for labor remains constant.

What does change if an effective minimum wage is raised is the expense to the firm; hiring a lower skilled worker at an artificially high wage necessarily increases labor costs. As the price of labor increases, firms can only afford to hire a smaller quantity, so the quantity of labor demanded falls. At the same time, the number of workers willing to work at the higher wage increases. There is often an increase in the high school dropout rate quickly following an increase in the minimum wage as some teenagers leave school in the hopes of finding a job at the higher wage.

Suppose that there are two workers, Jacob and Emily. Jacob's DMRP is $7 an hour and Emily's is $8.10. If the market wage is $6.00 an hour and there is a free market, both workers will be hired because both will be profitable to their employers. Both of their DMRPs are greater than the market wage rate. Now suppose that, due to political agitation, Congress increases the minimum wage to $8 an hour. Jacob is no longer employable, because his firm loses $1 for every hour it employs him. Emily will remain employed at the higher wage. We see that as a result of the higher wage the quantity of labor demanded decreases. The quantity of lower skilled workers that are actually hired is less than there would be without the minimum wage.

What happens with an effective minimum wage is illustrated in the labor market graph in figure 15.6.

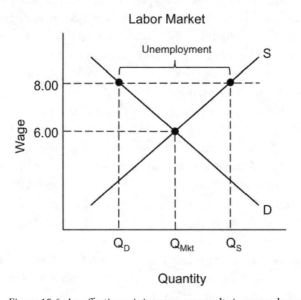

Figure 15.6. An effective minimum wage results in unemployment.

In a free market, the equilibrium wage of $6.00 an hour would be paid and the quantity Q_{Mkt} would be hired. At the minimum wage of $8.00 an hour, the quantity of labor demanded, Q_D, is less than the quantity supplied, Q_S, so an excess supply develops. Because it is illegal for the more eager workers to bid down the wage they will accept, the excess supply becomes a surplus that will persist until the minimum wage is removed. A surplus of labor is what we commonly call unemployment. The quantity of labor actually hired will decrease from Q_{Mkt} to Q_D. Whenever minimum wage is enacted above the market wage rate, those who remain employed will receive a higher wage. They do so, however at the expense of those who are laid-off or are not hired to begin with. Some people benefit while the number of people hired decreases and unemployment results. A minimum wage never benefits all lower skilled workers but harms the lowest skilled workers.

Not only does an effective minimum wage result in unemployment for the lowest skilled workers in the short run, but it also places them on a lower income trajectory over time. One of the benefits that low-skilled workers, especially those without a high school diploma, receive from even low paying jobs is the development of a work ethic and job skills. These are traits that are learned on the job. As people work their first jobs, they learn more and develop character so that they become more productive, so they get promoted or hired by different firms at higher wages. By keeping the lowest skilled workers from getting a job initially, however, these people are kept from developing job skills. They are forced to play catch up indefinitely in the future.

Additionally, effective minimum wages result in a decrease in general societal wealth. If entrepreneurs are forced to pay a wage higher than they otherwise would, this forces them to allocate the available land, labor, and capital goods inefficiently. In the free market, all factors will tend to be utilized so that the price of each factor will be equal to its DMRP. If the quantity of the factor used is such that its DMRP is greater than the price of its service, there is even greater profit to be reaped by using more of that factor until its DMRP becomes equal to its price.

If an effective minimum wage is in place, the price of lower skilled labor is held artificially high, which alters the combination of labor and other factors that are utilized. Firms can cut the hours of their entry-level workers and have managers pick up the slack. They can automate certain functions faster than would be economically feasible in a free market. They might simply reduce customer service. In any event, the ratio of capital goods and land to labor will be one that is suboptimal so that the total output that is produced is less than it could be with the same quantity of factors. For the economy as a whole, this means that we must pay higher prices for fewer goods. This harms everyone, especially the poor.

Furthermore, increases in the minimum wage do not reduce poverty. In fact, the minimum wage aggravates poverty for those sub-populations who are most affected by the minimum wage. One of the main contributors to poverty is unemployment. One is most poor if he does not receive an income. If a lower-skilled person is out of work, he is less likely to earn an income, and more likely to be poor.

What can we conclude about the effect of the minimum wage? At best it benefits some workers at the expense of other workers. Some lower-skilled workers will get paid more, while the lowest-skilled workers are left unemployed. Even higher-skilled workers can benefit from effective minimum wages, however. Unionized forklift operators, for example, benefit when stock boys are kept out of work because of above-market minimum wages. Those who are kept out of a job earn lower incomes over time because it takes them longer to develop job skills. Society in general is worse off because they have fewer goods available for which they must pay higher prices. Once again, the result of government intervention is exactly the opposite of what proponents of such intervention say they intend. An effective minimum wage does not reduce poverty. Rather, it places obstacles in the path of lower skilled workers in their quest to improve their lot in life.

INEFFECTIVE PRICE FLOORS

It is possible for price floors to be ineffective in the sense that they have no effect on the market. Such price floors do not alter the price actually paid by buyers and actually received by sellers.

An ineffective price floor is one which is set at or lower than the market price, as illustrated in the graph in figure 15.7.

Figure 15.7. A price floor fixed at or below the market price has no effect on the market.

We see there that the price floor P_F, is less than the market price P_{Mkt}. What is the consequence of such a price floor? To answer this question let us begin by supposing that the price in the market falls to the floor. If this occurs the market would be in a state of excess demand. We know this by comparing the quantity demanded Q_D, and quantity supplied Q_S, at the price floor. From the graph it is easy to see that the quantity demanded is greater than quantity supplied. Therefore, if the price would be at the floor price there would be excess demand resulting in frustrated buyers.

In a free market, when there are frustrated buyers due to excess demand, the more eager buyers will bid the price up until everyone who wants to buy can buy. As the price is bid up, the less eager buyers will drop out of the market for this good, searching out less expensive substitutes to buy instead. At the same time more sellers are willing to sell a greater quantity as the price rises. Therefore, as the price is bid up, the excess demand shrinks until it disappears completely at the price at which quantity demanded is equal to quantity supplied.

This is the pricing process that would take place in a free market. You might object, however, that this market is not free because of the government mandated price floor. You would be right. However, remember that a price floor is not the single price that must be charged but merely the lowest legal price. Buyers can pay any price greater than or equal to the price floor that they want. Sellers can charge any price greater than or equal to the price floor that they want. Therefore, if there would be excess demand at the price floor, the floor price would not be the price actually charged in this market. Because it is legal for the price to be above the floor, the price adjustment process is allowed to go forward in this case and the price actually paid and received is the market price. Therefore, the price floor is superfluous. It has no effect on the market and is therefore labeled ineffective.

I witnessed such a market in action the summer following my sophomore year in college. At that time that national minimum wage was $3.35 an hour. This was quite effective in the state of Iowa where I lived and attended college. In hopes of gaining a higher income during the summer, I took a job painting houses in Westchester County, New York. I was paid $5.00 an hour for my labor. When I got out there, I saw numerous signs in the windows of fast food restaurants, such as Kentucky Fried Chicken, Burger King, and McDonald's, recruiting new workers for the starting wage of . . . you guessed it, $5.00 an hour. What struck me is that in the lower skilled labor market in New York the national minimum wage was entirely meaningless. If fast food places and other employers of lower skilled labor tried to pay their people $3.35 an hour, they could not hire as many employees as they wanted. There would be excess demand. Therefore, in order to hire the optimal number of lower skilled workers from their point of view, businesses had to pay the market wage of $5.00 an hour.

This leads us to the conclusion that minimum wages lead to unemployment only if they are effective. In the Midwest the national minimum wage was effective and led to fewer lower skilled people hired. In New York at the time I was painting houses, the minimum wage did not lead to unemployment. A minimum wage either leads to unemployment of lower skilled workers or it has no effect. Neither are great options for a policy that is supposed to help the poor.

PROGRESSIVE INTERVENTIONISM

The case of price controls is also useful for illustrating a principle of political economy, meaning economic policy as it is actually worked out by politicians and bureaucrats. This principle is progressive interventionism and was explicitly brought to light in the work of Ludwig von Mises.

Effective price ceilings result in shortages due in part to the fact that many producers cannot profitably operate by selling their output at the ceiling price. In the 1970s poultry farmers killed their baby chicks and my father was laid off as his beef plant shut down. Because such measures taken by businesses are rather drastic and relatively unpopular with their employees and with the public at large, they might seek other ways to stay afloat.

One option is to lobby for even more price ceilings on the markets for factors of production. Suppose that there is an effective price ceiling placed on milk because politicians hope to ensure that all citizens, especially the poor, will be able to afford the milk necessary to form strong teeth and bones. This generates the expected shortage and milk producers begin shutting down their production facilities. Once it becomes apparent that the dairy farmers face substantial losses, they want to do what they can to avoid taking it on the financial chin. Farmers could merely shut down operations, but this only cuts their losses, it does not make it profitable to operate. They are still not gaining any return on the capital they have sunk in their dairy investments.

In light of this, the American Dairy Council lobbies Congress explaining that if something is not done, dairy farmers will have to cease milk production because they simply cannot afford to keep producing. Congress responds by placing effective price controls on the factors needed to produce milk. The prices of feed and automatic milking machines are held artificially lower. This allows dairy farmers to now maintain the same rate of return as before. But now the squeeze is placed on the manufacturers of feed and milking machines. Their producers lobby for help in the form of price controls on their factors of production. There is an incentive for this process to spread throughout the economy, until every market is regulated by effective price ceilings. If this scenario was fully played out, the economy would be in *de facto* socialism.

Like the economy of Nazi Germany, such a scenario would still have the appearance of a market economy, but because all of the prices would be constrained by the state, the entire economy would in effect be centrally controlled by the state. Although the policy makers start out with only a little intervention, its negative consequences result in pressure for politicians to apply even more intervention. As Ludwig von Mises explained, "Middle-of-the-road-policy leads to socialism."[1] The only alternatives in response to the negative results of price controls (or any intervention for that matter) are to remove the intervention and return to a free market, or move forward toward socialism. There is no stable third-way economy in between the free market and socialism. The economy is always moving toward one or the other.

THE ETHICS OF PRICE CONTROLS

Often people desire price controls for what can be considered laudable ends. Christians especially should feel compassion when others less fortunate struggle to feed and house their families because of high food prices and rents or because they earn very low wages. Recall from chapter 4 however, that Christian ethics require that we not resort to acts of violent aggression in order to alleviate the suffering of the downtrodden. Notwithstanding any good intentions on the part of government price regulators, price controls are not merely economically futile; they violate the Christian ethic of private property and hence should not be embraced as a Christian means to achieve desired ends.

1. Mises, "Middle-of-the-Road Policy Leads to Socialism," 18–35.

Both price ceilings and price floors prohibit people from voluntarily exchanging their property. Price ceilings on consumer goods prevent some consumers and producers from engaging in peaceful exchange of their property. Rent control keeps some potential renters from renting apartments, while others benefit. Price floors likewise, hamper the voluntary exchange of private property. Minimum wage laws prohibit some workers and entrepreneurs from making mutually beneficial work agreements. Some workers, who are not productive enough to warrant employment at the artificially high wage, are prohibited from exchanging their labor as they see fit. Some workers benefit at the expense of the least productive. Because price controls violate the Christian ethic of private property, they must be rejected as a means of achieving a better society.

The Christian property ethic does not allow for using the power of the state to keep people from entering into voluntary contracts. Christians are called to be charitable to those who need it and to be merciful to the poor and to see that widows and orphans are not oppressed. Justice demands, however, that people use their own resources to help those in need and to convince others voluntarily to do so. Lobbying for government price controls violates the right to property and therefore violates the Christian ethic.

THE FAILURE OF PRICE CONTROLS

When a government enacts price controls in an effort to solve various economic problems, it inevitably fails. If it mandates a price ceiling above the market price or a price floor below the market price, its controls have absolutely no effect on the market. Effective price controls, however, result in a plethora of negative consequences and even fail to achieve the stated goals of their proponents.

A price ceiling fixed below a good's market price will result in a shortage, not an ideal market where everyone can afford to buy the good in question. The quantity of the good demanded will increase, but the quantity of the good supplied will decrease. The amount of the good actually traded will fall. The shortage will be worse the longer the price control is in force as producers reduce production and buyers rush to make their purchases. In affected markets, the quality of the goods will tend to decline as manufacturers try to cut production costs. Goods increasingly will be rationed by waiting and even black markets may develop.

A price floor set above the market price likewise will have negative consequences. Such a price floor will cause a surplus as the quantity supplied will be greater than the quantity demanded. The quantity bought and sold will actually decrease. Resources will be wasted as factors of production initially are directed into those markets by the siren song of artificially higher prices. If the price floor is a minimum wage, unemployment will develop and lower-skilled workers will be hindered from gaining valuable job skills because it will be harder for them to enter the market. Some workers will receive a higher wage, but only at the expense of those laid off or not hired to begin with.

Price controls do not work to increase the number of ends people can satisfy. Price controls inevitably result in preferences being frustrated as voluntary exchange is hampered. If that were not bad enough, price controls violate the Christian ethic of private property. Therefore, such economic policy is neither economically nor ethically sound.

SUGGESTED READING

Beisner, *Prosperity and Poverty*, 161–74. An excellent exposition of the ethics and economics of price controls.

Block, et al., "Rent Control," 253–63. A very good article explaining the economics and some of the ethics of rent control.

Mises, *Human Action*, 752–73. Mises' definitive account of the economic consequences of price controls.

———, "Middle-of-the-Road Policy Leads to Socialism," 18–35. In this essay, Mises explains his theory of progressive interventionism.

Ritenour, Shawn, "What You Should Know About the Minimum Wage." An exposition of the nature and consequences of a government minimum wage above the market wage.

Rothbard, *Man, Economy, and State*, 892–900. Rothbard's excellent treatment of the economics of price controls.

16

Taxing and Spending

IN THE PREVIOUS TWO chapters we discussed the nature and consequences of two forms of government intervention in the market. The
first was inflation and the second was price controls. The state, however,
also has a great impact on the market through its budget. Government
taxing and spending have serious economic consequences.

Every year the state spends funds on whatever it decides is worthy.
There are two primary reasons that the government spends money. One
is to benefit particular people or groups of people. The other is because of
a belief that government spending can help generate and maintain economic prosperity and the public good.

In order to do anything, the state needs economic goods. Because the
state has a legal monopoly on the use of violence, it could just confiscate
the goods it needs from the citizenry. However this method of acquisition
is extremely unpopular and increases the likelihood of rebellion; it is also
somewhat inefficient because the state would have to know who had what
goods and some of the goods that it wants (such as tanks and anti-aircraft
artillery) may be very hard to locate.

Therefore, the state finds it advantageous to operate within the monetary economy. To function in the monetary economy, government needs
money. The state must have access to funds that it can spend trying to
achieve whatever ends it has. If the government wants to enforce price
controls, if it wants to provide military defense or police services, if it
wants to support artists or scholars by subsidies, or if it gives out educational welfare in the form of Pell grants and subsidized student loans, if
the state wants to accomplish anything, it must spend money.

In order to spend money, the government must first obtain revenue.
The path that the state takes to receive revenue is intervention in the free

market, obtaining funds through coercion. There are three sources of revenue available to the government: inflation, taxation, and borrowing.

We have already made a fairly extensive inquiry into the nature and consequences of inflation. We have seen that inflating the money supply does not provide a general social benefit. While people have more money, they must also pay higher prices. Inflation merely redistributes wealth to those who receive the new money earlier at the expense of those who receive it later or not at all and generates the business cycle, culminating in recession usually accompanied by unemployment. This chapter, however, will be devoted to deriving economic principles relevant to taxation, government borrowing, and government expenditures.

GENERAL EFFECTS OF THE GOVERNMENT BUDGET

Although we can conceptually distinguish between government taxing and spending, in practice the two are very closely linked. In fact, taxing and spending are different stages in the same total process that results in the government's burden on the economy. First money is taken from some people. This taking is always backed up by the threat of violence. Then the government uses the tax revenue to purchase whatever it wants. Both stages contribute to the total burden of the process on the economy. We will first focus on the general effects of the government budget on the economy and then proceed to a more detailed investigation of the more specific effects of taxation and then government spending.

To understand the nature of taxation, it is crucial to define what we mean by taxation. A tax is a coerced levy paid to the state. As such, all taxation is a forced exchange between the citizen and the government.

Some might object to the language of coercion when applied to taxation, arguing that some willingly pay their taxes as a moral duty. Christ's words, "Render to Caesar the things that are Caesar's" (Matt 22:21), and Paul's admonition to "Pay to all what is owed to them: taxes to whom taxes are owed" (Rom 13:7) come to mind. Notwithstanding people's willingness to submit to the powers that be, it is still proper to view taxation as coercive, because no one has the legal right not to pay what the state says is owed. As mentioned above, the government's demand for taxes is always backed up by the threat of violence. Try not paying your income tax and see what happens. You will be fined and perhaps imprisoned; your property can be confiscated and sold to pay your tax bill. No voluntarism here. The essence of the coercive nature of taxation is communicated by the etymology of the word *tax*. It is derived from the Latin word, *taxare*, which means, "to touch sharply."

All forms of taxation do what all forms of government intervention do: separate people into two classes. There are those who benefit from the intervention and those who are harmed. Taxation takes money from some citizens and gives it to others. Therefore, as noted by John C. Calhoun in his *A Disquisition on Government,* government taxing and spending divides people into tax payers and tax consumers. Tax payers are those who are net payers of government taxes. Tax consumers are net beneficiaries of the tax. Tax consumers benefit from taxation, while tax payers bear the cost.

Who are the tax consumers? They are mostly those who fully rely on government tax revenue. They include government officials and bureaucrats as well as those receiving government subsidies. Some might be wondering why we call government bureaucrats tax consumers when they must pay taxes too. Why are they not also considered tax payers? The reason is that we define tax consumers and tax payers in an economically meaningful way.

Suppose, for instance, that an associate commissioner of some relatively minor government agency is paid a salary of $100,000. Suppose that all of his taxes come to $25,000 or 25% of his income. While it is understandable that some might mistake him for a tax payer, he is really, a tax consumer. He is so because all of his salary is funded by tax dollars coercively taken from other people. He is paid $100,000 out of tax money and at the end of the year must pay $25,000 back to the government. The associate commissioner keeps a net income of $75,000. This income is funded entirely by tax dollars. Consequently, while the bureaucrat's net income paid from tax revenues is smaller than if he would not have to pay any tax, on net he is still a tax consumer to the tune of $75,000 a year. He is really merely reimbursing some of his tax income. He *receives* $75,000 in tax revenue.

At the same time, tax payers are those who pay more in taxes than they benefit from government payments. Suppose that a relatively successful artist receives an income of $50,000, $1,000 of which came in the form of a grant from the National Endowment for the Arts. Come tax time, we find that his total tax bill is $10,000. In this case, our artist is a net tax payer. The artist paid $10,000 and received a $1,000 subsidy that was funded by tax money. Therefore, his total tax bill is $10,000 - $1,000 or $9,000. While the arts subsidy partially offset his tax burden, he was still a net tax payer. The associate commissioner mentioned above benefits at the expense of our artist. Intervention through taxation necessarily creates castes that live in conflict with one another.

Taxation has two primary economic consequences. One is that it alters the way resources are directed by changing the allocation of scarce factors of production and consumer goods. Therefore, taxpayers are hindered from best satisfying their ends. The state forces them to give up a fraction of their income. Taxpayers' incomes become smaller so they are able to purchase fewer consumer goods. Their standard of living falls while resources are allocated away from the consumers and toward fulfilling the ends of state officials. Meanwhile tax consumers have wealth redirected toward them.

Suppose that the government makes the Masterpiece Record Club a taxpayer to the tune of $500. The Masterpiece Record Club pays a net tax of $500, which is money it would have spent purchasing Bach records that the club would have advertised to its members. The tax revenue finds its way into the budget of the Bureau of Labor Statistics (BLS), a part of the Department of Labor charged with collecting wage and price data, including the CPI. Believe you me; the BLS is a tax consumer. One of the goods the BLS buys with its funds is ink pens that it issues to its field economists who collect the raw wage and price information throughout the country.

What affect does such a process have on the producers of Bach CDs? Because the Masterpiece Record Club has its $500 taxed away, they are unable to purchase the Bach recordings. The demand for Bach CDs falls, which will result in a reduction in the market price for such recordings. This reduces the profit that can be reaped from Bach CD production. Investment in the Bach recording industry decreases along with incomes earned in that industry.

The results in the pen industry are exactly the opposite. Because the government is spending more on pens, the demand for pens increases. As we have already learned, if the demand for pens increases, the market price for pens will also increase. This higher price for pens will allow for greater profits in the pen industry attracting increased investment by entrepreneurs. As entrepreneurs spend more money on factors used to produce pens, incomes in the pen industry rise. The full consequence of the $500 tax is that the pen industry benefits at the expense of those in and closely connected to the Bach recording industry.

Consequently, we see that the taxing and spending that is inherent to the government budget necessarily distorts the allocation of resources, consumer goods, and incomes from what they would be in the free market. Because the free market is merely the network of voluntary exchange, the government budget process directs wealth in ways other than those actually wished by members of society. If people really wished that an additional $500 would be taken from record clubs and spent on pens by statistical agencies, they would voluntarily spend the money on pens and the governmental process by which this happens coercively would be entirely superfluous.

In fact, both taxing and spending by the government contribute to the total government burden on the economy. In our example, the recording industry essentially is forced to pay for the removal of pens from the private economy into the hands of government bureaucrats. Additionally, all pen consumers must pay higher prices for pens due to increased government demand.

The general effect of the government budget is to coercively redistribute resources, consumer goods, and incomes from the productive to the nonproductive, from tax payers to tax consumers. At the same time government taxing and spending interferes with the process of voluntary exchange, which is made more difficult for both buyers of Bach records and non-governmental purchasers of pens.

The larger the government budget is, the more redirection of resources, consumer goods, and incomes there is, and hence, the greater impact on the economy. As the government budget increases relative to free market activity, more and more resources are being directed toward government consumption. At the same time, the base of productive activity—the source of all government taxation—becomes increasingly narrow.

A second related general consequence of the government budget is that it severs production from distribution. In the free market there is no separate question regarding the *distribution of income*. In fact, in the free market the very term—distribution of income—itself is misleading. The very phrase income distribution implies that everyone's income is passed out by a central authority as if an income czar monitors the wealth that is contributed to the Society Chest and then doles it out as the he sees fit.

This concept is totally foreign to the free market, in which income is earned by providing productive services. In a free society, workers earn income by selling their labor services; landowners earn income by renting or selling their land; capitalists earn interest income by supplying present money in exchange for future money; entrepreneurs earn income in the form of profits by more successfully serving future customer demand better than anyone else. In order to receive an income in the free market, people must serve others.

The government budget process, however, cuts the link between productive service and income distribution, allowing some to receive incomes without contributing to society. Tax consumers live off tax payers. Government taxing and spending enable some to receive incomes without contributing anything to society. Those who do not make a net contribution to society get paid money coercively taken from others. This income is distributed to the beneficiaries. The extent of the misdirection of resources is directly related to the total government taxation and spending. Consequently, and as we will see later, the total level of taxes and government spending determines the extent of budget impact on the economy and not the type of tax or spending.

THE ECONOMICS OF TAXATION

Having discerned the general economic consequence of the two-stage budget process, we now examine in greater detail the effects of taxation on the tax payers. Because of its pervasiveness, we first will discuss the economics of the income tax. As the name implies, an income tax is a coerced levy on the taxpayer's income. All income taxes create a disincentive for productive activity. This is seen by examining how taxation affects the different sources of income.

Let us begin by examining the effect of a tax assessed on income generated by the sale of services by an original factor of production. What is the effect of taxation of income earned by laborers? As the income tax increases, the marginal benefit of working decreases because each hour earns less as more of the worker's hourly wage is taxed away. In general, fewer workers offer their services in the labor market and those that do are willing to work less.

Some have argued that higher taxes can actually provide a boost to the economy, because people will want to work longer hours to make up for their lost income that is taxed away. It is possible that some people may indeed work more hours in response to a higher income tax, especially if they are in debt for fixed amounts that do not decrease as their taxes increase. However, this does not mean that these workers are better off than before the tax was raised. Everyone when deciding whether to work at a job weighs several factors including the income that will be received versus the value of leisure time. Leisure time is a good that people desire as well as monetary income.

As a result of a hike in the income tax, if someone decides to work more hours in order to maintain a relatively stable monetary income, he is still worse off than before the tax increase. Suppose a lady living in a tax-free economy earns $50,000 for working 40 hours during a five-day work week, leaving her 80 hours of leisure time during the week. Then suppose that due to a tax increase, she now must work an extra two hours a day just to maintain her earlier income level.

The end result is illustrated in table 17.1.

Figure 17.1

Work Income and Leisure		
	Before Tax	After Tax
Income:	$50,000	$50,000
Leisure hours:	80	70

She brings in the same amount of monetary income but has less wealth, which includes leisure. If she works an extra two hours every day, she is working ten hour days resulting in 70 hours of leisure instead of 80. Before the tax increase she was earning $50,000 and consumed 80 hours of leisure during a work week. After the tax hike, she still earns $50,000, but has only 70 hours of leisure. Therefore, even if somebody does work longer in response to an income tax, she is still harmed.

In general, however, fewer workers will be willing to work and the supply of labor will tend to decrease. Because of a decrease in the supply of labor, the price of labor will tend to increase so that some of the cost of the increased tax will fall on the employer. Additionally, because the labor pool would be smaller, the quantity of production would tend to decrease so that fewer goods are produced.

Similar results are found when landowners are assessed income tax. An income tax on landowners will decrease the net land rents received by landlords. As land rents decrease, so does the capital value of land that is sold outright. If this is the case, less land will be brought into productive use. More land will be left standing idle. Again, if less land is brought into productive use, fewer economic goods will be produced and societal wealth decreases.

A tax on investment income tends to reduce saving and investment on the part of capitalists. Just like an income tax decreases net wages, a tax on investment income reduces the net return on a capitalist's investment. If a capitalist invests $1000 in a project that will bring in $1,100 a year from now, his net revenue will be $100 and his rate of return will be 10%. However, a 10% income tax on all investment income requires $10 of that $100 to be given to the government; the real return to the capitalist is only $90 or 9%. Capitalists who have time preferences that require a rate of return over 9% in order to invest will drop out of the time market. Consequently, taxation on investment income reduces saving and investment.

GOVERNMENT SPENDING

The other side of the fiscal coin from taxation is government spending. We already have seen the general effects of government spending for items necessary to run the bureaucracy, such as ink pens. Sometimes, however, the state spends money not merely on goods necessary to support its activity, but to achieve additional social purposes.

An important form of this type of government spending is a *subsidy*. A subsidy is a sum of money the state spends by giving it to particular people. An example of a government subsidy would be the money that is paid to a farmer if the price of his wheat drops below a target minimum. Another example would be the grant received by a symphony orchestra from the National Endowment for the Arts. Subsidies, therefore, are a way for some people to gain income without providing a reciprocal productive service.

There are only two ways of acquiring wealth: voluntary production and exchange or confiscation by force. In a free market, people gain wealth only to the extent that they serve others or are given a gift. Government subsidies, however, are not part of the free market. They are a coerced transfer payment from taxpayers to the recipients. Consequently, government subsidies allow people to gain wealth to the extent that they influence or get control of the state. The more influence in government people have, the more they can direct the distribution of wealth at the government's disposal.

Not surprisingly, government spending greatly affects the allocation of income. Government subsidies distort the allocation of income away from those who are most efficiently serving their customers, prolonging the life of inefficient firms at the expense of efficient ones.

Imagine entrepreneurs in two different industries: shrimp production and barbecue sandwich production. The rate of return on shrimp production is 4%. The sole proprietor of a fish business invests $40,000 in his business for the year and takes in a total of $41,600, so his annual net revenue is $1,600 or 4%. Suppose, however, that in the barbecue industry entrepreneurs reap a rate of return equal to 6%. If this is the case, what would tend to happen in a free market? Entrepreneurs would tend to exit the shrimp industry and enter the barbecue industry because the available factors of production have their highest valued use, as determined by members of society, in barbecue production.

Suppose, however, that due to successful lobbying on the part of the ASP, the American Shrimp Producers, Congress passes a law providing subsidies designed to keep struggling shrimp fishers afloat. Our lone shrimp man receives an $800 subsidy increasing his total revenue from $41,600 to $42,400. That minus his $40,000 initial investment results in a net revenue of $2,400 for a 6% rate of return on his investment. Because of the subsidy, the rate of return in the shrimp industry is equal to that in the barbecue industry even though societal preferences remain the same as before the subsidy.

Consequently, there is no more incentive for entrepreneurs to exit the shrimp industry or for entrepreneurs to enter the barbecue industry. Factors of production will remain in a less efficient industry and less productive use because of government intervention. There will be more shrimp than society wants produced and less barbecue than society wants available.

Government subsidies, therefore, necessarily create class conflict. This conflict is not between the Marxist classes of labor and capital but between the beneficiaries of government largesse and those who are harmed. One person benefits at the expense of the other. The shrimp man benefits at the expense of the efficient barbecue producer. This class conflict necessarily politicizes the marketplace. As government subsidies increase and become more widespread, people have a greater incentive to influence and control government because more and more income is determined by the policies of politicians and bureaucrats.

Production and overall living standards become further diminished because time, resources, and energy are diverted away from productive activity to political activity. Those who succeed in the political arena are those with a comparative advantage in politics, gaining control of the state apparatus themselves or winning favors from those who do. These are not necessarily the same people who have a comparative advantage in the profitable production of anything.

WELFARE PAYMENTS: AID TO THE POOR

One popular form of government subsidization is what used to be called poor relief and now is more commonly termed welfare. In the United States welfare to the poor derives mainly from two programs: Temporary Assistance for Needy Families (TANF), and Women, Infants, and Children (WIC).

Although taxing people to pay for subsidies to others designated as poor is a violation of private property rights, welfare programs tend to be quite popular with Christians because they recognize the numerous times in the Bible that God calls for his people to care for the poor. Many seem to draw a straight logical line from the commands to care for the poor to government mandated welfare subsidies. God wants us to be caring toward the poor therefore we need the welfare state.

What are we to make of this argument? How do we reconcile the mandate to provide charity to the poor without violating property rights and the prohibition of theft? First, we need to discern what are our duties regarding the poor. Then we will be able to gain a better understanding of how we minister to the poor without doing evil to someone else.

God does make it clear that we are to help the poor. We are to be imitators of God and he tells us that he cares for the poor (Ps. 35:10). God tells us that the poor and orphaned are to be defended from would-be oppressors (Ps. 82:3). We definitely should not turn a deaf ear to the cry of the poor. In fact, God tells us that whoever ignores the plight of the poor himself shall not be heard when he calls for help (Prov. 21:13). God tells us that in times of trouble, he will deliver the one who has consideration on the poor (Ps. 41:1). Whoever is charitable to the poor lends to the Lord and God will repay him for his generosity (Prov. 19:17). The mandate to minister to the poor even includes our poor enemies (Prov. 25:21).

We receive similar instruction in the New Testament. When the rich young ruler asked Jesus what last thing he needed to do to be perfect, Jesus told him to sell all his possessions and give the money to the poor (Matt 12:21). In the early chapters of Acts we find the Apostolic Church ministering faithfully to those in need. Additionally, James clearly teaches that it is not enough to feel compassion on the poor, but we are mandated to provide them with real material help when they are in need (Jas 2:15–16).

© United Feature Syndicate, Inc.

What has caused confusion regarding how we are to fulfill God's requirement to show charity to the needy is a change in definition of *poverty* over the past century. Throughout the twentieth century, a linguistic shift occurred so that poverty has become defined in a relative, not absolute sense. The Biblical view of poverty, which is the view that informed much social thought into the 1900s, defines the poor *not* as those who are not rich, but as those who are so lacking in the material necessities of life, they cannot survive without charity.

As intellectuals became increasingly egalitarian in their thinking, however, the modern definition of poverty evolved to a relativist definition. One is considered poor, not if they lack the necessities of life, but if they have *sufficiently less* than other people (who are by default considered middle class or rich). Note that this relativistic definition is at serious odds with the biblical definition of poverty. When God commands us to give aid to the poor, he is not calling us to give money to someone merely because they spent their money on a $200 pair of Nike sneakers instead of on food.

From a biblical perspective, we find that there are very few truly poor people living in the United States. How can this be when we hear the statistics about how many poor people and children in particular that live in poverty? Those poverty statistics are generated based on the number of people earning an income at or below the poverty threshold. This threshold is an income amount that says little about the actual wealth that a person possesses.

The U.S. Census reveals that most of those who are considered officially poor by the government are in fact living relatively comfortably. According to data from the U.S. Census Bureau, forty-six percent of those Americans officially designated as poor in 2002, owned their own homes. Ninety-seven percent of officially poor Americans owned a television and over half owned two or more color televisions. Sixty-two percent of poor Americans had cable or satellite television service. Poor adults are more likely to be obese than the non-poor adult population. The next time you hear someone ask "Why is it, in the wealthiest country in the world, so many people have to go to bed hungry?" you can respond, "It just isn't so." There are very few citizens of the United States that are indeed poor as defined by the Bible and historically understood to be truly needy.

To say that there are few people in the United States who are poor is not to say that there are none. Certainly there are many who are truly poor, especially in other countries. What about their plight? What can we do about them? We will discuss the issue of economic development in chapter 19. Drawing upon the economic principles explained throughout this book, you will find there are three engines of economic development: the division of labor, capital accumulation, and entrepreneurship. For all of these engines to produce prosperity, the institution of private property is necessary. Without private property there can be no exchange that makes the division of labor possible, there would be no saving and capital accumulation, there would be no technology bound up in those physical capital goods, and there would be no entrepreneurs directing capital toward its most highly valued uses. Consequently, the solution to widespread poverty is to establish private property and let loose the engines of development.

Often that solution is not palatable to those who want instant fixes and egalitarian ones at that. Consequently, income transfers to those designated as poor have been undertaken by most wealthier governments. Our present investigation of the economic consequences of welfare payments to the poor will show that, like price controls, income transfer payments to alleviate poverty end up doing the opposite of the expressed goals of those advocating welfare. Government poor relief in the form of TANF or WIC payments results in the subsidization of poverty. The trouble is we get more of whatever we subsidize.

The direct effect of welfare payments to the poor is that it reduces the opportunity cost of leisure time. While income taxation reduces hourly wages and hence the marginal benefit workers receive from their labor, welfare subsidies reduce the cost of not working. Such transfer payments make leisure more attractive. Suppose that in a free market a relatively lower skilled worker finds that his only job offer is for one in which he could receive $1,500 a month working full time. In actuality a free market would most likely provide a number of employment opportunities with differing salaries. But for the sake of argument, let us suppose that this one possibility is all that a person has. In a free market, his only other alternative is to not work and earn nothing for the month. This person has a considerable incentive to accept the job paying $1,500 a month, develop a good work record, learn more skills, and strive to get promoted. As he stays on the job, he will be developing character and job skills that enable him to increase his income over time.

The introduction of welfare payments changes this scenario. Suppose that a $1,500 per month job is not enough for our friend to earn an income above the official poverty threshold, and he qualifies for government income assistance. The government provides $1,200 a month and while this is not as much as he could earn working full time, it is a lot more than nothing. The ability to receive $1,200 in welfare reduces his opportunity cost of not working. If he chooses not to work, instead of receiving no income he receives $1,200, so not working costs him a total of $300 instead of $1,500. While not every lower skilled worker would immediately jump for the government dollar, some of the marginal workers will succumb to the temptation to opt for welfare instead of working full time. They might be willing to take a $300 a month pay cut if they can stop working all together.

Consequently, instead of winning Lyndon Johnson's War on Poverty, welfare transfer payments actually exacerbate the very problem they were intended to solve. Income assistance provides an incentive for people to engage in the very behavior that results in poverty to begin with. Because of the incentives inherent in such transfer payments, idleness and poverty actually increase. Because welfare reduces the opportunity cost of not working, the number of poor people who decide not to work is higher than it would be without welfare. Consequently, their incomes remain lower than would be otherwise, and such recipients do not develop the work ethic and job skills necessary for occupational and salary advancement.

This analysis has been born out in the economic history of the United States during the twentieth century. While official poverty rates declined throughout the first half of the 1900s, the drop in poverty ceased and the rate became relatively stable soon after the United States began its War on Poverty as part of Lyndon Johnson's Great Society legislation.

Not only do institutional income guarantees exacerbate the very problem they purport to solve, they also has grave social consequences. One of the most popular justifications for the welfare state is that we must do something for the children. The repetition of the number of children in poverty successfully plays on the emotions of voters and notwithstanding the effects of welfare on employment, people acquiesce to growth of the welfare state. Voters reason it is not the children's fault if the parents do not work or do not know how to manage a budget. Is it the child's fault if a husband and father abandons his family or if a young couple has children with no intention of marriage and family?

Consequently, many welfare programs exist to provide monetary payments specifically to help support children born out of wedlock because of the recognition that children born into such circumstances are more likely to live in poverty. This policy, however, is just as prone to failure as the war on poverty in general. Remember that we get more of what we subsidize. Subsidized poverty begets more poverty. Subsidized illegitimacy, begets illegitimacy. The economic results of welfare payments tied to children born out of wedlock is to reduce the opportunity cost of unwed pregnancies. A young woman who knows that the government will provide a certain amount of guaranteed income will be less likely to be chaste and will be more likely to engage in reckless sexual behavior, because the financial risk associated with such behavior is reduced.

This is not to say that all unwed mothers are lightning fast calculators estimating the monetary profit and loss associated with having a child out of wedlock. The incentive structure set up by welfare payments to unwed mothers operates on a much more subtle basis—perhaps almost unconsciously—merely because it reduces the cost and financial risk if an unmarried young women does end up getting pregnant.

Even if welfare transfer payments do provide some financial relief to some of those officially designated as poor, such a policy is doomed over the long run. This is because of the effects of taxation we have already studied. Remember that all of the money the government transfers to the poor has to come from somewhere. If poor relief is financed by taxation, the productive taxpayers have an incentive to reduce their burden through both legal and illegal means.

As taxes increase to pay for the expanding welfare state, production tends to fall. Because of higher taxes, people have less of their income to save and invest. There is a decrease in the capital stock to the extent that the state is successful in collecting the increased taxes. Less capital results in lower labor productivity, lower wages and incomes, and results in fewer goods available to society. At the same time that people have less income to save and invest, they also have less incentive to save and invest what disposable income they have left because interest income from such investment will be taxed as well.

Additionally, increased taxes provide an incentive for the development of what is called the underground economy. The underground economy is the network of voluntary exchanges that take place in ways that are unreported and undocumented to the taxing authority. As taxes on incomes increase, more and more transactions are undertaken for cash or barter, so that no record of a sale and no record of income are evident to the government. If the government does not know of income earned, it is difficult to tax.

The bottom line is that over the long term tax revenues will tend to decline. The longer taxes are raised to help pay for income maintenance of the poor, the more production will decline and the larger the underground economy will grow. Consequently, the tax base will become smaller, reducing tax revenues.

Another policy designed to help people in financial need is unemployment insurance. Unemployment insurance is a transfer payment the government gives to people who have lost their job. The payments are designed to tide one over until he finds a job; therefore, the payments are limited in duration.

Although unemployment insurance is designed to alleviate the stresses associated with unemployment, not unlike much government intervention in the economy, it actually encourages the very problem it is supposed to alleviate. Just as welfare to poor people subsidizes poverty, unemployment insurance subsidizes unemployment. If we subsidize unemployment, we will get more unemployment. We have already seen that in a free market there will be no involuntary unemployment that persists for any length of time. In every labor market a wage will be established that will clear that labor market. At that wage the quantity of labor supplied will equal the quantity demanded. At that wage there will be no frustrated suppliers of labor. Everyone who wants to work can work.

We have seen that the only reason unemployment will persist is if a wage is set above the market wage rate. If union activity successfully negotiates a wage above the non-union wage, there will be unemployment in the unionized firm or industry. If the government mandates a general minimum wage for the entire economy, there will be general unemployment.

Unemployment insurance results in prolonging unemployment by allowing wages to remain above the market clearing wage for longer periods of time. Unemployment insurance reduces the opportunity cost of remaining unemployed.

Suppose that, because the United Food and Commercial Workers Union is successful in a campaign to raise wages in meat packing plants, some workers in that industry are laid off. The laid off workers were earning $15.00 an hour. After looking for work for a week, they are all offered jobs at a manufacturing plant starting at $13.50 an hour. They can either accept this job or decline it and keep looking in hopes that a better paying one will become available. In a free market, every day they refuse to accept the job for $13.50 an hour they earn no income. Every day their choice is between a job paying $13.50 an hour and getting paid zero.

In an interventionist economy that features unemployment insurance, the choice is different. Suppose that for twelve weeks after someone loses their job, the government will pay unemployment insurance in a sum that averages $9.00 an hour. For twelve weeks, if these unemployed meat cutters decline to accept the manufacturing job for $13.50, they will not receive zero but $9.00 an hour. Admittedly $9.00 is not as much as $13.50; but it is much greater than zero. While some laid-off workers might take the first job that comes along, many others might hold off on accepting the first job available in the hopes that a better one is offered to them soon.

The fact that they are getting paid unemployment insurance makes not accepting a job easier. Because they are less likely to accept the job offer for the lower wage, unemployment insurance helps keep wages above the market wage prolonging the time that people remain unemployed. Once again we are left with the conclusion that society gets more of whatever is subsidized. If we pay people to remain unemployed, it should be no surprise that more people remain unemployed. In the name of helping people overcome the hardship of being without work, unemployment insurance results in more unemployment.

INCOME EQUALITY

Some of the proponents of the welfare state are not driven merely by the desire to help the poor. Increasingly it seems they are driven by the ideology of *egalitarianism*. The word egalitarianism is derived from the French word meaning equal. Egalitarianism is the ethic that all people should be equal. In the economic sphere, this ideology demands that the allocation of income be made such that everyone earns the same or at least roughly the same income.

The only way to attempt to do this without confiscating all private property and establishing a full-blown socialist system, is to use the government budget. Tax from the productive who earn larger incomes and redistribute the money to those who earn less or no income. This policy is essentially the same as the policy designed to help the poor we have just described above.

Such a scheme to bring all incomes into equality is doomed to failure. In the first place it is a complete denial of reality. People earn different incomes because they have different comparative advantages in the economy. These comparative advantages are determined by the relative capital, location, and labor skill endowments given to them by God. Any attempt to use force to create equal incomes from an economy made up of different people with different skills living in different places and with different stocks of capital is an attempt to thwart the sovereignty of God; it smacks of materialism at its crassest.

Additionally, economic theory demonstrates that a policy of establishing a welfare state in order to create equality of incomes contains the seeds of its own destruction. Such policies put in place the very incentives that result in more poverty while at the same time reducing both the incentive and ability to engage in productive activity necessary to fund the welfare state via tax revenue. The result would be a progressive reduction in production, incomes, and societal wealth.

Jesus said that the poor would always be with us (Matt. 26:11). This is partly so because poverty is not merely a condition that one catches like the common cold. Poverty is often the result of destructive choices. The Bible includes a number of cases in which poverty abides in the person. People can become poor because of laziness (Prov. 6:10-11), keeping close company with lazy people (Prov. 28:19), hedonism (Prov. 21:17), and drunkenness and gluttony (Prov. 23:21).

In his first letter to Timothy, the Apostle Paul sets forth principles that inform a Christian view of charity to the poor. The first is that charity literally begins at home. A widow in need of assistance should be provided for by her family (1 Tim. 5:4). Additionally, Paul instructs that one who does not provide for his own house is worse than an unbeliever (1 Tim. 5:8). He then restricts the number of widows who may apply for permanent church assistance to those who are truly destitute and most likely unable to remarry. He tells Timothy not to provide material aid to young widows because that encourages them to be idle. Instead, Paul says that he would rather have younger widows marry again, bear children, and productively guide the household (1 Tim. 5:11, 14).

The picture that Paul paints is very instructive. Notice that there is no mention of any role for the state in caring for poor widows. The responsibility is placed squarely on the family first and then on the church. He calls for private—not state—charity.

What if the church fails in its responsibility to care for the truly poor? Should not the state step in? This is the position of many well-meaning Christians. They understand that we have a calling to minister meaningfully to the poor. They also think that the Church has failed in the calling. Therefore, to care for the poor, another large institution with enough money must take up the slack. By default that institution is the government.

What is the proper response to this position? First, we must remind our well-meaning brothers and sisters of the biblical definition of poverty. For the vast majority of those considered officially poor by the U.S. government, the biblical mandate to care for those in poverty does not apply because they are not genuinely poor. They merely have fewer (sometimes a lot fewer) goods than average.

For those concerned that private charity cannot take up the slack because it is not now doing so, it is important to recognize that this is partly the result of the very welfare state that is supposed to be filling the gap left by the church. When the state begins assuming responsibility for charitable action, private charity tends to diminish. One reason for this is that people simply have fewer resources. People who find their incomes shrinking because more is taxed away to pay for welfare programs, have less income at their disposal and have less ability to make charitable contributions to poverty-assistance organizations. At the same time, once people become accustomed to the state taking care of the poor, they decreasingly view charity a personal responsibility. They can tell themselves that they have already made such contributions through their tax bill.

GROWING THE ECONOMY WITH OTHER PEOPLE'S MONEY

Sometimes the state embarks on spending money, not to benefit specific people or groups of people such as the poor, unemployed, or starving artists, but rather in the hope of expanding the general economy. Government officials have become accustomed to view themselves as responsible for growing the economy, as if they can give it just the right amount of water, sunlight, and Miracle Econogrow and watch the GDP rise as high as an elephant's eye. This turn of events is certainly unfortunate because these same politicians and bureaucrats greatly misconstrue both the nature and causes of a progressing economy. Efforts to provide prosperity via increased government spending are doomed to failure.

As has already been explained in chapter 11, an expanding economy is the result of increased capital accumulation that makes labor more productive and allows for the expansion of the division of labor. For capital accumulation to progress there needs to be increased savings and investment made possible by lower time preferences and private property rights. In chapter 13 you saw what happens if government seeks to grow the economy by increasing the money supply. Entrepreneurs are led astray in an inflationary boom that resolves itself into a recessionary bust as soon as the inflation stops.

Another hoped for route to prosperity is increased government spending. The issue of government spending and gaining revenue for such spending through taxes is often referred to as fiscal policy. Fiscal policy is economic policy related to the government budget. Keynesian economists have often advised increased government spending because they have the notion that aggregate demand really drives the economy. If their view was correct, then it makes sense to boost government spending because this necessarily will result in an increase in aggregate demand, national income, and GDP. The problem is that their view is wrong.

We have already discussed many of the problems with this approach, but the one we want to focus on presently is the fact that if the government spends money, they must get it from somewhere. If the state wishes to boost the economy by increasing spending without the negative consequences associated with monetary inflation, they have two options for funding their spending: taxing and borrowing.

We have already seen that taxation has a decided negative impact on production. Higher taxes reduce the incentive for laborers to work, for land owners to rent land for productive uses, and for capitalists to save and invest. In other words, taxation hinders the use of land, labor, and capital goods for productive use. Economy-wide output decreases, real wealth falls, and people see a drop in their standard of living. This is hardly the way to expand the economy. Funding increased government spending by increased taxation is like pouring weed killer on your garden, all the time thinking that it is Miracle Grow.

Another source to which the government can turn to finance its spending is borrowing. The government must borrow money if it wants to spend more than it receives by taxation. If the state spends more than it receives in tax revenue, the government budget runs a deficit, so such spending is referred to as *deficit spending*. Keynesian economists believe that an effective way of getting an economy out of recession is to engage in such deficit spending. However, not only is government borrowing to fund spending of no value to an economy in recession, it is essentially poisonous.

There are two sources from which the government can borrow to finance deficit spending: the banking system and the non-bank public. If it borrows from the banking system, the practice is essentially inflationary because banks will lend newly created fiduciary money resulting in an increase in the money supply and all of its associated negative consequences.

Borrowing from the non-bank public is not inflationary but does have a serious economic impact. Government borrowing diverts savings from private investment to government consumption. Some savings that would have been invested in productive activity will instead be spent on whatever government officials desire. This shortens the structure of production, reducing the economy's stock of capital, the productivity of the labor force, and the general standard of living.

At the same time, government intervention in the time market in search of funds competes with private citizens for present money in exchange for future money. Consequently, government borrowing in the time market results in an increase in demand for present money in exchange for future money, raising interest rates. The diversion of savings out of the private time market and into government hands makes the rate of interest entrepreneurs must pay to borrow funds from capitalists higher than it would be otherwise.

Additionally, government borrowing has an even greater impact on savings and investment than taxation. Some, perhaps most, of the money taken by the government in taxes would have been spent on consumption. Therefore, for every dollar taxed away from the citizenry, only a fraction of it would have gone toward productive investment. Every dollar lent to the government by the non-bank public comes wholly out of private savings, which would have completely been spent on investment. Consequently, government borrowing has a greater dollar for dollar impact on savings and investment than taxation.

You may understand the above analysis, but still be wondering why we treat government borrowing as coercive intervention on the part of the state. No one forces citizens to loan the Federal Treasury money by buying government bonds. They do this of their own free will. True enough. But this only considers part of the time market transaction that constitutes government borrowing.

Why does the state, which is essentially a consumer and not a producer, have the ability to convince people that if they lend to it they will be paid back at a handsome rate of interest? Because the government always has the ability to back up its revenue stream with coercion. The state hardly has to worry about not being able to pay back its debt because it has access to taxation and inflation, both of which are clearly interventionist.

Since the government's ability to pay back its debt is relatively certain, it can pay lower interest rates than other private borrowers. This lower rate is possible because the risk premium attached to government debt is lower than for debt borrowed by private citizens who can neither force others to pay taxes to them nor legally counterfeit money. The state's ability to intervene in the economy via taxation and inflation, also allows it to intervene in the time market by offering savers an investment that is relatively safer, providing an offer many cannot refuse.

But wait, there's more. Because of the nature of the time market, government borrowing results in a distortion of savings and investment twice. First, when the loan is made, savings are taken out of the private economy and diverted toward government consumption. Second, when the loan is paid back it must be paid back with either tax revenues, new money resulting from monetary inflation, or further borrowing. Each of these sources of government funds necessarily further distorts savings and investment in the economy. Hence, savings and investment are allocated against the wishes of members of society both at the time of borrowing and at the time of repayment. Both episodes of distortions result in a shrinking capital stock, lower productivity, lower incomes, and a general decline in wealth.

During the New Deal days of the Franklin Roosevelt Administration, the President's staff began justifying the supposed benign nature of deficit spending by assuring everyone that they could relax about the government debt because, "We owe it to ourselves." These propagandists were crying "Peace! Peace!" when there was no peace. While somewhat successful as a public relations device, the statement suffers from the fallacy of viewing society collectively instead of personally.

It is not true that we owe the government debt to ourselves. When the government borrows money from the non-bank public to fund its deficit spending, it is not borrowing from everyone, but borrowing from certain individuals. At the same time, when the government taxes money or inflates or borrows against the funds to pay off the debt as it comes due, it does not take wealth from everyone uniformly. A statement describing the nature of the debt more accurately than, "We owe it to ourselves," is "Some citizens owe it to other citizens." No matter how politicians try to popularize government borrowing, it will always have the same economic effect.

Recently, it has been noticed that an increasing amount of new debt issued by the U.S. government is held by foreigners. More and more investors from other countries have purchased U.S. government bonds. Consequently, some policy makers have tried to assuage popular concerns over deficit spending. Because foreigners are buying more of the newly issued debt, this debt is not a drag on our economy they argue. These people have exchanged "We owe it to ourselves," with "We owe it to other people."

The reason this is supposed to be soothing is that if the new debt is being financed by foreign savings, deficit spending by the United States will not be a drain on domestic savings. Interest rates in the American time market will not be affected and domestic investment will not be artificially lowered. Economic analysis reveals that this view is the result of glasses that are a bit too rose-tinted.

In the first place, if foreigners have savings they are eager to invest in U.S. government bonds, it is just as likely that, without government debt issues, these foreigners would have invested in private production. Therefore, if foreigners are willing to invest in U.S. Treasury bonds, it is reasonable to conclude that those investors had savings that would have been supplied in the American time market in any event. An increase in foreign lending to the U.S. Government and in foreign investment in American enterprises indicates an increase in the supply of present money in exchange for future money in the American time market. The increased supply is due to an increased quantity of savings available for American investment. Without the increased demand resulting from government borrowing, more savings and investment would be available for productive activity in the United States. The interest rate would be lower. Entrepreneurs would find it easier to borrow savings from capitalists. The structure of production would be longer. The stock of capital would increase. Productivity and wages would increase. The standard of living would be higher over time.

However, this increase in prosperity is retarded because of government borrowing. Even if government borrowing is funded by foreign savings, the interest rate will be higher than it would be otherswise. Nevertheless, government borrowing still leaves society with less wealth than it would have if the foreigners directed their savings toward private production in the United States.

At the same time, we need to remember that U.S. dollars that foreigners lend to the U.S. government cannot at the same time be spent on any other goods produced by Americans. If a foreigner wants to buy a product from an American producer, he must pay the producer with U.S. dollars. If, however, foreigners spend their dollars on government debt, they have fewer dollars to purchase American made goods reducing American exports. Consequently, increased U.S. government debt funded by lending from foreigners also comes at the expense of lower incomes for American exporters.

What is the bottom line? Because the government is essentially consumptive and not productive, the only way it gains funds is by taking from others—not by successfully serving customers. Therefore, the only way the state can fund spending is by taxing its citizens, inflating the money supply via fiduciary money, or borrowing. All three have negative consequences on savings and investment. All three result in a misallocation of capital. All three work to shorten the structure of production, reducing the stock of capital and productivity of labor. All three methods result in lower real incomes and lower standards of living. We simply cannot grow the economy into prosperity by resorting to government spending. It cannot be done.

GOVERNMENT TAXING AND SPENDING

"I'm from the government and I'm here to help." In our popular lexicon this has become a recognizably laughable anti-truism. Whenever the state attempts to make things better through intervening in the economy, it either falls noticeably short or actually makes things worse. Economic theory teaches us why this is the case when the government attempts to solve social problems by taxing and spending. Government subsidies, taxation, and activist fiscal policy all fail to achieve their goals. When they try, their efforts are frustrated because they cannot overturn economic law.

In general, government taxes and subsidies distort the division of labor away from the wishes of citizens and toward the wishes of rulers. Taxation takes wealth out of the hands of those who earned it, reducing producers' incentive to save, invest, and engage in production. Subsidies reward people for being less productive than they have to be in a free market and provide a great incentive to spend their time, effort and resources in influencing and controlling the political apparatus rather than in productive activity.

Economics teaches us that we cannot easily remove a problem merely by erecting a government wealth transfer program. Society gets more of whatever it subsidizes. Using subsidies in an attempt to solve the poverty problem merely results in more poverty. Recipients have an incentive not to work and taxes required to pay for the subsidies reduce the incentive to save, invest, and engage in productive activity. Capital is consumed, leaving labor less productive and earning lower wages. Subsidizing illegitimacy, as charitable as it may seem, results in more illegitimacy.

© United Feature Syndicate, Inc.

Attempting to stimulate the economy via government spending will likewise fail. It may be that activity will be generated in those industries where the government spends its money. However, government spending must be funded somehow and all three sources of funding—taxing, inflating the money supply, and borrowing—all produce negative economic consequences that more than offset any benefit received from increases in government spending. Increased taxation results in capital consumption by reducing citizens' ability and incentive to save and invest in capital maintenance and accumulation. Monetary inflation destroys money's purchasing power and sets in motion malinvestment creating business cycles that result in capital consumption and unemployment. Government borrowing also consumes capital by drawing it out of the hands of private capitalists and directing it toward government consumption, leaving workers less productive than they would be otherwise, reducing output and incomes.

The moral of the story is that a country cannot tax and spend its way to prosperity. The best thing a state bureaucrat can say upon his arrival is not that he is here to help, but "I'm from the government, and I am staying out of the way."

SUGGESTED READING

Beisner, "Poverty," 111–30. Beisner's excellent essay on what the Bible does and does not mean when referring to the poor.

———. *"Prosperity and Poverty*, 191–98. An outstanding exposition of the nature and causes of poverty from a Christian perspective.

Garrison, "The Trouble with Deficit Finance." A brief and succinct discussion of the negative economic consequences resulting from each of the ways a government can finance deficit spending.

Rector and Johnson, "Understanding Poverty in America," 1–17. An excellent documentation of the myth of widespread poverty in America and the chief causes of children living in lower income households in the U.S.

Rothbard, *Man, Economy, and State*, 878–85, 907–61. Rothbard's outstanding economic analysis of government taxing and spending.

Schlossberg, *Idols for Destruction*, 39–87, 102–21. Schlossberg's devastating Christian critique of the ideology fueling income redistribution policy as well as his analysis of the economic and moral consequences of such redistribution.

17

Voluntary Exchange and Regulation

Economic theory explains how, in a free market economic cal-culation allows entrepreneurs to direct scarce factors of production toward the ends most valued by consumers. There are occasions, however, when people who are not happy with a particular market outcome put enough pressure on politicians to enact some form of regulation of the market. Some argue that the free market is fine and generally very efficient in delivering the goods, but if allowed to operate freely it can too easily careen off the tracks, leaving a trail of wounded souls in its wake. The state, it is argued, needs to be ever watchful and ready to steer the market back toward serving the public good. Unlike the ideology of socialism, which seeks to bring all factors of production under the ownership of the state, the regulatory ideology seeks to engage in *ad hoc* social engineering intervening in the market when it is politically necessary.

The ideology of interventionism results in a hampered market economy. The market is not abolished as it is under socialism but it is hampered from operating how it would in a world of complete voluntary exchange. In an interventionist society the state does not restrict its activity merely to protecting private property, but engages in other activities such as price controls and monetary inflation, the effects of which we have already analyzed. Government intervention in the economy can also take the form of state regulation of commerce, the subject of this chapter. Such intervention is always backed up by government coercion.

DIRECT EFFECTS OF INTERVENTION

Before we can understand the consequences of government regulation of the marketplace, we need to revisit the nature of voluntary exchange. Remember from our study of human action that people act in order to increase their future state of satisfaction. They hope that by taking an action their future states will be better than it would be had they not acted.

We have seen that the free market allows everyone to achieve more of their ends than otherwise. They do so via voluntary exchange. Both parties to a voluntary exchange perceive themselves to be better off after the exchange than before because what they receive is valued more highly than what they trade away.

A simple example will remind us why this is the case. Suppose that Nova runs a bakery. Her end is not to eat all of the bread that she bakes but to trade that bread for money that she can then spend on other goods that she wants. The more money she makes, the more she has to spend and the more ends she can achieve. On the other hand, Casey already has some money and seeks to use that money to relieve the hunger she will feel in the evening just before supper. She cannot eat the money so she looks to exchange it for some bread at Nova's shop. Suppose that Nova is willing to sell a French baguette for $3.00 and Casey is willing to pay $3.00 to buy one baguette. Their respective value scales are as illustrated below:

<u>Casey</u>	<u>Nova</u>
(1 Baguette)	($3.00)
$3.00	1 Baguette

As Nova sells one baguette to Casey for $3.00 and as Casey buys the baguette from Nova for $3.00, both walk away from the transaction happier than before. Both receive something that they value more highly than what they give up.

What is true in this simple example is true for all voluntary exchanges. Each time people engage in exchange, they do so because they think they will be better off. In this way the free market tends to maximize the satisfaction of society. Intervention in the market, however, hinders this process and necessarily creates conflict. Instead of an exchange which is mutually beneficial, one party benefits at the expense of another party.

GOVERNMENT REGULATION OF BUSINESS

In addition to mandating price controls or attempting to control the economy via monetary inflation and the government budget, policy makers often resort to intervention in the market in the form of product controls. Instead of mandating the price or prices at which buyers and sellers can legally trade, rulers mandate what types of goods can be sold by which sellers under what conditions. An example of such a regulation is the European Union law requiring sellers of class A eggs to have the egg shells stamped with a code identifying their establishment. The code is not to be stamped on the crate or carton but on *every egg*. To obtain the code the egg producer must register with his national government. Those who do not register and obtain a valid code cannot legally sell class A eggs.

Economically, production controls amount to a grant of special privilege to particular owners. If certain egg producers simply do not have the time to stamp the eggs themselves or the money to hire people to do it for them, they are kept out of the market for class A eggs. The egg producers that have the cash to absorb the extra costs are to a certain extent shielded from would-be competitors. In this sense, product controls always provide monopolistic privileges. If only one seller is able to abide by the government's product controls, he becomes a legal monopoly. As is more often the case, a group of sellers able to work in accordance with state regulation are conferred by the product control a quasi-monopoly. This allows them to act as a legal oligopoly—a group of sellers somewhat shielded from competition by law.

Before discussing specific types of product control, it is possible to apply economic principles to discover the general economic effects of legally mandated product regulation. We will first focus on the most extreme case where one seller is granted a legal monopoly so that he is the only legal producer.

In the first place, such intervention benefits the seller receiving the special privilege. Whoever has the legal right to produce is protected from potential competition. Potential competitors are barred from entering the regulated market, reducing the number of substitutes for the good made by the privileged seller. As the number of substitutes decreases, the demand for the product made by the legal monopolist becomes more inelastic, as shown in figure 17.1.

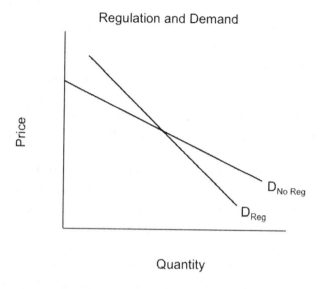

Figure 17.1. If an established firm faces less competition due to government regulation of the market, its demand curve (D_{Reg}) will be less elastic
than its demand curve ($D_{No\ Reg}$) would be without the regulation.

If the monopolist's demand becomes inelastic, he can raise his total
revenue by raising his price above the free market price. He is able to
command a higher price for his product as he decreases his output. Not
surprisingly, legal privilege in the form of product controls benefits the
recipient of the privilege.

Now it is true that monopoly profits—like all profits—are never permanent. They dissipate over time as the market tends toward the evenly
rotating economy. However, the gains to the monopolized factor used by
the monopolist are permanent. The benefits of the legal monopoly are
imputed to the monopolist's right to sell in the regulated industry. As long
as this privilege remains, so do the monopoly returns. The factor to which
the monopoly return is imputed is the monopoly privilege itself.

This legal privilege to a firm allowing it to sell its product makes the
firm's assets more valuable than other firms. If the owner of the privileged
firm sells his business to someone else, the monopoly privilege will be
capitalized into the value of the firm. Any subsequent buyer will be willing
to pay a price for the firm that is equal to the value of the assets in a free
market plus the additional value. The buyer will not reap a net gain on the
monopoly privilege because he will be paying for it up front. This buyer

will only earn the same interest on his investment that any other capitalist earns. The chief beneficiary of the regulation is the seller to whom the monopoly privilege is first conferred.

At the same time, product control intervention harms those potential competitors who would otherwise be actively selling in the regulated market. These potential sellers are kept from entering profitable industries. They are forced, therefore, to accept lower revenues in other industries in which they are less efficient and the value of their output is lower. Lower revenue translates into lower profits. Additionally, workers who would be employed by potential competitors are harmed because they will also be forced to work in lines of production where their discounted marginal revenue product is lower than it would be in the regulated market.

Consumers are also harmed by product controls. To the extent that government production controls keep them from buying goods they would buy if no regulation was in place, they are prevented from making mutually beneficial exchanges. At the same time, the price they must pay for the goods sold by the monopolist increases. Consumers end up paying higher prices for fewer products.

In general, society is less prosperous. When legal privileges are given to some sellers, factors are allocated away from their most valued uses. From the perspective of people in society, too few producers are in those industries where entry is hindered and too many are in those other industries where the value of production is lower. Resources are directed away from their higher valued and more efficient uses. Consequently, resources are wasted and fewer goods are produced with the same quantity of land, labor, and capital goods. People are left with fewer means with which to attain their ends so fewer ends are satisfied.

This is exactly what the World Bank found when it undertook a large empirical study comparing regulation and economic performance across a multitude of countries.[1] In countries where regulation was historically higher, productivity was lower, costs of production were higher, and production delays were longer. The greater the government regulation of the economy, the poorer that country tended to be.

Another social consequence of government regulation is the rise of corruption. Whenever the state intervenes and makes it harder for voluntary exchange to take place, the incentive increases for some to skirt the rules. Some sellers will simply ignore the regulations taking a sort of don't-

---

1. The World Bank, *Doing Business in 2004*, xi–xvi.

ask-don't-tell approach in the hopes that their production and selling will slip undetected under the legal radar screen. Likewise, some buyers will pay a premium to purchase the goods they want, but which are forbidden by law. If buyers or sellers are discovered, they will often bribe government officials in order to continue in their illegal trade. This puts corrupt government bureaucrats in the position of passing out favors arbitrarily and extracting relatively large sums of money by allowing certain buyers and sellers to violate the law. Government regulation also creates an incentive for certain producers and consumers to spend resources in an attempt to change the regulations and also creates incentives for politicians to favor interest groups who provide the most campaign contributions and votes.

Indeed, historical studies indicate that the level of government regulation in a society is positively correlated with the amount of corruption. With more intervention comes more corruption; with less intervention we find less corruption. The same relationship between regulation and corruption has been documented in studies that compare the level of regulation and corruption across many different countries. Higher regulation is associated with a larger underground economy—the black market—and more corruption.

The difference between product controls that provide one seller a legal monopoly and controls that provide a group of sellers legal privilege is one of degree, not of kind. The same principles that apply to legal monopolists apply to a group of quasi-monopolists. A group of sellers who receive a legal privilege is shielded from competition from sellers who are not in the group. The number of substitutes for the group's products decreases making their demand curves more inelastic, increasing their ability to raise prices and increasing revenue and profits. The effects are similar to what we find in a cartel except that in the case of government regulation the cartel is legally enforced, mitigating the threat of competition from outside the cartel.

Each privileged seller, however, does face competition from within the group so there are more substitutes than there is when there is only one legal seller. Therefore, the demand curves for the privileged sellers become less inelastic than it would for a single monopolist. Therefore, the sellers would be able to increase the price by a smaller amount when maximizing their revenue and profits.

Nevertheless, while government regulation does not always result in a monopoly in the sense that there is only one legal seller, it is always monopolistic. The results of regulation granting special privilege to a group of sellers are the same as in the case of a single seller. Only the magnitude of the effects is smaller.

TYPES OF MONOPOLISTIC PRIVILEGE

Rarely does the state grant special privilege to a seller or group of sellers as directly and explicitly as did kings and parliaments in the heyday of European mercantilism during the sixteenth and seventeenth centuries. Today, such privileges are usually granted to groups of sellers indirectly in the form of various product regulations promoted as ways of benefiting society. The economic results, however, are no different. We now take up an examination of a number of the various forms in which such monopolistic privilege is granted.

Government License

A very common way government privilege is extended today is a government license. A government license is a legal document allowing only the possessor of the license to offer his product for sale in the marketplace. State licenses deliberately limit the number of sellers who may produce in the regulated industry. In order to obtain a license the government usually requires monetary payment in addition to a number of requirements that must be met. For example, in order to legally cut hair for a beauty salon, every state in the union requires that beauticians successfully complete training at a certified cosmetology school, pass an official examination, and pay several hundred dollars just to have the right to take the test. Anyone who does not meet the government requirements is prohibited from legally cutting hair as a trade.

Such licenses actively prevent some amount of mutually beneficial exchanges from taking place, leaving prospective exchangers worse off than they would be without the licensing requirements. A good example of how government licensing restricts trade gained notoriety in the spring of 2003. An Arizona teen started a successful rat-proofing business covering homeowner's exterior pipes and vents with steel mesh for a flat fee of $30 per house. Most of his clients were elderly and happy to pay for his services rather than for the much higher priced services of licensed pest controllers that they could not afford. The day after the local newspaper ran a large story about his business, the young entrepreneur received nu-

merous requests for his service plus a personal visit from a state bureaucrat informing him that he was violating the law by providing pest control
without a state license. After much public uproar, the state commission
that oversees pest control revoked their prohibition. But what would be
the case if they had not? This entrepreneur would have been forced to do
other work at which he either was less fulfilled, earned less, or both. His
clientele would have had to settle for a pest control service that was much
higher in cost or merely forgo such service completely, surrendering to
the rats. The Arizona teen was able to continue only because he was not
attempting to exterminate rodents, which requires a two-year apprentice-
ship and acquisition of a state license.

Quality Standards

Often the state explicitly mandates that products to be sold in the market
must meet some official standard of quality. The primary reason given
for such laws is that the government has a responsibility to establish and
enforce quality and safety standards. If the state does not provide such
regulation, it is argued, the numerous rogues that populate free-market
capitalism will be all too able to cheat people by charging exorbitant prices
for inferior or dangerous products. Protecting society is the stated goal of
politicians favoring quality and safety standards.

Bureaucratically ensuring that all products sold are of a particular
quality grade is extremely difficult because quality is a relative term.
Remember that economic value is subjective to the actor. Likewise, the
standard for product quality also varies from buyer to buyer. Buyers decide according to their own tastes and preferences what they mean by
quality and they also subjectively determine how much they are willing to
pay for different goods with different perceived levels of quality.

By refusing to allow buyers to freely pay for goods that do not meet
official standards, the government harms those buyers who have decided
such a purchase is beneficial. Buyers routinely make price-quality trade-offs
when making purchases. Many would like to drive a Lexus, for example, but
know they can only afford a Ford Focus. They often opt for the best quality
they can afford. At other times they are willing to buy a much lower quality good if the price is quite low. Government mandated quality standards,
however, forestall these type of rational trade-offs. Buyers are prohibited
from purchasing goods that are deemed below standard, even if they desire

to buy them. They are unable to legally pay for lower quality goods that fit their budgets and instead are forced to do without them altogether.

At the same time such regulation benefits officially certified sellers by protecting them from competition. They do not have to worry about other sellers undercutting them on price, and as a result can charge higher prices for their product.

Additionally, if the quality standards are for physical goods as opposed to services, they can actually have a perverse affect on the quality of the goods in the regulated industry. If the government establishes official definitions of products and forces producers to follow these definitions in order to sell their products, the state actively restricts innovation. With fewer legal competitors, the officially approved sellers do not have to worry about other sellers in the industry offering new products that do not adhere to the official standards.

Even if state bureaucrats finally recognize increased quality products that are comprised of different components than those mandated by regulation, it takes a long time for regulations to change. There is a tremendous amount of inertia in the bureaucratic culture. Any improvement is very slow to occur because bureaucrats have no incentive to react quickly to changed consumer preferences. Bureaucrats do not reap profits by successfully serving consumers and do not incur losses by failing. They do not have to worry about keeping ahead of the competition in terms of improving product quality. Therefore, by the time bureaucrats recognize the political expediency of updating official product regulations, technology has already passed them by again. Such bureaucratic drag serves to hold back increases in product quality by restricting innovation rather than encouraging it.

Safety Standards

A very popular justification offered for government regulation is consumer safety. No one wants to be harmed as the result of making an exchange. People are rightly concerned with safety. Therefore, governments have seized the opportunity to regulate particular markets by establishing safety standards that buyers and sellers must maintain.

One market that has strict safety regulations is the labor market. In fact, the U. S. Department of Labor houses the Occupational Safety and Health Administration (OSHA), an entire agency charged only with the task of establishing and enforcing workplace safety standards. Again, however, the result of such regulation is to reduce the quantity of mutually beneficial exchanges in labor markets. Like life itself, no job is completely safe. Prospective workers make subjective trade-offs between the desirability of working conditions and wages. Some are willing to receive relatively little for perceived comfort and safety. Others are willing to accept less comfort and more risk for more pay. Labor safety regulations prohibit employers and workers from voluntarily negotiating acceptable terms of trade unless they abide by government standards. Workers who would be willing to work in conditions that do not meet OSHA guidelines are forced to accept work elsewhere for less income. Employers are forced to allocate factors of production in a way that is contrary to what consumers want and contrary to the most highly valued use of the factors. Production is less efficient and society is able to produce, exchange, and consume less.

Similar principles hold true for safety regulations that apply to goods produced. They mandate certain safety qualifications that must be met and prohibit the sale of goods that do not adhere to the rules. This proscribes the safety-price trade-off that consumers make similar to that made by workers described above. The end result is the same. Buyers are prevented from purchasing goods that would increase their satisfaction. Some sellers are forced to earn lower revenues because they are forced to less successfully serve customers. Those sellers who do meet government standards enjoy protection from competition and are thereby able to charge higher prices for their output.

One might be concerned that if a free market would persist in the areas of quality and safety, we would be left with an anarchic free-for-all with no constraints on potentially dangerous conmen fraudulently plying their worthless wares like Yakov peddling his Elixir in *The Inspector General*. However, to say that a market is not regulated by the state does not mean that it is unregulated.

In the first place, all transactions are voluntary in a free market. In order to make a profit, a producer must successfully give consumers what they want. In order to reap profits over time, he must be able to successfully keep his old customers while winning new ones. An entrepreneur will never maintain fabulous wealth if he becomes known for killing off his clientele. He will also find it hard to maintain customers if he becomes known for passing off junk for high prices. Consumers will simply not tolerate such practices for very long.

I remember standing in line in academic regalia for my first graduation as a university economics professor. Thankfully I had purchased my regalia so I was assured that everything fit. A history professor friend who was also a new hire had to rent his. The cap they gave him was supposedly a one-size-fits-all contraption with elastic at the back designed to fit every head. Not only did it not fit securely, but it looked rather goofy from the back. My friend, who had certain interventionist leanings, tried to goad me by pointing out the inferior quality and stating, "Someone should see that this is regulated." I retorted, "Well, *you'll* never rent from this company again." "*Touché,*" he replied. A seller who does not cater to his buyers, but who routinely sells products that they perceive as being a poor value will not survive long as a producer.

Additionally, when customers are concerned enough about product quality and safety, entrepreneurs will find it profitable to offer them independent information. Magazines and internet sites provide product ratings for enquiring buyers. Private organizations test and evaluate products for both quality and safety and only certify products that meet their standards. The difference between certification from private entities and the government is that if consumers do not find private certification relevant or helpful, producers will cease submitting their product for evaluation. The certification organizations will subsequently not be able to turn a profit and cease operations. Private certification firms have every incentive to provide exactly the sort of information that producers and buyers find useful.

Examples of such private organizations are probably more plentiful than you realize. Most television weathermen, for example, advertise that they have received the American Meteorological Society's seal of approval, indicating that they are knowledgeable in their field. Many automobile mechanics post signs informing customers of the training that they have received from respected mechanic schools. A very influential organization that tests and certifies various products is Underwriters Laboratories. Underwriters Laboratories is a privately owned firm that tests products like toasters, space heaters, and televisions for safety. Producers pay Underwriters Laboratories a fee for putting their goods through the rigors of quality testing. Only products that pass muster are able to display the UL symbol. Countless producers and consumers rely on this private firm to aid them in selling and buying products that are relatively safe.

Finally, if sellers are guilty of misrepresenting goods as being in top condition or of selling goods that actually injure the buyer, the seller can be brought to court and sued for damages by the victims. If a seller advertises a hamburger patty made of 100% ground beef but in actuality is 20%

soy, he is guilty of fraud and can be required to pay monetary compensation to his customers who essentially had their money stolen. In a free market, if a pharmacist sold a drug to someone and the drug made them seriously ill, the victim is allowed to sue for medical bills. If the patient died, the seller of the drug could be prosecuted for manslaughter. Sellers have no incentive to expose themselves to civil or criminal suits. As a result, they have incentive to provide products that have as much quality and are as safe as consumers demand.

A good example of the problems associated with government safety standards is the regulatory activity of the U. S. Food and Drug Administration (FDA). This government agency has the legal mandate to ensure that pharmaceutical drugs are safe and effective. The FDA also has the authority to monitor and control the methods by which drugs are tested and control the way in which drugs are marketed. On the one hand, the FDA routinely demands long, drawn out evaluation procedures, contributing to increases in drug prices and a delay in availability for drugs that appear safe and effective. Rather than increasing societal well-being, it is estimated that such restriction of drugs reaching the market has resulted in the premature deaths of hundreds of thousands of people over the past four decades. On the other hand, the FDA sometimes rushes to approve politically popular drugs. It is not uncommon for the FDA to approve drugs with seriously negative side effects and therefore make doctors and patients think that they are safe, when they may not be. Additionally, recent studies have estimated that tens of thousands of people die in American hospitals every year due to adverse reactions to drugs approved by the FDA. In light of these figures, it is clear that the FDA is not succeeding in ensuring the safety and well-being of our citizenry. The FDA arbitrarily increases costs to some drug manufactures while providing monopolistic privileges to those firms that meet all FDA requirements. Such companies face less competition and fewer substitutes for their products and can thereby raise the price of their drugs and increase their revenues and profits. They do this, however, at the expense of consumers who could benefit from the availability of the prohibited drugs or are harmed by taking drugs the FDA mistakenly labels safe and effective.

At the same time, private, market-driven drug evaluation is possible. In fact, it is already occurring. Many alternative, private suppliers of quality evaluation have arisen in the current highly regulated market. The ECRI Institute provides evidence-based assessment of drug quality and safety. TÜV Product Service in Europe and Underwriters Laboratories provide quality assurance testing and ratings for medical devices. These and other organizations would likely do more if not for the FDA crowding out private evaluation.

Anti-trust Laws

Another very popular justification for business regulation is keeping the market free of monopolies. The state uses anti-trust legislation to ostensibly restrain monopolistic business practices. As observed in chapter 10, justification for anti-trust laws stems from the ethic of consumer sovereignty. The argument is that if a firm becomes so dominant that it obtains monopoly power, that company can reduce its output and raise its prices and thereby increase its revenue and profits.

Most economists recognize that in a free market, competition between entrepreneurs provides consumers with the goods they want the most at the lowest possible prices and in the most efficient manner. Poorly managed firms and entrepreneurs that incorrectly anticipate and respond to demand are forced to quickly adjust or exit the market. Resources are directed toward their most valued uses and kept from being wasted.

Opponents argue that often the free market fails to provide such results whenever successful firms come to dominate a market and gain monopoly power. Such power is seen as the ability of a firm to reduce its output and thereby gain the ability to increase its price. Interestingly, this is exactly what firms can do that receive monopoly privileges from the government. Consumers are left with fewer goods from which to choose and must pay higher prices for those goods.

Monopolistic behavior on the part of the dominant firm is seen as uncompetitive and harmful to societal welfare resulting in all sorts of negative consequences. Possession of monopoly power is said to result in the discouragement of entrepreneurial initiative and efficiency, lower output, higher prices, exclusion of competitors from the marketplace, and the misallocation of resources. Because monopolists can charge a monopoly price higher than the competitive price by restricting output, some charge that potential transactions that could benefit society are not undertaken. Additionally, they argue that some firms will expend resources merely in attempting to obtain and maintain monopoly power rather than produce goods to meet consumer demand. Because they can get away with charging a higher price, they can afford to operate less efficiently and thereby waste resources.

In our earlier discussion of monopolies we have seen that there are several problems with these claims. If any seller successfully reaps profit in an industry he always faces the possibility of other entrepreneurs entering that market and offering a product that consumers prefer at a lower price. In a free market, the concept of a monopoly price as distinguished from a competitive price is meaningless. In reality, the only monopolies that are a cause of concern are those that exist due to government granted privilege.

In the free market how can a firm increase its market share? That company must serve customers better than anyone else. The entrepreneur therefore is encouraged to innovate, resulting in improved products and more efficient and less costly production. Consequently, the seller can charge lower prices. All of these things better serve the customer.

If a successful firm with a large market share attempts to exploit its position by restricting output and raising its selling price, the firm will not be able to maintain its position. Higher profits resulting from higher selling prices attract investment into that industry from other entrepreneurs. This new investment tends to result in increases in the supply and more of the good being sold at lower prices. If the original firm seeks to keep competitors at bay by lowering its selling price, it will increase its output. The original firm has no so-called market power to maintain high profits by restricting quantity produced and charging a higher monopoly price. The only way to successfully maintain a market share is for a firm to become more efficient so that it can charge lower prices for better products than competitors. Again, this benefits consumers instead of harming them.

Rather than maintaining competition, antitrust laws actually work to restrict competition by hampering voluntary exchange. Such laws tend to enforce the status quo preventing the best entrepreneurs (from the point of view of the consumers) from taking advantage of their efficient operations. Antitrust laws, therefore, reduce incentive to improve efficiency and better serve customers.

The roots of the antitrust laws in the United States are found in several state legislatures. In the 1880s, farmers in Missouri lobbied their state government to pass a law designed to keep large businesses at bay because of concerns that land ownership was concentrating into the hands of a few capitalists. During that time, however, the price of agricultural products fell instead of rising. This is precisely contrary to what we would expect if in fact monopolists were running rampant in agricultural industries.

The most prominent early antitrust case was that of Standard Oil. John D. Rockefeller became a successful producer and seller of petroleum products by becoming more efficient than other sellers. He was able to reduce production costs in several ways. He engaged in shrewd bargaining for crude oil and undertook good research and development. He maintained effective control of the transportation of both crude and refined oil and made managerial innovations that boosted efficiency. Consequently, Standard Oil was able to sell more petroleum products at lower prices. Kerosene prices, for instance, fell from thirty cents per gallon in 1869 to less than six cents per gallon in 1897.

Other sellers sought to follow suit. Standard's market share fell from eighty-five percent in 1890 to sixty-four percent in 1911. That same year Standard Oil was prosecuted by the U. S. Government for violating antitrust regulations. Standard was in essence punished for its innovation.

The most famous case of antitrust enforcement in recent history is the Microsoft case. Microsoft was declared to be a monopolist in the personal computer operating system market. The company was found guilty of using its monopoly to leverage market power into the Internet web browser market by providing Internet Explorer for free as part of its Windows software package.

In fact, Microsoft never controlled the entry into the computer operating systems industry. Like Standard Oil, Microsoft achieved a large market share through successful innovation and promotion of an operating system that could relatively easily integrate several software applications. Microsoft never prohibited computer retailers from installing other web browsers such as Netscape on their computers as well. Also, the users were free to download other web browsers. Clearly an operating system with a web browser included free of charge is better than an operating system without a web browser. Microsoft was hardly harming the public.

By hampering the free market process, antitrust regulations ultimately create monopoly privilege. In doing so, they promote inefficiency, reduce voluntary exchange, and thereby reduce social welfare. Antitrust laws do the very things advocates of the law worry about: reduce competition and harm consumers. They protect firms that are less successful at serving customers from competition from those that are more adept at satisfying them.

Child Labor Laws

Another popular form of regulatory intervention is child labor laws. These laws forbid the labor competition of workers below a certain age. When I was fifteen, I was employed by a local fast food restaurant. I had worked there only a month and a half when I was called into the office and notified that I was not to come in to work the following week. They had failed to notice my age on the application and said that it was illegal for anyone less than sixteen years old to work in front of a hot grill, something employees did regularly in serving up hamburgers red hot. Though perfectly capable of cooking hamburgers to the satisfaction of customers without burning myself, the law was the law. The restaurant had to look for another worker and I had to look for another government approved job if I wanted income. As indicated by my own personal experience, child labor laws create compulsory unemployment. The state prohibits some workers who are willing to work at the prevailing wage rate from working, resulting in a surplus of labor.

Economic theory tells us that such labor prohibitions benefit the remaining workers by allowing them to earn artificially high restrictionist wages. Wages in labor markets covered by child labor laws are higher than they would be in a free market because such regulation reduces the general supply of labor. As I am sure you already know, if the supply of something decreases, the market clearing price increases.

The immediate social impact of such labor regulation is to lower production for the entire economy. Child labor laws reduce the working population without directly reducing the consuming population. Fewer people are able to work at production, while the same number of people need food, clothing, shelter, and other goods. There are fewer goods per person in society so the general standard of living is reduced.

Additionally, child labor laws tend to harm families with children because the income of families in which children could work is lowered. Childless families gain at the expense of families with children because the labor restrictions lengthen the time during which children remain net monetary liabilities. Over time, increasing the financial liability of children puts in place the incentive to reduce birth rates. Rather than being fruitful and multiplying, married couples have less economic incentive to have children.

Populations in wealthier, more developed countries find it difficult to understand how children working could be desirable at all. Indeed I am unaware of anyone who would want their young children to work full time. However, in poorer economies many families see the extra income their children bring in as vital to the entire family's very existence. We cannot improve the family's financial situation by merely saying the children should not have to work. As a former professor wisely pointed out, if people are poor because they have few opportunities, we cannot improve their situation by reducing the number of their opportunities still further. In such circumstances, prohibiting child labor would work to keep the child's family in poverty for a longer period of time.

REGULATION AND DOING THE RIGHT THING

Some proponents of the regulations discussed in this chapter are well-meaning and it is easy to sympathize with them to a large extent. What mother wants her children to take unsafe medicine? Who wants to be taken to the cleaners by purchasing a lemon? Who enjoys paying a high price for a good merely because there are not very many suppliers of that good? No one likes the thought of children working ten hours a day in a factory. And there certainly are abuses. There are greedy and unscrupulous sellers in the market place, just as there are greedy and unscrupulous politicians, ministers, college professors, and students. What can be done to ensure that people are not the victims of greedy sellers? The easy thing, it seems, is to pass laws prohibiting or regulating these abuses.

However, in discussing regulatory policy, we quickly move beyond pure economics and into the field of ethics. At this point we no longer describe how things are, but what we *should* do to achieve *right* results. A Christian must look to biblical ethics in addition to economics in order to derive wise policy.

A Christian ethic recognizes that greed is sin. The Tenth Commandment forbids coveting those things that are not ours. Luke records that immediately before telling the parable of the rich fool Jesus warns, "Be on guard against all covetousness" (Luke 12:15). In fact, in the third chapter of the letter to the Colossians, the Apostle Paul refers to covetousness as idolatry. In his first letter to Timothy, Paul says that the love of money is a root of all kinds of evils (1 Tim. 6:10). There is even a Proverb that indicates the popular disdain generally heaped on a seller who restricts his supply with the purpose of charging a higher price. "The people curse him who holds back grain, but a blessing is on the head of him who sells it" (Prov. 11:26).

Christian doctrine also teaches that fraud is a sin. In the Old Testament Law, God specifically commands his people to use honest weights and true measurements when doing business (Lev. 19:35–36; Deut. 25:15). In Proverbs God tells us that "a false balance is an abomination, while a just weight is his delight" (Prov. 11:1). The prophet Isaiah had strong words for those who attempt to deceive customers (Is. 1:22). Christ includes the commandment "do not defraud" in his response to the rich young man's question, "What must I do to inherit eternal life?" (Mark 10:19).

We know that greed and fraud in business activity are sins. Both violate the second great commandment, "Love your neighbor as yourself" (Mark 12:31) and the Golden Rule "As you wish that others would do to you, do so to them" (Luke 6:31). The question remains, however. What should we do about it? What economic policy should be pursued in light of the reality of greed and fraud?

One principle that should guide all policy is not to resort to evil in order to do good. We have seen that economic regulation violates a person's right to his own property by forbidding exchanges seen as beneficial by all of the parties involved. As such, it is a form of state thievery and violates the commandment against theft.

Additionally, the Bible informs us that the civil magistrate is given the responsibility of defending life and property, not maintaining general holiness. We are to do God's will in the way he ordains. In the twelfth chapter of Romans we are commanded to, "never avenge yourselves, but leave it to the wrath of God, for it is written, 'Vengeance is mine, I will repay says the Lord'" (Rom. 12:19). Just as God does not ordain that we avenge evil that is brought upon us, neither does he charge the magistrate to serve as the military wing of the church punishing all wickedness.

A local magistrate, let alone a federal bureaucrat, cannot know precisely what is in the heart of any person. Suppose the bureaucrat observes a seller raising prices and some buyers complain. The bureaucrat is not given supernatural powers to read minds. He has no way of knowing for sure if the seller raised his selling price in an attempt to feed his family or if he is merely greedy.

While it may be wholly proper for the civil ruler to use force to punish manslaughter, theft, and its stealthier cousin fraud, it is improper for the civil magistrate to attempt to enforce commands against greed. This is an area that is between man and God, not between man and the government. As long as a seller is not violently forcing a person to buy from him, he cannot be guilty of violating a buyer's rights. He may be guilty of covetousness, but he is not stealing. Likewise, Scripture provides no warrant for the state to regulate or prohibit things that are merely undesirable occurrences.

On a more practical point, in a free market even the greediest of sellers can reap profits over the long run only by successfully giving consumers what they want at prices they are willing to pay. If they try to sell their product at prices people think are too high, they will not sell enough goods to maximize their income. If they gain a reputation of selling goods that are unsafe or of generally inferior quality, they will be regulated not by the state but by consumers who no longer purchase their goods. In other words, one of the practical benefits of the free market is that it requires even greed to be actualized in acts of production designed to serve others. This does not mean that greed is good. It merely means that in a free market, a greedy entrepreneur can only be successful if he serves others. In interventionist and socialist economies, greed can be actualized in gaining control of the state and its regulatory apparatus thereby allowing the greedy person arbitrarily to commit theft and murder.

In light of both biblical and economic principles, it seems that the best economic policy is to eliminate product controls. This allows for all mutually beneficial exchanges to take place. This is not a merely *buyer beware* economy. To be sure, buyers would have an incentive to gather and process relevant information before making purchases. However, sellers would be responsible to deal without deceit with customers and called to account for unscrupulous business dealings. The magistrate would prosecute crimes of theft, fraud, and murder while allowing buyers and sellers to voluntarily negotiate the quality and prices of products sold.

Some Christians might fear that such a free market economic policy will result in an unbridled capitalism that produces a society characterized by harsh, greedy, unrestrained industrialists who stop at nothing as they increase their fortunes. This worry misconstrues the nature of the free market. In a free market entrepreneurs cannot force anyone to buy their products. To receive revenue, firms must convince people to voluntarily purchase from them. The action of a profit-seeking entrepreneur is far from unregulated. In a free market, the entrepreneur may not be regulated by the state but he will be regulated by his conscience and especially by consumer preferences. If people do not want to buy from an entrepreneur with a reputation for wrong-doing they are free to refrain. The accounting firm Arthur Andersen went into bankruptcy at the mere allegation of improper accounting.

In this way that the church can properly act to regulate the economy. The Christian ethic of private property does not allow them to use the coercive state to achieve their ends for a better society. Instead Christians are called to evangelize and disciple converts in the paths of righteousness. As the church does what it is called to do, people will change their preferences. They will begin to be more loving and kind to their neighbors. If Christians really want different market outcomes, they should be obedient in their calling and have faith that God can transform the hearts and minds of men and women.

SUGGESTED READING

Armentano, *Anti-trust*. The case for repeal of antitrust legislation based on the argument that such regulation actually hinders free competition instead of preserving it.

Beisner, *Prosperity and Poverty*, 175–82. A discussion of the ethics and economics of government regulation of commerce from a Christian perspective.

Boudreaux and DiLorenzo, "The Protectionist Roots of Antitrust," 81–96. A revealing account of the origins of American antitrust regulation.

Friedman, *Free to Choose*, 189–227.
    Milton and Rose Friedman's explanation of the economic consequences of various types of government regulation of commerce.

Higgs, "The U.S. Food and Drug Administration," 59–73. A provocative exposition of the political economy of the FDA. Higgs argues that this regulatory agency reduces social welfare.

Rothbard, *Man, Economy, and State with Power and Market*, 1086–44. Rothbard's outstanding economic analysis of economic regulation.

18

Socialism

THIS BOOK IS DEVOTED to explaining the foundations of economics from a Christian view of both the universe and man. The first twelve chapters featured the development of economic principles of voluntary exchange. In order to analyze the nature and consequences of various government interventions in the market, we must first understand how the market works in the absence of such intervention. Chapters 13 through 17 examined the nature and consequences of government intervention in the market economy. In this chapter we will investigate socialism—the polar opposite of the free market.

WHAT IT IS

The defining characteristic of socialism is that the means of production are owned by the state. In a socialist economy there is no private ownership of land, labor, or capital goods. Consequently a central authority must direct all production. This authority can be one person—a sort of production czar—or a group of people like the central planning boards that called the economic shots in the Soviet Union and its Eastern European satellites.

Two main types of socialism have existed in history. The most famous is Soviet-style socialism. This is the communism that was tried and ultimately failed in the USSR. All of the means of production were explicitly owned and directed by the central planning board. Everyone understood there was no market for factors of production.

The other type is fascist-style socialism. This is the form of socialism that was tried and failed in Nazi Germany. In fascist-style socialism the terminology of the market economy is retained, but there is no real buying and selling because all exchange is directed by the central authorities. Owners keep the legal deeds to their firms, but they are told what factors to buy and what prices to pay for them. Although the law says they still own their assets they cannot use them as they see fit; in actuality the central authorities own the assets.

MARX AND SOCIALISM

Many people think that socialism originated with the communistic writings of Karl Marx. This is understandable. Marx is the most famous socialist writer of all time. His notion of a future communist utopia took hold of the hearts of many people in the nineteenth and twentieth centuries. There are still pockets of Marxists scurrying about, mainly on American college campuses.

Marx, however, did not originate either socialism or his particular brand, namely communism. The origins of the socialist idea go back to Plato and it was later embraced by Christian heretics. Marx secularized it and transformed it into the popular doctrine that it became. What Marx did develop was a type of polylogism, the notion that different classes possess minds with different logical structures. Not only do people place different emphases on certain facts and not only do they arrive at different conclusions, but Marx claimed that somehow, different classes actually followed different laws of logic. Marx alleged that the proletarian class was subject to a different logic than the bourgeoisie.

In the case of Marx, the alleged polylogism that mattered was the difference between proletariat and bourgeoisie logic. This polylogism was very convenient for Marx because he could deflect any criticism of his economic theories by merely claiming that those economists who did not agree with him suffered from bourgeoisie thinking. Marx rationalized that we should not expect these economists to think like proper proletarians because they are using different logic; rather they are merely shills for the dominant capitalist class. Marx's appeal to polylogism protected him from dealing with logical economic criticisms of his system because instead of having to deal with the economic arguments, he could simply pass off criticism as stemming from logic that was irrelevant to the problem at hand.

Another major contribution, if we want to call it that, was his development of what Mises called the *socialist creed*. This creed has two essential doctrines. The first is that society or its agent the state is omnipotent and omniscient and essentially perfect. The other article of faith is that the coming of socialism is inevitable. This doctrine was the product of Marx's dialectical materialism. He used Hegel's dialectic of a thesis/antithesis/synthesis to explain the unstoppable march of history. The Hegelian dialectic proposed that history progressed as some proposed thesis is opposed by an antithesis. The conflict resolves itself in synthesis, being neither identical with the thesis or the antithesis but affected by both. This synthesis then becomes the next period's thesis, which is opposed by another antithesis and the cycle continues to repeat itself.

Marx used this dialectic to explain world economic history. Marx thought that what he called the *material productive forces* are what propel all historical events and changes. These material productive forces are the sum of our systems of technological production methods. These forces create all of the relations of production (such as legal and property institutions) independent of people's wills. These relations of production, according to Marx, make up what he called the *economic structure of society*. The economic structure is the base which causally determines a society's *superstructure*, its natural science, legal doctrines, religion, and philosophies. Consequently, in Marx's view of the world, technology determines the modes of production, which determine the relations of production, which determines our ideals, religious beliefs, values, and art. Social change can take place then, only through a change in technology.

While the dialectic is the engine that propels economic change, the class struggle is the power that animates the dialectic. According to Marx, the privileged ruling classes are living incarnations of the social relations of production and its corresponding property system. They are the Hegelian thesis. At the same time, another rising economic class embodies oppressed modes of production and serves as the antithesis. The contradiction between the oppressed material productive forces and the oppressive social relations becomes embodied in the class struggle between the ruling and rising classes.

The inevitable dialectic of history resolves itself in the triumphant revolution of the rising class. The revolution brings the relations of production and technology into peaceful harmony again until further technological development gives rise to new contradictions. From Marx's perspective, the first two great social conflicts of history were between political classes. These conflicts produced first oriental despotism and then feudalism.

When Marx described the movement to capitalism, however, he switched his categorical framework from *ruling versus ruled* to *capitalist versus laborer*. The exploitation is no longer the product of violent coercion on the part of the political rulers but the exploitation supposedly inherent in the wage contract.

Consequently, for Marx the key to the class conflict inherent in capitalism is the labor theory of value—that relic from classical economics. The labor theory of value explains the value of any economic good by appealing to the amount of labor it takes to produce the good. The theory tells us that the entire value of any good is solely the product of the labor it takes to make it. If the value of a good is determined by the labor used to make it, profits earned by the businessman are derived only by exploiting laborers. Marx argued that businessmen earn profits as the surplus value of goods over and above the wages necessary for labor to subsist. Because the value of a good is determined by labor, the surplus value pocketed by the businessman—the profit—rightfully belongs to the workers but is reaped instead by the capitalist. Finally, Marx posited what he called *laws of motion* that are pushing history toward the communist revolution. In doing so, he attempted to demonstrate the bad conditions that were allegedly brought about by capitalism.

The first of his laws is the accumulation and centralization of capital. As capital is accumulated by capitalists, the uniform rate of profit supposedly is doomed to keep falling. This is because in Marx's system the profit resulting from production has nothing to do with the quantity of capital accumulated but is only due to exploiting labor. Therefore, as capital investment increases, the rate of return capitalists reap by exploiting laborers in the same production processes falls.

At the same time that profit rates are falling, impoverishment of the working class will take place. Because capital will be increasingly centralized and the ranks of capitalists will be continually decreasing, the size of the proletariat class will continually increase. The proletariat will become increasingly poor as they are pushed harder by the capitalist. They will be increasingly alienated from their output due to the specialization inherent in the division of labor. Their misery will increase so much that eventually the proletariat class will rise up in revolution.

The final law of motion is that there will be increased business cycles. Marx thought that capitalism naturally produces business cycles that will get progressively worse. The primary cause for the increasingly severe business cycles is underconsumption. Marx asserted that increased production made possible by capital accumulation eventually outpaces the ability of exploited workers to consume. As businessmen overproduce, they will experience a falling rate of profit. When profits fall below a certain rate, the growth of capital ceases, and the economic crisis ensues. Capitalism therefore leads to over-accumulation of capital. Because in Marxist thought the market has no effective coordinating mechanism, all production and exchange is chaotic. There is no effective way to deal with the problem of underconsumption. Business cycles that are spawned by capitalism will become more and more volatile until the crack-up comes and the revolution ensues.

When the crack-up comes, the proletariat masses will have simply had enough and revolt against the capitalist class. According to Marx this proletariat revolution will begin to usher in the communist utopia. Private property is abolished. In this socialist workers' paradise each works according to their ability and each receives according to their need. Finally, this first stage of raw communism will be transcended by the final stage of history—full communism—in which there is superabundance and where almost no work is needed because everyone will be so much more productive. Such is Marxism in a nutshell.

As an economic system, socialism is open to a dual critique. We will first evaluate socialism from an economic point of view by answering the question of whether a socialist system can operate as a system of division of labor. We will then provide an ethical critique of socialism by addressing whether it is morally right.

THE ECONOMICS OF SOCIALISM

There are numerous problems with Marx's philosophy of history in general and his view of capitalism in particular. In the first place, when ascribing all economic change to material technology he never explains from where technology comes. He fails to recognize that technology is the product of ideas and not the other way around. Machines are the embodiment of ideas. The institution of private property that allows for the savings and investment that result in the accumulation of capital goods is the result of ideas. The proper causal path is from ideas to the alleged technological base.

Additionally, there is no logical link between technology and a class. There is no way for an economic class to embody either technology or productive relations. People from any social class can take advantage of the technology available if they act in suitable ways. In a modern economy, there is no strict distinction between the capitalists and the proletariat. While we have distinguished between capitalists, entrepreneurs, laborers, and consumers, we are distinguishing between economic functions and not necessarily people. Often the entrepreneur is also a capitalist in that he invests his own saved capital. Do not overlook, however, that because the goal of every producer is to gain income to spend on consumption, everyone is a consumer. Also, every laborer who saves part of his money and invests it in a retirement IRA or merely deposits some in a savings account is a saver and investor. He is a capitalist who exchanges money in the present in return for an interest payment in the future.

On top of all this, we have already seen that the market prices of economic goods are determined by the subjective value of all market participants. Economic value is not an objective entity that can be measured like feet and inches. Economic value is subjective. Consequently, we cannot use units of labor to measure or calculate the value of a good. The quantity of labor used to produce a good does not determine its price. There is not, therefore, an unfair exploitation of labor inherent in profit. Laborers receive an income determined by the DMRP of the marginal worker. Capitalists receive interest for satisfying time preference. And entrepreneurs reap profit as a reward for forecasting future market conditions better than their rivals.

In light of these problems, and many others we do not have time to explore, it does seem that the best explanation for the popularity of Marx's vision of socialism is due to a religious faith in the coming utopia. Some people want so desperately to live in an egalitarian socialist workers' paradise that they look past all of the theoretical problems with Marx's specific description of the dialectic and focus on the abundant life to be had by all in the socialist end of history. They interpret all of life in terms of this vision.

One thing that Marx and his followers left out, however, was any detailed explanation of how the economy is supposed to be organized in a socialist society. Merely asserting that everyone will produce according to their ability and everyone will have according to their need is no substitute for a working economy. Suppose that we overlook all of the theoretical problems with Marx's description of history. Suppose we graciously turn the other intellectual cheek in the face of his dialectical materialism and grant that we can all look forward to socialism coming soon to an economy near you. Still we are behooved to investigate the workability of such a system. So many people have tried and continue to try to impose their own version of socialism on their contemporaries. Considering whether socialism will indeed deliver the goods is a worthy exercise. Will socialism give life and give it more abundantly? Can social-ism even operate as a social system of division of labor?

Economic analysis shows three fundamental problems that social-ism cannot overcome. The first is the incentive problem, the second is the knowledge problem, and the third is the problem of economic calculation.

THE INCENTIVE PROBLEM

In a private property environment, people bear the full costs of their ac-tions and reap the full benefits. Therefore, they have an incentive to engage in productive activity. They have an incentive to only do those actions for which the benefit outweighs the cost. Producers have an incentive to make only those goods that return them a profit. Therefore, they also have an incentive to use resources carefully. They are encouraged to maintain and accumulate capital and invest that capital wisely.

In a socialist economy, however, the situation is very different. Because there is no private property, people do not bear the full costs or reap the full benefits of their actions. If Melody Singer, a producer of clas-sical records, is able to manufacture and sell five hundred thousand box sets of Sibelius symphony recordings in one year and that she could earn a two dollar profit on each set in the free market, her profit would one million dollars in one year. Suppose that she lives in a socialist country, however, and the central planning board decides that, even though she has great ability to produce interesting Sibelius compilations, she only needs $25,000 a year to live on. What does that do to her desire to produce Sibelius records? Her desire dissolves faster than you can say *Finlandia*.

Suppose on the other hand, that the central planning board decides that a family of four needs $100,000 on which to live even though neither the father nor the mother engages in productive employment. If they do nothing, they still receive their income. What does this do to their incentive to work? Their incentive to work likewise withers away faster than the grapes of wrath left on the vine to spoil.

The incentive problem is exacerbated by the pattern of incidence of the costs and benefits of shirking. In a socialist economy where income is not tied to the value of productivity (as it tends to be in the free market), the costs of shirking tends to be dispersed while the benefits of shirking tend to be individualized. If Joe Socialist Worker discovers that he will get paid no matter his performance, the cost of his shirking in terms of reduced output is spread throughout the economy. Everyone pays a relatively small increase in price. Because the increased price due to lower output is spread throughout a large number of consumers, people do not notice it as much.

The benefit to Joe of shirking, however, is entirely his. If he decides to only work half the time he is assigned while still receiving his government income check, the entire extra leisure time is his to do with as he pleases. He gets all the benefit of shirking, but bears only a very small fraction of the cost.

Every time people have attempted to put some form of socialism into practice the incentive problem has been observed. A dramatic example of this is the economic arrangement of the Pilgrims who landed at Plymouth Rock in 1620. Their first year was extremely hard with a large number of their party dying due to illness and not enough food. In his account of Plymouth Colony, William Bradford explains how he and others began to realize that the main reason their colony was not producing enough food to sustain the population is because of its socialistic arrangements. All produce was brought to a common storeroom and then distributed according to people's perceived need. Because of the above described incentive problems, the total quantity of food available for distribution was not enough to feed the colonists.

Bradford made the life-giving decision to allow people to work for themselves and keep their produce. Almost immediately this had a very positive impact on the total output of the colony. Food production increased and illness and starvation abated. The colony survived and then thrived and the rest, as they say, is history.

The incentive problem of socialism was also recognized by the truest of true socialist believers in the Soviet Union by the middle of the twentieth century. There is presently quite a market for motivational and propaganda posters from the USSR as a sort of souvenir of the Cold War. Many of them merely have images of Lenin or Stalin, but many are more narrowly focused on labor. They sport slogans such as "Work as if you are working for yourself," and the ominous, "We will not let you do a bad job."

Many socialists including Marx thought that the incentive problem was no problem because, they argued, incentives rule and corrupt only in a capitalist economic system. Under socialism, it is claimed that since everyone has an equal stake in the commonwealth of society, people will be motivated by their felt brotherhood with all mankind. More fanciful socialists such as Charles Fourier thought that socialism would be so productive that no one would have to work more than a few hours a day freeing up the majority of hours to pursue leisure and cultural activities. In such a society the temptation to shirk would have no force.

Surely experience and Scripture argues against such wishful thinking. People naturally care the most about those things closest to them. They are simply more inclined to be productive if their efforts benefit themselves and their families. A system that allows for such provision also tends to generate a larger total stock of goods available for charitable giving.

The Christian doctrine of fallen man also works against the socialist theory. Because of sinful human nature it is difficult for Christians (indwelt with the Holy Spirit) to always do their duty. There is no reason to expect an unregenerate person to persistently avoid shirking.

THE KNOWLEDGE PROBLEM

Another very important problem for an economy is the practical issue of dispersed knowledge. In order for a central planner to be a successful director of a society's economic affairs, he must make sure that all of a society's factors of production that are available are put to their best uses. Such decisions about the use of means are not simple. They require knowledge of which resources are available where, for what sort of production, and how these factors can be combined in the way that provides for the most output that people actually want or need.

The problem for socialism is that such knowledge is never centralized and wholly made available to a single economic czar or central planning board. Such knowledge is dispersed among the vast number of participants in economic society—the market division of labor. The best way to accumulate the right sort of factors and the best way to combine those factors toward the production of a specific product is known only to a relatively small number of people inside a particular industry. Magnify this fact times the number of different products that are produced in modern industrial economies found in more developed countries and you have an idea of the problem that would be faced if a single central planner attempted to direct all economic activity alone.

Economic progress is not merely a matter of advances in technology and scientific knowledge. More important is what F. A. Hayek called, "the knowledge of the particular circumstances of time and place."[1] One of the things that makes the division of labor more productive than a society where everyone produces only for his own use is that different people have different advantages over others regarding the unique information necessary to make wise decisions at specific times in specific places. That is why so much knowledge that is valuable in the performance of any occupation is learned via on-the-job training. Textbook, scientific knowledge gained in academic training is helpful no doubt, but to make such training truly operational it must be adapted to suit the specific circumstances in which decision makers find themselves. Again, no socialist central planner is in the position to accumulate and process all of the important requisite knowledge of specific time and place.

The only practical way for such knowledge to be harnessed for the most good to society is to allow for the market division of labor to develop free from government hindrance. In a free society where the factors of production are by owned private entrepreneurs, every participant has the incentive to be as productive as he can be and hence, put his unique knowledge to use for his own good as well as the good of society in the form of more goods that are desired by others.

THE PROBLEM OF ECONOMIC CALCULATION

These observations regarding human nature and the practical problem of centralizing dispersed knowledge have led some to the mistaken notion that socialism is a nice idea that works in theory but simply does not work very well in practice. A large number of people seem to long for a socialist utopia where all mankind treats each other like brothers on equal footing, but they also reluctantly grant that it is very difficult for socialism to work because people just do not have the incentive to work hard and produce for the entire community and because it is too difficult for central plan-

1. Hayek, "The Use of Knowledge in Society," 521.

ners to accumulate all of the knowledge necessary for a socialist economy to be very productive.

This notion is wrong. Socialism does not work in practice to be sure. The reason it does not work in practice, however, is because it does not work in theory. Socialism is both wrong in theory and a failure in practice. When we say that socialism is wrong in theory we mean that even if there were no incentive problems inherent in a socialistic economy and if the knowledge problem could be overcome it still would not work. We come to this conclusion by applying economic theory specifically regarding economic calculation.

The Achilles heel of socialism it turns out, is neither the incentive problem nor the knowledge problem but its failure to allow economic calculation. The impossibility of economic calculation under socialism is the reason why socialism always has and always will fail. The lack of economic calculation is why socialism can never work as an economic system.

To fully understand the consequences of the impossibility of economic calculation in a socialist economy we must keep in mind the chief economic characteristic of socialism. One entity alone chooses for the entire economy. Be it a lone economic czar or a central planning board, all economic decisions are determined centrally. In order to organize the economy, that person or group needs to determine whether the execution of various possible projects will increase the well-being of people. One does not after all live in a workers' paradise if the workers are left in worse shape than when they started.

Suppose for example that the central planning board wants to build more new houses to provide shelter to a growing population. There is more than one way to build a house and each requires different quantities and combinations of factors of production. Carpenters could use wood, steel, or cement for the structure; it could be sided with brick, wood, vinyl, or aluminum. The interior floors could be wood, carpet, or tile. The contractor could put in electric or gas heat. The heat could be in the floor or via airflow. What determines just how the carpenter will build the house? Which method should he choose?

Now multiply these decisions for every single good produced and you get an idea regarding the immensity of coordination that must take place in any economy. Not only would the central planning board have to make a plethora of decisions regarding the particular factors of production used in making each good but it also must determine the best location for each industry and best size of each plant and piece of equipment.

We cannot solve such problems by looking to technology. Merely because something is technically possible does not mean that it is a wise thing to do. Why do we not produce water synthetically in factories? We have the technology to do it. We know how to manufacture water but we do not do so. Why? Because it costs too much. Sellers would have to charge such a high price for water that not enough people would buy it to make it profitable.

How do entrepreneurs calculate profit? They do so by comparing the revenue they expect to reap from a production project with the total costs involved. How do they calculate their expected revenue? They take the quantity of goods they expect to sell and multiply it by the market price for which they think they can sell their goods. How do they calculate their costs? They sum the prices of all the factors of production necessary to make their good. They compare the price of the product with the sum of the prices of the factors of production used to make the product. If the price of the product is greater than the sum of the prices of factors of production then that project earns a positive rate of return. If that rate of return is larger than the price of time—the interest rate—that project is profitable and they will continue with it.

Did you notice one common tool used to calculate profit and to calculate the revenue and costs associated with producing and selling economic goods? Entrepreneurs calculate profit and loss using market prices. If the price of their product increases and the prices of the factors used to produce that product decrease, the expected profit resulting from producing that good increases, and entrepreneurs would be encouraged to produce that good. On the other hand, if the expected profit from producing and selling a good decreases because the selling price of the product decreases and the sum of the prices of the factors necessary to produce that good increases, the expected profit received after production decreases, so entrepreneurs are discouraged from producing that good.

A nifty characteristic of the free market is that the prices of both the product and the factors of production are determined by people's subjective value scales. The prices of the products are constrained by how much people really want them. The prices of factors of production are determined by their discounted marginal revenue product (DMVP) and the amount of income the services of the factor could generate in its highest valued alternative use. The DMVP of a factor is determined by the marginal physical product of the factor, the price of the goods it is used to produce, and social time preferences. The price of this good is constrained by how much people really want that good.

Because factors of production are scarce goods, if they are used in one production process any other production process must be foregone. Therefore, to reap a factor's MRP in one line of production means that the owner of the factor must sacrifice any other MRPs that he could receive if he employed his factor in any other use. The highest alternative MRP that a factor could reap but must be forgone is that factor's opportunity cost. Therefore, the prices of factors of production must be high enough to cover that factor owner's opportunity cost.

In a free market, entrepreneurs are encouraged to use factors only in ways that are profitable, and at the same time are being encouraged to use factors only in ways that are most highly valued from the point of view of the people who make up society. In a free market, entrepreneurs have the incentive to not waste resources, but to produce exactly what society wants in the least costly way possible thereby maximizing the wealth available with the scarce resources in existence. In socialism the necessary tool to calculate profit and loss—market prices—is gone. There are no real prices for factors of production in a socialist economy so there is no way for the central planning board to steer the production process.

The defining economic characteristic of socialism is that all of the means of production are owned by the state. Because the factors of production are not private property, there is no voluntary exchange for the factors. The state cannot exchange with itself. There are no entrepreneurs who own the factors of production and who, therefore, are free to buy and sell them if they so choose.

Additionally, without private property there is no real money. We learned in chapter 5 how the medium of exchange emerged as a result of voluntary exchange as people traded away less marketable goods in exchange for more marketable goods. The process continued until people traded the most marketable good in all markets. This most marketable commodity became used as the medium of exchange and is called money. If there is no private property, then there can be no voluntary exchange and therefore there can be no process by which a most marketable commodity becomes the general medium of exchange. Whatever so-called money there is in a socialist economy is merely artificial and more like ration coupons because in a socialist economy there are no market prices because there is no voluntary exchange. Prices are determined and fixed by the central planning board as are workers' incomes. If the economic authorities in a socialist economy determine both prices and incomes, they *de facto* determine how many of each good a household can purchase. In any case, there can be no prices for factors of production.

If there is neither voluntary exchange nor money in an economy there can be no market prices for economic goods. If there is no general medium of exchange in an economy, there is no single good that is traded against all other goods. Consequently, the central economic planners cannot reduce the market value of units of various goods to a common denominator. A carpenter cannot compare the economic value of bricks and wood, vinyl, or aluminum siding. He cannot compare the economic usefulness of a power nail gun and a hammer. Without money prices the central planning board cannot compare the market value of all of the different factors of production that could be used to produce each good. Additionally, without money prices, definite numerical expression cannot be attached to either waiting time or the duration of serviceableness of various factors of different qualities.

Some have argued that socialist managers could use labor hours as the unit of calculation. The root of this suggestion is the labor theory of value which was prominent in the writings of Adam Smith and David Ricardo. We have seen that it was swallowed whole and made a key part of his theory by Karl Marx.

The labor theory of value, however, fails on at least three counts. In the first place, the labor theory of value alleges value to be an objective quantity. We know, however, that value is subjective. The labor theory of value alleges that if it takes twenty hours of labor to produce both a Cadillac Escalade and a Ford Taurus, then both of them possess the same value. This does not happen because value is not rooted in objective amounts of labor, but in personal preferences.

Additionally, even if value could be reduced to objective units of labor, the labor theory of value disregards differences in the quality of labor. Obviously different people have different skill levels and are able to provide different quality services. If there exist differences in the quality of labor, which quality level should be used to calculate the value of a good? If it takes one person twenty hours of labor to produce a men's suit and another person ten hours of labor, what unit should be used to calculate the value of the suit? Twenty hours? Ten hours? Fifteen hours? There is no scientific way to decide.

Finally, the labor theory of value takes no account of the value of original factors of production. If all economic value is determined by the labor necessary to produce a good, then land and labor (neither of which is produced with the use of labor) should have no value. Land and labor services obviously do have value, so the labor theory of value is a failure.

The lack of market money prices and any other effective method of economic calculation greatly hinders any central planning board's ability to effectively plan the economic activity in a socialist economy. In a socialist economy, the central planners are charged with directing production toward best satisfying the needs and desires of the citizenry. However, because they cannot resort to market prices to calculate profit and loss, there is no guide, no rudder, no compass, no (insert your favorite navigation metaphor here). As Ludwig von Mises aptly quipped, "What is called a planned economy is no economy at all. It is just a system of groping about in the dark."[2] Such groping about in the dark necessarily results in waste at best and starvation and death at worst.

If socialist central planners do not use market prices to guide their decisions, what do they do? They often resort to quotas. A former Soviet economist who now teaches in the United States once told the story of how the central planning board in the USSR sought to cope with economic reality without the aid of market prices. In an attempt to successfully regulate nail production, central planners first tried quantity quotas. They charged managers placed over nail factories with producing a certain number of nails per month. These managers soon figured out that the easiest way to meet the quota was to produce a relatively large number of small nails. A surplus of small nails soon resulted, while a shortage of larger nails also became apparent. Builders had more than enough shingle nails, but they had little use for them because there were not enough larger nails to erect building structures that needed to be roofed and shingled.

After the nail imbalance was detected central planners changed their orders and required weight quotas instead of quantity quotas. They required nail factory managers to produce a target weight of nails. Managers soon decided that the quickest way to meet their monthly quotas was to produce a relatively small number of large nails. Not surprisingly, the nail imbalance pendulum swung the other way. Shortages of smaller nails appeared, while the USSR was glutted with larger nails. New structures went up more rap-

2. Mises, *Human Action*, 700.

idly, but they remained unfinished and without shingles because there were insufficient amounts of smaller nails to complete the job.

On another occasion Soviet central economic planners mandated weight quotas for manufacturers of chandeliers. Like nail manufacturers, chandelier makers discovered that the fastest way for them to meet their quotas were to make fewer, relatively large light fixtures. Such large chandeliers became a real problem because they were too heavy for the average ceilings in which they were hung and began to fall out seriously injuring several people.

The central planning board's reliance on weight and quantity quotas opened the door for much corruption. All along the chain of the bureaucratic command false numbers were passed along to superiors as evidence of meeting productivity targets. This culminated in the reporting of artificial output statistics. When in doubt Soviet economists were known to simply make up numbers. While stories of made up numbers and shortages and surpluses of large and small nails are amusing and tempt us to take the economic problem inherent in socialism lightly, the economic chaos that ensues under pure socialism is no laughing matter. In truth, such chaos is literally a matter of life and death.

The nail incidents occurred during a period of Soviet Union history after it had already abandoned full socialism. The USSR had made allowances for some semblance of market activity because it quickly became obvious that a socialist economy simply could not deliver even a minimum of the goods.

Whenever full-blown socialism has been attempted, history demonstrates its disastrous consequences. While it is commonly thought that the USSR possessed a socialist economy throughout its entire existence, pure socialism was in place for only four years—from 1917 through 1921. Soon after the Russian Revolution, Lenin established a socialist economy for the production and distribution of goods throughout the country. In doing so Lenin was way too optimistic about the ease with which socialism could produce abundant living. He thought, "The accounting and control necessary for this [socialism] have been simplified by capitalism to the utmost, till they have become the extraordinarily simple operations of watching, recording and issuing receipts, within the reach of anybody who can read and write and knows the first four rules of arithmetic."[3]

3. Lenin, *The State and Revolution*, 120–21.

Lenin mistakenly thought that the key to economic coordination was merely effective management. He believed that if managers could simply read reports, keep good records, and observe patterns of production, the economic decisions inherent in any economy could easily be made.

For socialism to work, Lenin knew this required putting in place a detailed and precisely calculated plan. Labor was assigned to various branches of production. Lenin nationalized all industry, which was overseen by the Supreme Council of National Economy, the central planning board.

The economy over which the Supreme Council governed was fully socialized. All banks were seized and controlled by the state. The government confiscated all industrial enterprises. All foreign trade was managed by the government and all labor strikes, even peaceful ones, were outlawed. Real estate could not be owned by private citizens. All mortgages of more than 10,000 rubles were annulled. Lesser amounts were converted to state loans. All output was delivered to communal warehouses. From these warehouses citizens were to draw goods out in accordance to their self-defined needs. Because of a lack of a market price system, it was very difficult to determine the location where the food should be distributed. As economic theory teaches, chaos occurred because there were no market prices that could adjust in response to changing market conditions. Drastic rationing became necessary as a result.

The economic plan of the Supreme Council was further hindered as the incentive problem kicked in. Peasants who were not eager to go along with all of the socialist policy began refusing to yield their produce to the communal distribution centers. They reduced their sowing acreage and buried some of their produce in hiding. Black markets arose placing further obstacles in the path of the central planners.

The result of Lenin's socialism was disaster. The economy came to a standstill. A great famine spread throughout the Russian countryside even as the government stepped up enforcement of the plan. The aggression against the peasants and the dearth of food resulting from socialist economic policy resulted in approximately six million deaths.

Even Lenin could see that the socialism that he and the Supreme Council had established was horribly destructive in terms of human life. Lenin understood that something had to be done and he developed his New Economic Policy which was in place from 1921 to 1928. The New Economic Policy was Lenin's compromise with the market system and an admission that the monetary incentives inherent in a free market were necessary to some extent—an implicit admission that centralized distribution was a failure.

Agriculture policy was de-socialized. Farmers were able to keep their surplus grain and instead pay a tax based on the amount of the surplus grain. Peasants were allowed to hire more labor and rent more land if they desired. In 1923 the USSR made legal again a farmer's right to farm for himself. Industry was also decentralized. While the most important industries making up approximately nine per cent of all production remained under full central planning, others were allowed to base operations on incentives with private/state cooperative partnership management. Along with the most important industries, the state still retained a monopoly on foreign trade and there were many regulatory constraints that were kept in place.

Nevertheless, the move back toward a freer market had many positive economic results. There was a decrease in starvation as more people began getting enough to eat. During the 1920s there was output growth in industries such as steel, coal, oil, electricity, copper, and cement. There were increases in employment and wages. A new bourgeois arose that alarmed orthodox Marxists and prompted a socialist reaction once Stalin achieved power.

The Stalinist economy, which began in 1928, took the USSR back to full central planning. Farms and industry were collectivized and sometimes relocated. Rural peasants were again forced to give up their produce to feed industrial workers in the cities. Production, consumption, wages, and prices were all directed toward the goal of the plan instead of consumer demand. Labor was entirely directed by the central planners, even if this required forced labor.

The consequences were similar and in some cases more acute than the USSR's first experiment with full socialism. There resulted a massive famine and chronic food shortages in urban areas. Approximately fourteen million people had died by the end of 1933.

The economic history of the USSR is the quintessential example for what economic theory suggests will happen under socialism. There was a disastrous decline in economic output under pure socialism. The USSR economy managed to progress under the relatively more market-oriented New Economic Policy. Growth slowed again during and after Stalin. The industries in which there was growth tended to be in the wrong industries in the sense that they continued to produce certain goods like steel and concrete far beyond the demand for those goods. Much of the increased statistical growth in these industries turned out to be economically wasteful. Output was heavily skewed toward military and industrial commodities like steel, oil, coal, and tractors and away from consumption goods. The USSR experienced chronic shortages of housing, schools, hospitals, and roads.

We see a similar decline when looking at Russian agriculture. From 1909 to 1913, annual grain exports from Russia totaled 10.5 million tons. In 2005 Russia was the world's largest importer of grain.

The overall consequences of socialism in the USSR were exactly what economic analysis indicates. Instead of generating life, prosperity, and a thriving culture, socialism produced starvation, death, and cultural malaise.

SOCIALISM AND CHRISTIAN ETHICS

We have seen that socialism fails in both theory and practice. Economic analysis tells us that pure socialism is unworkable because of the problem of economic calculation and the practical knowledge and incentive problems. Additionally, wherever pure socialism has been tried, many people have starved to death and their survivors suffered a lower standard of living.

However, material goods are not the be all and end all of existence. Man cannot live by bread alone. Some argue that while socialism is not as productive as free market capitalism, it is more moral than the free market. Because the market indeed does not determine what is true and right, we must examine the ethical claims of the advocates of socialism. We must submit socialism to a biblical ethical critique.

This is important in light of several different defenses of socialism or socialistic economic policy allegedly based in Christian ethics. One of the more radical schools of Christian thought advocating socialism is liberation theology. This is the theory that because of the Church's mandate to act charitably toward the poor and oppressed, the entire political and economic system of historic capitalism must be overthrown. Usually liberation theologians advocate that Marxist socialism replace the free market. If necessary, the revolution resulting in the removal of capitalism may include violence and indeed, the Church is implored to participate in this violence as positive Christian action.

Besides liberation theologians, others advocate a kinder, gentler socialism because it is thought that socialism is generally more Christian.

Socialism is allegedly based on motives of goodness, generosity, sharing, gentleness, and compassion. Two passages in The Acts of the Apostles are pointed to as the Scriptural foundation affirming socialism as the ethically right economic system. Both describe the Church in its infancy in Jerusalem. "And all who believed were together and had all things in common. And they were selling their possessions and belongings and distributing the proceeds to all, as any had need" (Acts 2:44–45). Two chapters later we read again of a seemingly communal arrangement among believers. "Now the full number of those who believed were of one heart and soul, and no one said that any of the things that belonged to him was his own, but they had everything in common. And with great power the apostles were giving their testimony to the resurrection of the Lord Jesus, and great grace was upon them all. There was not a needy person among them, for as many as were owners of lands or houses sold them and brought the proceeds of what was sold and laid it at the apostles' feet, and it was distributed to each as any had need. Thus Joseph, who was also called by the apostles Barnabas (which means son of encouragement), a Levite, a native of Cyprus, sold a field that belonged to him and brought the money and laid it at the apostles' feet" (Acts 4:32–37).

How are we to understand what the Scripture teaches about economic systems? Because economics is a science of human action, we must keep the Christian view of humanity at the forefront of our thinking. God has made mankind in his image to have dominion over the earth by being fruitful, multiplying, subduing, and filling the earth. God created mankind and the world in such a way that certain rights are the necessary means of fulfilling human destiny in God's universe. In order to fulfill God's mandate to have dominion in this fallen world of aggravated scarcity, we must take advantage of the division of labor and capital accumulation. An economy with a well-developed division of labor is also an economy with a very complex structure of production. In order for economic activity to be well-coordinated, entrepreneurs must be able to use free market monetary prices to calculate profit and loss. Money prices in a free market are only possible in an environment of private property. Consequently, our study of God's created order reveals that in order to follow God's command to have dominion, we must live in society that embraces private property. Therefore, socialism is not an option.

Additionally, Scripture itself contains many passages that together teach that a just society respects and protects private property, which disallows any sort of state-mandated socialism. God makes very clear in his word that rulers are not autonomous. They are not a moral law unto themselves. God calls rulers to act justly. They cannot kill their citizens at will. They cannot steal from them at will. God holds rulers to the same moral law as he does everyone else.

An obvious historical example of this truth is the account of Ahab, the evil king of Israel. Among the deeds of Ahab that are recorded are murder and the thievery of a vineyard. In 1 Kings 21 we read that Ahab had occasion to admire the vineyard of Naboth the Jezreelite. Ahab tried to buy it from Naboth but Naboth refused to sell. Ahab's wife, Jezebel arranged for Naboth to be charged with cursing God and the king, and as a result to be stoned. Upon hearing of Naboth's death, Ahab promptly took possession of it and claimed it as his own. For this action, Ahab was roundly condemned by God through the profit Elijah who told him: "You have sold yourself to do what is evil in the sight of the Lord. Behold, I will bring disaster upon you. I will utterly burn you up, and will cut off from Ahab every male, bond or free, in Israel" (1 Kings 21:20b–21).

One of the moral laws for all of God's creatures is his prohibition of theft. Throughout Christian history the commandment, "Thou shall not steal," has served as the moral foundation for private property. In his magisterial *Systematic Theology*, Charles Hodge wrote a lengthy commentary on the eighth commandment in which he concluded that this commandment communicates a divine right to private property. He argued that private property forestalls any form of socialism as ethically inferior. Similar conclusions were reached by R. L. Dabney. People individually do not have the right to take other's property. Likewise, the state has no moral right to collectivize the property of others.

If this is true, what are we to make of the two passages in Acts previously mentioned? In the first place, we must be clear regarding what the text is teaching. Clearly those early Christians were engaged in *voluntary sharing* of their own property. Their property was not confiscated and collectivized by either the civil or ecclesiastical rulers. Annanias and Saphira, who were struck down after failing to give to the church all of the revenue from selling some property, were not punished for *having* private property. The text makes it abundantly clear that they were punished not for keeping some of the revenue *per se* but for lying about their contribution. They represented the amount of money they gave to the church as the entire amount for which the property was sold. They lied and the Bible tells us that *lying to the Holy Spirit* is why they were both struck down in quick succession. Their condemnation actually assumes private property. Peter said to Ananias, "While it remained unsold, did it not remain your own? And after it was sold, was it not at your disposal?" (Acts 5:4). Peter here clearly presupposes the practical and ethical legitimacy of private property. Therefore the historical event recorded in Acts 5 can in no way be used as a justification for socialism.

At most, the social arrangements described in Acts 2 and 4 describe a community of believers voluntarily sharing in consumption, not centralized socialism of production. Each person brought their own consumption goods and shared them with those among the brethren who had need. They may have had a community of consumer goods but there is no indication that these Christians pooled and collectivized the means of production at their disposal in order to engage in centrally planned production.

Finally, even if we accept what is described in Acts is an example of socialism of consumption, there is no indication that God demanded that as an ethical requirement. There are to be sure, commands against greed and warnings against loving mammon and caring too much for the things of this world. However, there are no commands to socialize economic arrangements. Nowhere does scripture teach that private ownership is condemned explicitly or implicitly. In fact, commands against theft imply the ethical superiority of private property, which also implies that socialism violates Christian ethics.

THE ECONOMICS AND ETHICS OF SOCIALISM

Socialism, regardless of its specific variety, is both ethically unsound and an economic disaster. Socialism violates God's law regarding private property. The only way for the state to own all factors of production is to coercively assert itself as owner of all land, labor, and capital goods. This necessarily requires the state to seize other people's property, which clearly is theft and a violation of God's commandments.

Additionally, socialism attempts to violate the economic laws God has built into creation. Without ownership, there is no real exchange and without exchange, there are no market prices for consumer or producer goods that are a reflection of the subjective preferences of people in society. Therefore, economic decision makers have no way to calculate expected profit and loss from various production projects. They are left, consequently, groping about in the dark. The necessary result is waste, capital consumption, poverty and death. Regardless of the motives of those pushing for collectivistic economic reform, socialism is both an economic and ethical evil.

SUGGESTED READING

Hayek, "The Use of Knowledge in Society," 519–30. Hayek's classic article on the "knowledge problem" inherent in any social system.

Hodge, *Systematic Theology*, 421–37. Hodge explains that the commandment against theft mandates private property and rules out communism as a Christian social system.

Hoppe, *A Theory of Socialism and Capitalism*. This is an excellent treatise comparing the free market with the various forms of socialism.

Marx, *Capital*. The classic ideological root for the plethora socialistic experiments of the twentieth century.

Mazour, *Soviet Economic Development*. A brief history of the disaster of pure socialism in the USSR from 1917–21 and the extremely hampered Soviet economy from 1921–65.

Mises, *Human Action*, 685–711. Mises' excellent treatment of the economic problems of socialism.

———, *Socialism*. The greatest and most devastating analysis of the economics and sociology of socialism ever written. It contains a full explanation of why the economic calculation problem cannot be overcome in a socialist economy.

Rothbard, "Karl Marx," 123–79. Rothbard's outstanding analysis of the intellectual roots of Marx's communist ideology as well as an excellent exposition and critique of Marxist economics.

19

Fulfilling the Cultural Mandate

THE BOOK OF GENESIS is rightly called the book of beginnings. Not only does the word *genesis* itself mean beginning, but the foundation of knowledge dealing with so many issues of our lives is found there as well. This foundation includes the basis for all scientific inquiry including economics. From the very first chapter we find that we are in a teleological universe. God made all things for a purpose and gave man the first commandment: to have dominion over the earth. We are called to do this, however, in our present, fallen, and finite world featuring a scarcity of goods. Since banishment from the Garden of Eden, man has faced a central cultural dilemma: how is God's creation mandate fulfilled in a world of aggravated scarcity without people either starving to death or killing one another? This is not at all a moot point. Whether they recognize it or not, different societies seek to answer this question with every change of economic institutions and policies they make. History is full of stark examples revealing that different attempted solutions to our dilemma have resulted in widely different consequences. We will find that economic theory rooted in an understanding of man as a rational actor created in God's image teaches that to materially fulfill God's cultural mandate, we must take advantage of the division of labor, capital accumulation, and entrepreneurship. These three pillars of economic development require societal maintenance of the biblical ethic of private property.

THE CULTURAL MANDATE

The first command given by God to mankind is what has been variously called the creation mandate, the cultural mandate, or the dominion covenant. Even before sin and the fall of man, God told our first parents, "Be fruitful and multiply and fill the earth and subdue it and have dominion over the fish of the sea and over the birds of the heavens and over every living thing that moves on the earth" (Gen. 1:28). The cultural mandate is further developed in Genesis 2: "The Lord God took the man and put him in the Garden of Eden to work it and keep it" (Gen. 2:15). This mandate requires that man fill the earth, rule over creation, work at productive activity, and keep it from waste and spoliation.

One of the more controversial aspects of God's cultural mandate is the call to fill the earth with a growing population. In general there have been two broad perspectives on population: the pessimistic view and the biblical perspective. The pessimistic viewpoint at best sees population growth as something to be very concerned about and at worst a curse that will eventually doom us to poverty and starvation.

The modern pessimistic perspective on population has its roots in classical economics. The Reverend Thomas Malthus became very famous for his *Essay on Population*, first published in 1798 and revised and enlarged several times through 1826. The main thesis of Malthus is that fixed laws of nature will keep the population at the level of the means of subsistence. In other words, the laws of nature are such that the wages of people in general tend toward a level just high enough for them to have their basic needs met and no more.

Malthus based his conclusion on two postulates. Food is necessary for the existence of man, and passion between the sexes is necessary and will remain relatively constant over time. Malthus further argued that the power of population is indefinitely greater than the power of earth to produce subsistence for man. This is because he thought population left unchecked, increases geometrically while food increases arithmetically. If this is indeed the case, it would not take long for the population to vastly outstrip the ability of the earth to support everyone.

Given his assumptions, he was able to describe the process by which population and food adjust to maintain merely a subsistence level. Suppose that the population increases faster than food. The price of food would rise. This would mean those who are the poorest must live in much worse conditions and some in severe distress. Marriage would decrease because men would realize that they could not support a family, which will work to decrease population over time. At the same time, because of the increased present population the supply of labor increases, decreasing wages. The quantity of arable soil put to use increases, increasing the supply of food. The process continues until, as more food is produced and the population shrinks over time, the wages paid to labor are once again just enough to subsist.

In 1968, Paul Ehrlich famously updated the Malthusian story in *The Population Bomb*. He repeats Malthus' claim that population increases geometrically using the concept of compound interest to bolster his claim. He argued that the key relationship that explains population growth is the difference between the birth rate and death rate. The development of medicine was the final straw that tilted the scales toward a great reduction in the death rate resulting in the population bomb. People did not die as fast as new children were being born and the population exploded.

His book in 1968 was pessimistic at its very core. He claimed that because of the demographic changes already taking place, mass starvation would be inevitable. Ehrlich continues to warn about too many people and advocates comprehensive population control programs—using coercion if necessary—to curb our population woes.

© United Feature Syndicate, Inc.

The perspective of Ehrlich and others like him is markedly different from the biblical perspective. From the very beginning of his Word, God's cultural mandate indicates that he favors bountiful life. After creating the animals, "God blessed them, saying, 'Be fruitful and multiply, and fill the waters in the seas, and let birds multiply in the earth'" (Gen. 1:22b).

In verse 28, God repeats the same words to Adam and Eve. "And God blessed them. And God said to them, 'Be fruitful and multiply and fill the earth and subdue it'" (Gen. 1:28a). Knowing the connotations behind a few of the Hebrew words in this passage is helpful. The word translated, "Be fruitful," is the Hebrew *pârâh*, meaning to blossom or to bear fruit. The Hebrew *râbâh* is the word translated in English as multiply. *Râbâh* means become many or numerous, become great, grow, and increase. The word translated *replenish* in the King James Version of the Bible is the Hebrew word *male*, meaning to fill or overflow. The Septuagint, the Greek translation of the Old Testament, uses the word *plero'o*, which in the Greek means fill completely, so nothing is left over. God was telling Adam and Eve that he wanted the earth to be filled with people.

In the ninth chapter of Genesis, we find that the mandate to fill the earth continues after the flood. "Whoever sheds the blood of man, by man shall his blood be shed, for God made man in his own image. And you, be fruitful, and multiply, teem on the earth, and multiply in it" (Gen. 9:6–7). The Hebrew *sharats* translated as *teem* can also be translated as to swarm. Again, the idea is that of a tremendous number of people densely populating an area.

Not only is filling the earth with people a direct mandate, but God's Word views population growth as a blessing. Notice the command to be fruitful and multiply always comes in the context of God's blessing. God blessed the fish and birds, Adam and Eve, and then Noah and his sons as he commanded them to fill the earth (Gen. 1:22, 28; 8:17; 9:1, 7). A teeming population is presented as a blessing, not a curse.

God's language of blessing continued when he made his covenant with Abraham. God promised Abraham, "And I will make of you a great nation, and I will bless you and make your name great, so that you will be a blessing" (Gen. 12:2). God promised Abraham that part of the blessing involved Abraham having numerous descendants.

In Deuteronomy, population growth was promised as part of the blessing the nation of Israel would receive if they obeyed God. "He will love you, bless you, and multiply you. He will also bless the fruit of your womb and the fruit of your ground, your grain and your wine and your oil, the increase of your herds and the young of your flock, in the land that he swore to your fathers to give you" (Deut. 7:13).

God also revealed in Deuteronomy that population decline was one curse that God would send to his people for disobedience. "Whereas you were as numerous as the stars of heaven, you shall be left few in number, because you did not obey the voice of the Lord your God" (Deut. 28:62).

Not only does God view population growth a blessing on a national scale for Israel, God's Word also treats children as an individual blessing to parents. "Behold, children are a heritage from the Lord, the fruit of the womb a reward. Like arrows in the hand of a warrior are the children of one's youth. Blessed is the man who fills his quiver with them! He shall not be put to shame when he speaks with his enemies in the gate" (Ps. 127:3–5). "Blessed is everyone who fears the Lord, who walks in his ways . . . Your wife will be like a fruitful vine within your house; your children will be like olive shoots around your table. Behold, thus shall the man be blessed who fears the Lord" (Ps. 128:1, 3–4).

In the New Testament, God's promises of numerical growth for Israel broaden to include the extension of the people of God as believing Gentiles are grafted into the olive tree described in Romans 11. The conclusion we can take from the testimony of Scripture is that God has given us the mandate to be fruitful, multiply, and fill the earth with people. If God instructs us to fill the earth, it seems that we should not be filled with angst about some future population disaster. We are to view the birth of children as a blessing and not tinder poised to set off a catastrophic bomb.

The cultural mandate further requires that we cultivate creation. We are not called to live merely in an egalitarian harmony with nature, but to transform nature as we develop civilization. We are to subdue the earth and have dominion over it. The Hebrew word translated as dominion communicates the idea of dominion *for use*. Additionally, we find the cultural mandate reiterated in Genesis 2, as we see that Adam is to keep and till the garden. This is a command to cultivate both nature and himself. Note that the Garden contained all sorts of vegetation that was both pleasing to the eye and good for food (Gen. 2:9). We are to cultivate both what is beautiful and what is useful.

The cultural mandate is the basic, integral calling of mankind, and was given to man before the fall. The economic activity necessary to fulfill this mandate is not the result of sin, the fall, or the curse, but is organically linked to our nature as beings made in the image of God with moral and rational faculties. Furthermore, there is no evidence that God revoked his mandate after the fall. In fact, God repeated much of it to Noah and his sons *after* the flood. Therefore, fallen humanity continues to respond to the cultural mandate.

Mankind's response to the cultural mandate is similar to Adam's. In the second and third chapters of Genesis we find that Adam responded to God's command by having children. He demonstrated dominion over the created order in part by naming the animals (Gen. 2:19–20). He did this after observation and analysis, and thus we have the beginnings of science. He also grew crops, which necessitated the plowing of land (Gen. 2:5, 15). Faithfully responding to God's mandate requires humans to make concrete changes in nature. We also find Adam composing history's first poem after his initial glimpse of his newly created wife (Gen. 2:23) and in so doing Adam's response marks the beginning of artistic activity.

What we call civilization is essentially the outworking of man being the image bearer of God. We saw in the first chapter of this book that part of the image of God is our ability to think logically and act purposefully. Additionally, the image of God includes our desire and ability to act creatively, taking elements that God has created and combining them in new ways to achieve certain ends. The Christian view of man recognizes that he has a special relationship to the earth. Mankind was, after all, formed out of the dust and is in turn called to work the earth and keep it. Furthermore we are commanded to use our skills and abilities to develop the potentialities that are in the earth waiting to be discovered and realized.

Fulfilling the cultural mandate requires that we use our skills and abilities to develop the potentialities in the created order in a way that is glorifying to God. Fulfilling God's dominion mandate, therefore, requires wise ruling, working, and keeping. The possibility exists that we rashly try to draw too much from creation too quickly, make changes too abruptly, or do so without replenishing the earth. We could conceivably cut down all forests without replanting. We could dump dangerous chemical by-products of production into the waterways. Doing so will seriously diminish the earth's fruitfulness, spoil part of creation, and abrogate our responsibility to keep the garden.

We can, however, err on the other extreme and act as if nature is a museum and we are its curator. We can attempt to keep the earth in its pristine natural state by refusing to alter it in any way. We could prohibit all development, outlaw all new housing construction, and refuse to allow the production of any new factories. If we act in this manner we will fail to reap the resources necessary for our continued existence and the very cultural development mandated by God.

Our call to fulfill the cultural mandate wisely is made more challenging because of the fact of scarcity. Even in the Garden of Eden, Adam and Eve had to deal with the scarcity resulting from their finiteness. They could only do one thing at a time and therefore had to allocate their labor, land, and time to their most preferred ends. However, things were made much more difficult as a result of what we read about in the third chapter of Genesis.

As the result of Adam's disobedience, God cursed all parties involved. This curse is not only spiritual in nature. Much of it is also involves material creation. God told Eve, "In pain you shall bring forth children" (Gen.

3:16). He tells Adam, "Cursed is the ground because of you; in pain you shall eat of it all the days of your life; thorns and thistles it shall bring forth for you; and you shall eat the plants of the field. By the sweat of your face you shall eat bread," (Gen. 3:17–19). This obviously implies that in order to produce goods necessary to sustain and enjoy life, he had to work much harder than before the Fall. Instead of the ground naturally bringing forth bountiful fruits and vegetables, it now naturally brings forth weeds. In order to produce food, man must first expend much effort clearing weeds off the land and remain ever vigilant against weeds once crops are planted. As the result of the fall, man was banished from the Garden of Eden into a world of greatly magnified scarcity, where it became much more difficult to obtain one's daily bread.

In light of the cultural mandate and the nature of the cursed ground, two questions come to the forefront. How do we develop God's creation wisely? How do we fill the earth with people without our starving to death or killing one another in a barbaric struggle for survival? These are questions that must be dealt with and answered if we are going to fulfill God's cultural mandate in a way that glorifies him.

ECONOMIC EXPANSION AND DEVELOPMENT

Multiplying the population and subduing and exercising dominion over the earth requires economic progress. First and foremost, the cultural mandate requires survival. We cannot have dominion over anything if we are dying of starvation. The cultural mandate further requires that all of us individually develop our potential and the development of the potential of nature. Doing these things requires economic development. The cultural mandate requires an economic system that can support large numbers of people each possessing different skills and abilities. It requires the production of the right tools and other man-made objects that allow us to transform nature. It necessitates the building of buildings, the fashioning of clothing, the writing of books, the cultivation of the arts, and the production of a myriad of goods necessary to sustain and enjoy life. Fulfilling the cultural mandate, therefore, requires us to know something about the nature and causes of economic expansion and development.

Economic progress has been one of the perennial subjects for macroeconomics and indeed for the history of modern economics. We get an indication of this by examining the title of that economic best-seller of

1776, Adam Smith's *The Wealth of Nations*. Smith's title seems to indicate that the book is an investigation into the macro economy because it surely focuses on the national economy. However, we get a deeper insight of Smith's interest by looking at the full title of the book, *An Inquiry into the Nature and Causes of the Wealth of Nations*. The full title implies that Smith did not want to merely focus on the *national* side of national wealth, although he obviously thought that an important topic. He attempted to provide a better understanding of the *nature* of the wealth that nations possess and he further wanted to investigate the *causes* of national wealth. In other words, what causes nations to grow in prosperity?

The question of the source of prosperity is not merely academic. From a societal point of view the question is often a matter of life and death. People who live in the United States and other industrialized nations often forget this fact. They tend to take prosperity and relative wealth for granted.

As explained in chapter 16, even those classified as officially poor by the U. S. Census Bureau are relatively wealthy compared to those living in less developed countries. To be considered poor in America is not the same as being seriously destitute.

Because of relative material comfort, people in more developed countries too often assume that wealth is the norm and that poverty is the anomaly. In fact, historically, the opposite is true. For millennia, poverty was the norm for the majority of people. Conditions did not really change much from the fall of our first parents until the late middle ages.

For most of history, death came much faster than it does today. The average life span for a person born in twelfth century Europe, for example, was less than thirty years. Half of all people born then did not survive childhood. Famine and hunger were common. In the fifteenth century, the average French farmer produced approximately 3.25 pounds of wheat per man-hour of labor. To get an idea of how low that is, consider that today the average farmer produces 857 pounds per man-hour. Because of the limited amount of food produced, malnutrition was common. There was little rational medicine. In the fourteenth century, Europe had to contend with the plague. There were no antibiotics, but a lot of spurious treatments such as blood-letting. The vast majority of people were illiterate and ignorant. Education was only for the very rich. Few people had more than five or six years of schooling. Housing was crowded and people had few opportunities to better their material lot in life. The struggle for mere survival was a high priority.

How we got here from there presents itself as an interesting question. Why is it that for centuries most people lived in relative poverty and then experienced great increases in wealth? Economics helps to answer that question. The Bible makes it clear that ultimately it is God who brings prosperity and takes it away. Adam and Eve, when they were banished from the Garden presumably left with only the skins on their back. They had to make their way the best they could with the help of God's Providence and by productive activity. Psalm 75:6–7 tells us that it is the Lord who decides who prospers and who does not. However, God usually acts through the laws that he has established in creation. Just as God heals people through the means of medicine used in conjunction with the biological laws he has set in place, God providentially brings prosperity to those people whose actions are in harmony with the economic laws that he has established. Therefore, we can apply our understanding of economics in answering the question of why some people and nations develop economically and some do not.

Economic theory identifies three components necessary for economic progress: the division of labor, capital accumulation, and entrepreneurship. Understanding the nature of these sources of development helps us wisely fulfill the cultural mandate. Societies that fail to heed economic truth do so at their own peril.

THE DIVISION OF LABOR

From the economic principles established in this book, we can identify three engines of economic expansion and development. The first of these is explained in detail in chapter 4 where it is noted that one of the societal benefits of exchange is that it allows for specialization and the division of labor, which opens the door to increased productivity and hence, economic development.

The division of labor and human cooperation is the fundamental social phenomenon. Society is what makes fulfilling the cultural mandate possible. Society is the cooperation of people with various functions, abilities, skills, and gifts. The differentiation and specialization in tasks are inevitable as societies mature. As people discover the unique aptitudes that God has given them and as they develop these aptitudes, they become relatively more suited for some tasks and less suited for others.

The biblical record reveals that the division of labor has been part of human history from the very beginning. Eve was made for Adam as his helper. God said it was not good for man to be alone (Gen. 2:18). Adam needed help from a complementary person to achieve his calling. We find the division of labor in the second generation of humans as well. Abel was a keeper of sheep, but Cain was a farmer. We find further specialization and division of labor among Cain's descendants. In the fourth chapter of Genesis, Moses records that Jabal was a cattle rancher, Jubal a musician, and Tubalcain specialized as a metal smith. In the book of Ezekiel we find a list of the many different goods that were brought to the commercial center Tyre from many different countries. Charles H. Dyer identified thirty-seven different types of goods brought from twenty-three different cities or regions to be sold in Tyre. For example, from Tarshish was brought silver, iron, tin, and lead. From Rhodes was brought ivory tusks and ebony. Producers in Damascus brought wine and wool to Tyre and producers in Arabia brought lams, rams, and goats to the market there.[1] In 1 Corinthians we even have an exposition by Paul explaining that a well functioning church also takes advantage of the division of labor (1 Cor. 12:12–31).

The reason that the division of labor is so prevalent among society is that cooperative action is more efficient and productive than the isolated production of self-sufficient individuals. The division of labor allows us to obtain more economic goods using the same quantity of resources. There are many reasons for this.

God has created mankind with unequal abilities in human laborers. Some people are naturally gifted with manual dexterity and specialize in mechanical pursuits. Others are naturally gifted intellectually and become doctors, lawyers, scientists, clergy, or such. Others are gifted artistically and so forth. God has also providentially distributed natural resources unequally over the earth. Some regions contain large quantities of oil that are relatively easy to extract. Other regions have plentiful deposits of iron ore. Some countries have neither, but have a climate perfect for agriculture. Others have waterways that are excellent for transporting manufactured goods or for generating electricity. Capital goods are also unequally distributed. Different people have different quantities and kinds of tools, build-

---

1. Dyer, "Ezekiel," 1281.

ings, and machines with which to work. Because different people possess different human abilities, natural resources, and capital goods, they tend to be relatively more productive at certain tasks and relatively less productive at other tasks. These differences and the ability to produce more in certain lines of production create an incentive to specialize in production.

People find it in their interest to specialize in making those goods at which they are relatively better at producing. They will produce those goods at which they have a comparative advantage. People have such a comparative advantage if they are the low-opportunity cost producer of a particular good. As everyone specializes in producing goods they can make at the lowest cost compared to anyone else, the total output of all members of society increases. As people exchange their output with others who have participated in the division of labor, they are also able to consume more than they could without the division of labor.

As the division of labor develops, it produces a number of social effects. Geographic regions become further differentiated. As people of different regions recognize the comparative advantages they have as a result of their natural resource and labor endowments, they begin investing in capital goods that are suitable for the tasks at which they are relatively more efficient. Because these capital goods are more suited for these same lines of production rather than others, regional differences become even more pronounced. Suppose for example, that people dwelling in ranging grassland recognize that their land is more suitable for cattle ranching than growing corn and soybeans. These people will begin ranching and investing in capital goods such as trucks, barbed wire, branding equipment, and water tanks all used for cattle ranching. Because these capital goods are specifically useful for ranching and not for growing corn, the people become even more suited for raising cattle and even less suited for raising corn or soybeans.

Additionally, the division of labor intensifies the differences in the abilities of human labor. As a person specializes in one task for a period of time, his skills in that task increase. A person who spends his time raising cattle and none of it growing corn becomes even more suited to raise livestock and even less suited to grow grain. As the differences become more acute, people become even more productive, thereby further increasing the total goods that can be produced and consumed by society.

Finally, and very importantly, taking advantage of the division of labor makes it possible for the use of machinery in the production process.

Without the division of labor everyone would have to personally produce everything they consume. All production processes would be very short. There simply would not be enough time to produce a machine and then produce food, clothing, and shelter for oneself. The division of labor makes it possible for some people to specialize in growing food, tailoring clothes, and building houses and for other people to specialize in producing combines, sewing machines, hammers, and power saws that make it easier for their respective work. In other words, the division of labor makes it possible for people to take advantage of the use of capital goods, which greatly increases the productivity of labor. Again, the development of the division of labor opens the door to increased productivity, raises real incomes, and increases societal wealth.

By now, you should be getting the picture that if our desire is to increase the number of economic goods available to a society, the division of labor is a great thing. So how do we get it? We can only benefit from the division of labor if people are free to exchange the goods that they produce. As Adam Smith said, "The division of labor is limited by the extent of the market."[2] This means that being specialized in the production of a particular good only pays if there is a market for that good. What do we mean when we say there is a market for a good? A market for a good exists when people want it enough to pay for it. In fact, they must want a good enough so that they are willing to pay a high enough price to cover all of the production costs, including interest. On a desert island with only four inhabitants, it will most likely not pay for one of the four to specialize in doing rocket science. There is insufficient demand for rocket science on that island. In the civilized world, however, there is a large enough demand for rocket science that a good number of people can earn a very comfortable living specializing in the field.

Understanding how the extent of the market affects the division of labor allows us to identify part of the economic development process. As the division of labor expands, productivity increases. As people are able to produce more, their wealth increases and so does their ability to trade for other goods. As people's ability to trade increases, so does the market for other goods. As the market for these goods expands, it makes it possible for people to develop the division of labor even further increasing productivity even more, increasing incomes even more, increasing the market

2. Smith, *The Wealth of Nations*, 17.

for other products even more. This makes it possible to even more fully develop the division of labor, which continues the process even further. The more developed the division of labor is, the more productive society becomes and the more the economy expands and develops.

Remember, however, that we can only benefit from the division of labor if we can exchange what we produce. Carpenters cannot eat their houses. They are able to live only as they are able to sell the houses that they build so that they will have money to buy food and clothing from those who specialize in producing those goods. Remember that one of the conditions necessary for exchange is ownership. People cannot exchange what they do not own. We can only engage in exchange in an environment of private property. Therefore, in order to take advantage of the division of labor and benefit from the economic development that flows from it, society must have and defend private property.

CAPITAL

The sort of complex, highly developed division of labor observed in more developed countries would be impossible without capital goods. Many tasks in which people specialize would be simply impossible to accomplish without the aid of capital goods. Another source of economic development, therefore, is capital accumulation.

Capital is the whole complex set of goods destined for production evaluated in money terms. It is the sum of the monetary value of all a firm's assets minus the sum of the monetary value of all of a firm's liabilities that are dedicated to that firm's productive operations. These assets may consist of land, buildings, equipment, tools, goods of any kind and order, claims, receivables, cash, and so forth. Capital goods are produced means of production: tools, machines, buildings, and intermediary goods.

In chapter 7 it is explained that the use of capital goods increases the productivity of the user. They allow people to produce a greater quantity of output in the future. They allow people to produce some goods that could not be made at all without capital goods, such as watches, automobiles, or iPods.

You also found that producing capital goods requires a longer production process. In order to obtain capital goods, it is necessary to save. Saving is defined as the restriction of consumption. The transfer of saved resources in the formation of capital is called investment. Because capital goods are perishable, they must be replaced with further investment. At any moment in time, therefore, each producer has the option of accumulating capital, maintaining capital, or consuming capital. Accumulating and maintaining capital requires a certain amount of saving. Consuming capital requires only that producers use up capital.

The choice regarding whether a producer is going to accumulate, maintain, or consume capital depends upon how much that producer values present goods over future goods; it depends on his time preference. The higher a person's time preference, the more present-oriented he is. He tends to consume more and save less. If he saves less, he is able to invest less. If his time preference is high enough, he will consume capital, and over time have fewer capital goods with which to work. Fewer capital goods results in a shorter structure of production and less productive labor. Society reaps lower total output. Capital values and real incomes fall. With lower output and real incomes, society's standard of living falls as well.

The lower a person's time preference is, the less present-oriented he is. People with lower time preferences tend to consume less and save more. An increased pool of savings makes possible increased investment, which will result in increased capital funds. Over time, people will have more capital goods at their service. More capital goods allow for lengthening the structure of production. Labor becomes more productive. Aggregate output increases. Capital values and real incomes increase as well. With higher output and increased real income, society's general standard of living rises. We see then, that a chief determinant of whether an economy expands or contracts is the size of the stock of capital.

But what about technology? Isn't it true that much of our productivity gain is the result of new technology? To answer, we must first have fixed in our mind what technology is. Technology is the knowledge regarding how to do something. Because all production processes require a recipe or knowledge, it is true that technology does set a limit on our production. Technology limits how we can conceive of combining the various factors of production we have at hand in order to produce economic goods.

Capital, however, is a narrower limit. For technology to be usable, and hence relevant for production and economic expansion, it must be bound up in actual capital goods. For someone to benefit from technology it is not enough merely to know that a machine suitable for a task exists; it must be possessed. A secretary who knows about an electronic device to type and edit written documents if programmed correctly does not benefit if she has no have access to an actual personal computer. Consequently in order to take advantage of technology we must have capital investment. Therefore, capital is a more important engine for economic expansion than technology. Without capital investment, technology is of no use. With capital investment, technology will advance as entrepreneurs continually seek to use better, more efficient, capital goods.

In order for a society to experience economic development and enjoy prosperity, it has two choices. Society can build up capital over time by using its own savings, or it can accelerate the process of accumulating capital by taking advantage of the savings of others. Two different paths of channeling other people's money toward capital accumulation in less developed countries have been taken.

The general effect of foreign investment is the transmission of prosperity from more developed countries to less developed countries. As shown in chapter 8, in a free market investment flows from projects and industries of lower returns to projects and industries of higher returns. This same principle holds true for geographic regions. In a free market with no barriers to trade and contracting between citizens of different countries, investment is allocated out of areas of relatively low returns to areas of relatively high returns.

Without foreign direct investment, those living in less developed countries are forced to accumulate their own capital. As history has amply demonstrated, this is a long, drawn out process. It requires laborers to work for many years in relatively harsh conditions until enough capital is accumulated to raise their productivity sufficiently to generate the wealth necessary for companies to update their facilities and make for a better and safer working environment.

If, however, foreign investment is allowed, less developed countries benefit from the saving and investment of capitalists from more prosperous, more developed countries. There are a number of ways capitalists can invest in productive projects in less developed countries. They can buy shares of stock or bonds issued by companies in those countries. They can directly invest in business ventures in less developed regions as well. Additionally, companies that are headquartered in a country that is relatively more developed can use some of their capital to set up operations in a less developed region. Usually, the reason for doing so is the relatively less expensive labor available in such areas. Such forms of business organization are called multinational or transnational corporations. A transnational corporation is a company that crosses international borders and, therefore, has operations in two or more countries at the same time. A clothing company for instance, may be headquartered in New York City, but have a substantial amount of their manufacturing facilities overseas in countries like China or Indonesia. Economically, this is no different than a company that has part of their operations in New York and part in Pennsylvania. Ultimately describing a company as a multinational corporation is a political distinction, not an economic one.

Economic history bears out the importance of foreign investment in economic development. From 1750 to 1850 Great Britain was the world's outstanding economic powerhouse. Great Britain's economic expansion was not due to the mental capacities of its citizenry or the Britain's endowment of natural resources. There were people just as knowledgeable living elsewhere and countries with even more natural resources than those found on British soil. The wealth produced in Great Britain was the result of savings and capital accumulation.

The rest of the western world caught up to the British economically after 1850 due to British foreign investment. As British capitalists invested in business ventures in other countries in Europe and North America, the capital stock in those areas rose correspondingly. By the latter half of the nineteenth century the capital cities of Europe had light because British gas companies supplied them with gas. Railroads and many other industries in the United States were partially financed by British investment. Railroads, harbors, factories, and mines in Asia were also the product of British capital. In all of these cases, British capital investment allowed for the accumulation of more and better tools, which increased the productivity of the indigenous labor population. The wages of local laborers, therefore, increased over time.

The other path of channeling foreign money toward economic development in less developed countries is the use of foreign aid. This is the way of interventionism. Foreign aid is the transfer of tax revenue from one government, usually that of a more developed country, to another government, usually that of a less developed country. Foreign aid is a specific manifestation of taxation and government spending. As such, all of the economic consequences of taxing and government subsidies identified in chapter 16 apply to foreign aid. Taxation levied in more developed countries results in capital consumption in these countries. People in more developed countries will be both less willing and able to save and invest in capital accumulation and over time their capital stocks will decrease leaving them less productive and relatively impoverished. This, of course, will make it harder for more developed countries to provide aid.

On the other hand, the subsidies received by the governments of less developed countries provide an incentive for such states to keep their countries poor. Too often state rulers have spent foreign aid money on self-aggrandizing projects that provided little or no economic benefits to their subjects. Often the money merely ended up in a dictator's Swiss bank account.

Even if rulers of recipient countries do hold economic progress as a high priority, foreign aid is administered by state bureaucrats, not private entrepreneurs. Capital invested in less developed countries by private entrepreneurs tends to be invested more wisely and productively than is foreign aid. Entrepreneurs who invest in the production of those goods most in demand by customers get to keep their profits. Those who invest in the production of those goods not in demand by customers bear losses. Therefore, there is an incentive for entrepreneurs to wisely invest capital. There is also an incentive for capitalists to provide capital to those entrepreneurs who they think will be profitable. These incentives do not exist in the case of foreign aid. Because foreign aid is being allocated by government bureaucracies, there is a great tendency for the money to be squandered on wasteful projects and sometimes even pocketed by the rulers of the governments receiving the money. The work of economist P. T. Bauer has ably demonstrated that foreign aid is neither necessary nor beneficial for achieving the goal of economic progress in less developed countries.

ENTREPRENEURSHIP

Because increases in the stock of capital result in increases in the productivity in land and labor, there is a positive relationship between the size of the capital stock and economic expansion. In order for economic development to continue over time, it is important not to waste the capital that has already been accumulated. Capital must not be needlessly consumed because it was sunk in unproductive uses. This is where the third engine of economic development, entrepreneurship, plays a major role.

If the previously accumulated capital stock is employed unwisely it will shrink over time resulting in economic decline. Merely because a capitalist sits on a pile of capital that he is ready to invest does not guarantee that he will have an equivalent or larger pile of capital to invest next year. It is entirely possible that he could invest his present capital on ventures producing goods that are unwanted, in which case he will not reap enough revenue to maintain his capital stock.

In order for the stock of capital to grow over time, it is necessary for time preferences to decrease. But it is also helpful for entrepreneurs to make wise choices. As economist Ludwig von Mises said, "Capital does not 'beget profit'. Profit and loss is entirely determined by the success or failure of the entrepreneur to adjust production to the demand of the consumers."[3]

Capital is accumulated out of savings. Savings is income held back from consumption. If income declines, then savings declines assuming there is no accompanying decrease in time preference (and there is no reason to think that there would be).

Capital accumulation, therefore, also depends upon the wise use of capital. Waste is possible because production decisions in the present are based on a forecast of uncertain future market conditions. If a producer forecasts incorrectly, he will use his capital to make something people do not want enough for him to sell his output at the prices he anticipated. His income falls and his capital is consumed. Successful entrepreneurship, therefore, results in the wise use of capital.

While entrepreneurship is necessary for any action, its importance is heightened because of the division of labor. People who participate in the division of labor are necessarily producing goods more for the market than they are for themselves. They must therefore forecast the subjective values of other people.

In fact, entrepreneurship itself becomes part of the division of labor as some people become better at forecasting future market conditions than others. It is hard to count the times, as an economist I have been asked what the future holds for the stock market or the economy in general. If economists, by virtue of their being economists could make accurate quantifiable predictions of future market conditions, most would no longer be at university posts, but instead would be wealthy business tycoons and investors. The fact that most economists (including myself) are not tycoons is evidence that not every economist is a good predictor of future market demand. The entrepreneur performs the valuable social

3. Ludwig von Mises, *Human Action*, 295.

function of dealing with uncertainty, coordinating diverse land, labor, and capital goods owned by numerous people, and directing production into the most socially valuable ends.

In order to direct factors of production toward their most valued uses from the point of view of society, entrepreneurs need a way to make objective decisions regarding the subjective values of those living in society. Money prices are what allow for making such decisions. Money prices make it possible for entrepreneurs to compare the desirability of different production projects, because monetary prices indicate the relative scarcity of economic goods. Goods with higher demand relative to supply are bought and sold at higher prices than those goods with lower demands relative to supply. Money prices are objective units of money that are voluntarily exchanged for all other goods.

The fact that money prices are concrete amounts of money allows entrepreneurs to make meaningful comparisons of social value between different consumer and producer goods. We learned in chapter 5 that money prices reflect the subjective value of buyers and sellers, but that these prices exist as objective units of money. The subjective values of buyers and sellers undergird the supply and demand schedules in every market. These supply and demand schedules determine the objective money prices for which goods are bought and sold. Prices of consumer goods are a reflection of how consumers value such goods given their supply. The prices of factors of production reflect the values of those factors in their alternative uses.

Entrepreneurs use money prices to calculate profit and loss. They compare the price of the final product with the sum of the prices of the factors of production used to make the final product. If the expected price of the final product is greater than the sum of the prices of the factors of production, the entrepreneur will produce that good. If the price at which the producer expects to sell his product is less than the sum of the prices of the factors used to make the good, he will refrain from producing it.

Entrepreneurs earn profits by best satisfying consumers. If an entrepreneur forecasts the future market conditions more correctly than others, he will make those products that have the highest buyer demand relative to their supply, thereby allowing him to charge a price higher than his costs of production. Only then will he reap a profit. If he forecasts incorrectly, however, he will invest capital and buy the services of factors of production in order to produce a product that will not bring a price high enough to cover costs. He will reap a loss. Entrepreneurs earn a profit then, only by best serving customers. Because the price of consumer goods reflects the social value of those goods and the price of the factors of production reflects the values of those goods in their alternative uses, entrepreneurs reap a profit only by providing those goods that people value the most in the least costly manner.

Entrepreneurs have an incentive to serve customers to the best of their ability because in a free market they are what economists call the *residual claimant*. They keep whatever profits they earn and bear whatever loss is incurred. Here we see the real beauty of the free market price system at work. The same money prices that allow entrepreneurs to calculate profit and loss give them the incentive to act on that calculation thereby encouraging them to produce just those goods that are demanded most by society. The free market price system encourages the productive use of capital and discourages wasteful use from the perspective of others in society.

This profit and loss process is continual. Entrepreneurs cannot rest on past success. As soon as the entrepreneur fails to best satisfy consumer desires, he will reap losses. Those who cannot consistently earn profits and avoid losses will eventually be forced to give up their roles as entrepreneurs because they exhaust their capital funds. Their own funds will dwindle if they continue to reap losses and no one will want to loan them anymore. Again we conclude that successful entrepreneurship guided by market prices keeps accumulated capital from being consumed and allows for productive capital investment to take place so that the capital stock can grow, as long as time preferences do not increase.

What then, is necessary for successful entrepreneurship to occur? Clearly we must have a monetary economy. Economic calculation is only possible if entrepreneurs have money prices for various goods to compare. Additionally, in order for the decisions of entrepreneurs to be aligned with societal wants, the money prices they use to calculate profit and loss must be free market prices because only such prices are manifestations of the subjective values of the buyers and sellers in society. Consequently, in order for entrepreneurs to perform their social function a system of voluntary exchange is needed. We saw in chapter 5 that money originated via a process of voluntary exchange and that money prices are a reflection of subjective values only if arrived at by voluntary action on the part of both buyers and sellers. Therefore, we conclude that in order for entrepreneurship to assist economic expansion and development, there must be private property—the institutional foundation for voluntary exchange. Without private property there can be no voluntary exchange. Without voluntary exchange there can be neither money nor market prices. Without economic calculation those directing the allocation of factors of production have no way to know how to allocate them wisely. Capital is consumed and social standards of living fall.

THE IMPORTANCE OF PRIVATE PROPERTY

Our survey of the engines of economic development allow for drawing some conclusions regarding conditions necessary for such development. Institutionally, there must be a market economy. Even in a market economy hampered by various government interventions there can be some amount of economic expansion. The freer the market is the more economic expansion it can achieve. In order for economic expansion to be realized, people must be able to engage in social cooperation based on private property. Private property is necessary for voluntary exchange, and it is voluntary exchange that opens the door for the division of labor to develop. Private property also reduces the risk of saving and investment because in an environment of private property rights investors can keep whatever positive return they earn without fearing that it will be confiscated. Additionally, private property allows for the development of money and results in money prices that entrepreneurs use to calculate profit and loss. Hence, private property makes possible the productive use of factors of production.

Given the importance of private property and its corollary, economic freedom, to the development of the division of labor, the accumulation of capital, and entrepreneurship, it comes as no surprise that freedom and economic growth generally are positively related. History bears out this reality. Countries that have more economic freedom experience higher per capita incomes and greater expansion than countries with less economic freedom.

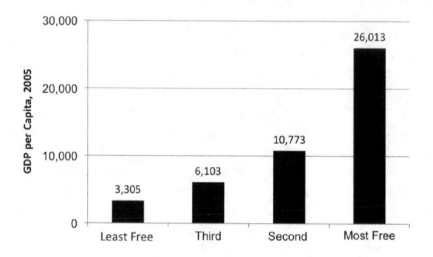

Figure 19.1. Economic freedom and prosperity are positively related. Those countries that scored in the highest quartile for economic freedom also had the highest per capita GDP. The Economic Freedom of the World Index is a statistic developed by the Fraser Institute and is designed to measure a country's level of economic freedom. Source: Gwartney, et. al, *Economic Freedom of the World*, 23.

What are we to conclude from all of the economic principles discussed in this book? God is clear in his Word that as his image bearers we are to have dominion over this earth. We are to be fruitful and multiply. We are to fill the earth and subdue it. This is the cultural mandate. Fulfilling the cultural mandate requires economic development. Economic expansion is the result of the division of labor, capital accumulation, and entrepreneurship. Neither the division of labor nor entrepreneurial activity is possible without capital.

The bottom line is that the social institution absolutely necessary for economic expansion and development is private property. Only private property makes possible voluntary exchange, the division of labor, the development of money, free market money prices, the economic coordination that results from successful entrepreneurship, and the accumulation of capital that makes the division of labor and entrepreneurial production possible.

Unclear or nonexistent property rights only hinder development. People are less willing to engage in voluntary exchange if they are uncertain whether they will be able to use what they trade for as they see fit. Uncertain property rights reduce the incentive to long-term capital investment.

Freedom promotes economic prosperity. Freedom also leads to a growing population, allowing us to obey God's command to fill the earth with his image bearers. The economic development that occurs in freer societies creates higher standards of living and makes it easier for parents to support more children. Freedom also allows for foreign direct investment that fosters global economic expansion. Consequently, the freedom that accompanies private property and the resulting growth of the market are both important for the development of each person and the diversity of individual differences with which God created us.

Unfortunately, economic policy in less developed countries often does not allow for economic output to increase along with population. Many governments in less developed countries boldly expropriate property of the productive. Many invoke draconian controls over foreign direct investment. Many impose high and discriminatory taxation. All of these policies prevent the formation of domestic capital and discourage the investment of foreign capital. Consequently, these policies retard economic expansion and result in social disintegration, poverty, and, in severe cases, starvation and death. Foreign aid is no panacea either because it manages to do little other than subsidizing interventionist and outright socialist economic policies of the governments that receive the aid.

Economic theory proposes that sustained economic expansion and development are possible. What prevents economic progress from occurring is not the depletion of resources, overpopulation, or capitalist exploitation in a world of free trade. The principle obstacle to economic development is an environment that penalizes individual initiative, is hostile to private ownership, discourages saving and investment, and severely restricts the operation of the free market.

Freedom provides the environment that fosters development. People need freedom to form, test, and act upon their own choices for the development of the talents, abilities, and desires with which they have been created. The freer the society, the more variety and diversity the population will exhibit. The more despotic a society, the less developed and diverse will be the personalities. That is why in totalitarian nations we also tend to see mass culture. People become less humanly unique and more like cogs in mass society.

To fulfill God's cultural mandate, society must be fully developed. As demonstrated earlier, people must have the opportunity to specialize if they want to develop whatever field for which they are called. Without specialization society must forego the service of the philosopher, scientist, builder, and merchant. Economic progress and social development are mutually reinforcing concepts.

A corollary of the security of private property is security in general. For the division of labor to develop and extend, society must enjoy peace. The division of labor is able to develop only because its participants expect lasting peace and have the ability to exchange that goes along with such peace. Conflict destroys the division of labor because it forces each group to consume only what it produces. Nations that are in conflict do not benefit from each other's comparative advantage because they do not exchange with one another. Even just wars that must be fought to defend one's social order from foreign aggression are inherently destructive. The division of labor that results in economic expansion requires the absence of violent conflict.

RELIGION, CULTURE, AND ECONOMIC PROSPERITY

Private property does not guarantee economic progress; it only makes it possible. Remember that one of the necessary conditions for economic expansion is the accumulation of capital. To increase the capital stock, it is necessary for society to work and to save and invest more than enough to merely maintain capital. Therefore, economic progress requires relatively low social time preferences. With higher time preferences people are less likely to put off present leisure and to engage in productive work. Likewise people with higher time preferences are less likely to put off present consumption. The lower people's time preference, the more they will save and invest and the more capital entrepreneurs will have available.

A society that maintains private property cannot, obviously force people to save and invest their incomes. In such an environment they are free to allocate all of their income to consumption if they wish. Therefore, in a free market, a society participates in just as much economic expansion as its members want. If people prefer more present consumption and less economic progress, they can achieve this. If people prefer instead to put off present consumption so they can consume more in the future as a result of economic development, they can do this too.

Because economic progress requires a minimum threshold of savings, private property is not a sufficient condition for economic expansion. Social time preferences are also very important, which implies that cultural and moral values play a vital role in economic performance over time. Cultures that are predisposed to highly value present consumption will not progress economically. Edward C. Banfield concluded in his landmark study of urban life, *The Unheavenly City Revisited*, that people making up the under-class tend to have relatively high time preferences. Consequently, they tend to place little value on work, self-sacrifice, or self-improvement. They simply do not have the long term future in view. Theodore Dalrymple reaches a similar conclusion in his more recent, *Life at the Bottom*. Societies that value thrift and foster longer time horizons will tend to experience more economic development.

A free and prosperous society will not thrive in every cultural environment. Certainly voluntary exchange will increase the welfare of the participants no matter what social values are like. However, certain cultural and moral values are more conducive to the development of the free market than others.

Because of its importance in shaping values, habits, and cultures, it is difficult to overemphasize the importance of religion as it affects economic prosperity. Ultimately, the problem of the cultural mandate and economic development is a spiritual and intellectual problem. Prosperity is a matter of capital investment, but it is not merely that. Economic development also depends on human attitudes. While economic expansion cannot guarantee spiritual, intellectual, moral, or aesthetic well-being, religious views do have a tremendous effect on general economic performance.

For example, religious error certainly places people in horrible danger for eternity, but such error also leads to destructive behavior in this present world. Habits of mind that follow from bad religion and that tend toward poverty include a lack of personal initiative, a platonic disinterest in material creation, a lack of personal responsibility, a high preference for leisure, an acceptance of determinism, the fear of change, belief in the occult, pantheism, and a mystical view of nature. The more primitive a society, the less civilized it tends to be in the sense that it tends not to take advantage of the social division of labor.

Additionally, bad moral decisions often tend to result in undesired economic consequences. Sexual promiscuity for example, tends to result in lives economically dependent on others. Increased out-of-wedlock births tend to inflate welfare rolls and foster an attitude of entitlement among those affected.

Metaphysical idealism and magic in many pantheist, Hindu, Buddhist, and animistic societies, undercut the rational development of knowledge and resources. Materialist humanism, such as what characterizes Marxism and much of the contemporary welfare state, undercuts motives for service necessary for development.

The Christian worldview, however, provides the best philosophical basis for rapid and long-term economic expansion and development. Christian doctrine provides the moral foundation for private property, a positive view of population, and a view of labor as a calling. A Christian worldview rejects centralized state planning in which brute force is used to impose the ruler's view on everyone else. Economic expansion and development requires the right to life, liberty, and property rooted in the Christian social ethic.

The Christian worldview also promotes lower time preferences. Christianity is forward looking and emphasizes planning for the future, thriftiness, and being diligent in labor. Behaviors that result from such attitudes promote prosperity. In his sweeping global economic history, *The Wealth and Poverty of Nations*, David Landes identifies the stress seventeenth century Protestants placed on redeeming the time as a significant reason for the flourishing of economic development in Northern Europe.

Cultural and moral values also affect the ease with which peace and private property are maintained. As market participants succumb to greed, they become more likely to lobby for state granted privileges via market regulation, reducing the scope of private property. Increased likelihood of fraud and theft increases uncertainty and results in more resources directed away from producing goods and more toward protecting property. Again, Christian faith and practice tend to constrain greed, fraud, and theft, and promote peace, honesty, and trust between our neighbors.

ECONOMIC DEVELOPMENT
IN LIGHT OF CHRISTIAN MORALITY

Not only are we called to rule, work and fill creation, but we are also called to keep and guard it. Consider then whether exercising dominion in a free market leads to waste, spoliation, or rampant consumption of the earth's bounty to the detriment of future generations. Private property and the free market are much more productive and, hence, seem to allow us to best fill the earth with people and sustain them through productive work. Does a free market do so, however, at the cost of leading us to abrogate our call to keep creation? Economic theory teaches us not to be overcome with worry about waste and spoliation in a society with an economy unregulated by the state.

Because all people are sinful, there is the possibility that they will seek dominion for their own glory instead of God's. This desire can easily manifest itself in the use and abuse of that part of creation over which God has given them charge. The relevant issue, however, is whether the proclivity to abuse creation is uniquely manifested or encouraged by a free market rooted in private property. Economic theory and history tell us that the answer is *no* on both counts.

For certain, people are fallen and may manifest sinful and destructive behavior in the marketplace. This, however, is not a function of the economic system but is a function of sin. People in socialist and interventionist economies are also sinful and can be quite prone to wasteful spoliation. The nuclear disaster in Chernobyl comes to mind.

One of the reasons that a free market does not encourage waste and spoilage is that in a society with private property, property owners are liable for any loss associated with the consumption of their capital. If they waste and destroy the productive capability of the land that they own, they do so at their own cost. They destroy a capital asset that could bring them income in the future. Therefore, owners have an incentive to maintain their capital and the productive capacity of their land.

When land is collectively owned the costs of use and abuse do not fall specifically on the user. Where no private property exists, it becomes less costly for people to overuse property because while they reap the full benefit of their use, they do not reap the full cost. There is an incentive, then, to overuse the property. This phenomenon is referred to as the *tragedy of the commons*, and regularly occurred in early colonial American towns and villages established by the Puritans. These small towns usually had a plot of land devoted to common pasturage open to all townspeople. As more outsiders moved to a community, the towns began assessing fees for the use of the commons. Users routinely over grazed livestock and over-harvested trees. Therefore, it is not a free market built upon private property that encourages waste and spoilage, but the *lack* of private property.

Additionally, economic theory teaches that as long as markets are free, we need not worry about a general lack of food or resources bringing economic development to a screeching halt. Food and natural resources are scarce like all economic goods. The demand for food and resources exceeds the supply of them freely available in creation. They must, therefore, be economized. Food is grown and marketed and resources are mined, refined, and sold by entrepreneurs. Entrepreneurs tend to invest in those lines of production that give them the largest profit.

If the demand for a good increases faster than its supply, the price of that good rises as the most eager buyers bid up the price. Those who are not willing to pay the higher price for the good abstain, searching for less expensive substitutes. At the same time, entrepreneurs direct more resources into the production of that good because it is now more profitable to produce. This increased investment often takes the form of capital goods that embody more advanced technology.

This is what happened regarding food. The availability of arable land increased and continues to increase with better technology and economic growth. The productivity per acre is increasing with better irrigation, planting, and harvesting made possible by more and better capital goods.

The Green Revolution of the 1960s and 1970s provided a big boost to world food production as it greatly increased yields. The Green Revolution occurred as farmers in less developed countries such as India shifted to high-yield crops, introduced irrigation and a better managed water supply, expanded use of pesticides and herbicides, and increased their own farm management skills. Biotechnology is poised to further increase yields with the development of plants resistant to drought, insects, disease, and salinity.

A similar economic process helps explain why, notwithstanding increased usage, natural resources appear relatively more abundant now than they were a hundred years ago. Suppose that due to global population growth and increased incomes in more developed countries, the demand for a natural resource increases. As is indicated in the graph in Figure 19.2, the market price for the resource will also increase.

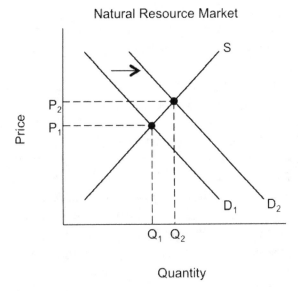

Figure 19.2. Increases in the demand for natural resources relative to their supply will result in an increase in their market price.

How do people respond to higher prices for the natural resource in question? As the market price for a good increases, potential demanders respond accordingly. The less eager buyers, those not willing to pay a higher price, will search for substitutes and work to mitigate some of the increased demand.

Additionally, as the price of a natural resource increases, users of the resource will be encouraged to increase resource efficiency. They will work at ways of getting more use out of the same quantity of resources. Today a single optical fiber carries the same number of telephone calls as 625 copper wires did just twenty-five years ago, freeing up copper for other uses. Newspaper technology has improved so that newspapers can be printed on thinner paper. Higher resource prices provide an incentive for technological development increasing resource efficiency. Higher prices for resources also encourage buyers to recycle, slowing the increase in demand for the resource in question.

At the same time that higher prices encourage more efficiency and recycling, they also stimulate investment that results in increased production. If the price of a resource increases, it becomes more profitable to produce. Consequently, entrepreneurs become more eager to invest in that industry. As they direct more investment into the production of that resource, the supply of that resource increases.

This increased investment includes increases in research and development. There have been for example, great increases in technology at the disposal of oil drillers and refiners. The quantity of oil that can actually be brought out of the earth increases as a result. Better technology allows previously uneconomical resource extraction to become economical, increasing known reserves.

We see that if markets are allowed to operate freely, market prices work to mitigate relative scarcity by encouraging both buyers and entrepreneurs to act in such a way that any crisis is avoided. If there is a noticeable increase in price, buyers are encouraged to economize and producers are encouraged to produce. Both actions work to lower the price of the good in question. In a free market, people are provided incentives to act to mitigate relative scarcity and are given free opportunity to do so.

We have already seen that the economic solution to all of these problems is to free the market. A free market encourages the development of the division of labor, capital accumulation, and entrepreneurial activity that results in economic prosperity. Remember, however, that a free market rooted in private property is also the biblical solution. The free market is the only system that is fully compatible with the Christian ethic of private property.

This should not be taken to mean that a free market guarantees a perfect outcome. Because the market is a network of voluntary exchange undertaken by men and women, the market is only as perfect as the people in it. If people demand corrupt popular music and pornography, then it is likely that some will supply that demand and earn a profit doing so. Also, while a free market is generally more productive than any other economic system, it does not guarantee freedom from all want for all people all of the time.

Some people are uncomfortable with the free market because they fear that it is essentially a system of unbridled capitalism ruled by greed. They view free enterprise as a system where big business roams about seeking which smaller businesses and neighborhoods it can devour. The concern is that capitalists get rich by turning out vulgar goods that corrupt the citizenry, while at the same time too many people must work extremely hard without managing to escape poverty. Problems do persist in a free society. We live in a fallen world. The issue is how we should respond.

Some argue that Christian compassion demands that someone step in and constrain capitalism while helping those in need. Surely, it is argued, Christian morality implies a certain amount of state regulation to assure some minimum outcome that is consistent with such morality.

Christian morality does demand compassion and assistance for the oppressed and less fortunate. The Christian ethic of private property, however, requires that we do not resort to acts of violent aggression in order to alleviate the suffering of the downtrodden. We are not allowed to force people to give to others. We are not allowed to use the power of the state to keep people from entering into voluntary contracts. We are called to be charitable to those who need it. We are called to be merciful to the poor and to see that widows and orphans are not oppressed. Justice demands, however, that we use our own resources to help those in need and to convince others voluntarily to do so. Lobbying for government intervention in the economy to force others to provide charity violates their right to property and therefore violates the Christian ethic.

Additionally, those who worry that a free society will result in an unbridled market that produces a society characterized by harsh, greedy, unrestrained capitalists who stop at nothing to increase their fortunes, misconstrue the nature of the free market. In a free market entrepreneurs cannot force anyone to buy their products. To receive revenue, firms must convince people to voluntarily purchase from them. As noted in chapter 17, the action of a profit-seeking entrepreneur is, therefore, far from unregulated. In a free market, the entrepreneur may not be regulated by the state, but is regulated by his own conscience and by the preferences of consumers. If people do not want to buy from an entrepreneur who has a reputation for wrong-doing, they are free to refrain.

In this way the church can properly act to regulate the economy. Christian property ethics forbid us to use state intervention to achieve a better society. Instead, God called his people to share the gospel and make disciples of all nations. As the church does this, people's values begin to change. People will begin to be more loving and kind to their neighbors. People will begin to demand more products that edify and fewer that are culturally destructive. If Christians really want different market outcomes, they should be obedient in their calling and have faith that God can transform the hearts and minds of men and women.

FREEDOM AND THE CULTURAL MANDATE

While our first parents were still living in the Garden, God told them to be fruitful and multiply, fill the earth, subdue it, and have dominion over every living creature. They were to cultivate both themselves and the created order. That mandate was reiterated after the fall. Fulfilling this mandate requires much economic expansion and development. The only way to cultivate and fill the earth without descending into a barbaric struggle for survival is to take advantage of the social division of labor, capital accumulation, and wise entrepreneurship. Allowing these sources of economic progress to flourish requires the security provided by the Christian ethic of peace and private property sustained by and combined with Christian virtues such as the forsaking of theft, honesty in our relationships, and low social time preferences.

Mere sloganeering about freedom will never be sufficient. Empty rhetoric about market economies coupled with interventionist economic policy will result in nothing except cynicism and misplaced anger fueled by an anti-capitalistic mentality that believes that the persistence of poverty is the result of a free market. Economics teaches that, in fact, poverty is the result of state intervention in the economy that constrains the division of labor, capital accumulation, and entrepreneurial activity. Only real peace coupled with private property and the cultural values that sustain it will enable people to more ably fulfill the cultural mandate God gave from the beginning.

SUGGESTED READING

Bauer, *Equality, The Third World, and Economic Delusion.* An outstanding collection of essays demonstrating the vast chasm between statist propaganda and economic reality as it relates to economic development.

Beisner, *Prospects for Growth*, 43–65. A biblical perspective on population in the context of modern demographic theory.

———. *Prosperity and Poverty*, 194–98. A brief introduction to the importance of religious belief as a contributor toward either poverty or prosperity.

———. "Sixpence None the Richer," 21–25. A brief, but informative discussion of economic development in the Western Civilization since 1000 AD.

Easterly, *The Elusive Quest for Growth.* An excellent documentation of the failure of foreign aid to produce economic development.

Hegeman, *Plowing in Hope.* An excellent primer on the Christian theology of culture in light of the cultural mandate.

Herbener, "The Role of Entrepreneurship in Desocialization," 80–86. An excellent concise explanation on the nature and importance of entrepreneurship.

Herbener and Gordon, "God's Mandate and Entrepreneurship," 90–99. An outstanding exposition of God's cultural mandate and the importance of entrepreneurial activity in fulfilling that mandate.

Hodge, *Systematic Theology*, 421–37. In this chapter, Hodge argues that God's commandment against theft mandates a divine right to private property.

Lomborg, *The Skeptical Environmentalist*. An outstanding empirical analysis of the state of various indicators regarding sustainable development.

Mises, "Foreign Investment," 75–91. An excellent introductory lecture on the importance of foreign investment for economic development.

———. *Human Action*, 157–64. Mises' excellent explanation of the characteristics and importance of the division of labor.

Osterfeld, *Prosperity Versus Planning*. An outstanding introduction to the political economy of economic development.

Rothbard, "Freedom, Inequality, Primitivism, and the Division of Labor," 3–35. A profound discussion of the importance of the division of labor for human development.

———. *Man, Economy, and State*, 47–70. An outstanding introduction to the economics of capital formation.

Schlossberg, *Idols for Destruction*, 39–139. A masterful Christian critique of modern ideology as it relates to a variety of issues connected to economic prosperity.

Bibliography

Armentano, Dominick T. *Antitrust and Monopoly: Anatomy of a Policy Failure*. 2nd ed. Oakland, CA: The Independent Institute, 1990.

———. *Antitrust: The Case for Repeal*. Revised 2nd edition. Auburn, AL: The Ludwig von Mises Institute, 1999.

Banfield, Edward C. *The Unheavenly City Revisited*. Boston, MA; Little, Brown, and Company, 1974.

Bauer, P. T. *Equality, the Third World, and Economic Delusion*. Cambridge, MA: Harvard University Press, 1981.

Beisner, E. Calvin. "Poverty: A Problem in Need of a Definition." In *Welfare Reformed*, edited by David W. Hall, 111–30. Phillipsburg, NJ: P & R Publishing, 1994.

———. *Prospects for Growth: A Biblical View of Population, Resources, and the Future*. Westchester, IL: Crossway Books, 1990.

———. *Prosperity and Poverty: The Compassionate Use of Resources in a World of Scarcity*. Westchester, IL: Crossway Books, 1988.

———. "Sixpence None the Richer," World, July 31, 1999, 21–25.

Block, Walter, et al. "Rent Control: An Economic Abomination." *International Journal of Value-Based Management* 11 (1998) 253-63.

Böhm-Bawerk, Eugen von. *Positive Theory of Capital*. Vol. II, *Capital and Interest*. South Holland, IL: Libertarian Press, 1959.

Boudreaux, Donald J. and Thomas J. DiLorenzo. "The Protectionist Roots of Antitrust." *The Review of Austrian Economics* 6, no. 2 (1993) 81–96.

Calvin, John. *Institutes of the Christian Religion*, edited by John T. McNeil. Philadelphia, PA: The Westminster Press, [1559] 1960.

Clark, Gordon H. *A Christian View of Men and Things*. Jefferson, MD: The Trinity Foundation, 1991.

———. *Thales to Dewey*. Jefferson, MD: The Trinity Foundation, 1985.

Dabney, R. L. *The Practical Philosophy*, Harrisonburg, VA: Sprinkle Publications, [1897] 1984.

Dalrymple, Theodore, *Life at the Bottom: The Worldview That Makes the Underclass*. Chicago, IL: Ivan R. Dee, 2001.

Dyer, Charles H. "Ezekiel." *The Bible Knowledge Commentary*, edited by John F. Walvoord and Roy B. Zuck, 1225–1317. Wheaton, IL: Victor Books, 1985.

Easterly, William. *The Elusive Quest for Growth: Economists' Adventures and Misadventures in the Tropics*. Cambridge, MA: The MIT Press, 2001.

The Editors of *Cook's Illustrated*. *The Best Recipe*. Brookline, MA: Boston Common Press, 1999.

Fiske, John. *The Beginnings of New England*. Boston, MA: Houghton, Mifflin, and Company, 1898.

Friedman, Milton, and Rose Friedman. *Free to Choose: A Personal Statement*. New York: Harcourt Brace Jovanovich, 1980.

Garrison, Roger. *Time and Money: The Macroeconomics of Capital Structure*. London, UK: Routledge, 2001.

———. "The Trouble with Deficit Finance, No pages. Accessed August 6, 2008. Online: http://mises.org/story/1158.

Gwartney, James, et. al. *Economic Freedom of the World: 2007 Annual Report*. Vancouver, BC: The Fraser Institute, 2007.

Hay, Donald A. *A Christian Critique of Socialism*. Bramcote, Nottinghamshire: Grove Books, 1982.

Hayek, F. A., ed. *Collectivist Economic Planning*. Clifton, NJ: Augustus M. Kelley, 1975.

———. "The Use of Knowledge in Society." *The American Economic Review*, 35, no. 4 (September 1945) 519–30.

Hegeman, David Bruce. *Plowing in Hope: Toward a Biblical Theology of Culture*. Moscow, ID: Canon Press, 1999.

Herbener, Jeffrey M. "The Role of Entrepreneurship in Desocialization." *The Review of Austrian Economics* 6, no. 1 (1992) 79–93.

Herbener, Jeffrey M. and T. David Gordon, "God's Mandate and Entrepreneurship." In *A Noble Calling: Devotions and Essays for Business Professionals*, edited by David Wesley Whitlock and Gordon Dutile, 90–99. Eugene, OR: Wipf and Stock, 2008.

Heyne, Paul. *The Economic Way of Thinking*, 9th ed. Upper Saddle River, NJ: Prentice Hall, 2000.

Higgs, Robert, "The U.S. Food and Drug Administration: A Billy Club Is Not a Substitute for Eyeglasses." In *Against Leviathan: Government Power and a Free Society*, 59–73. Oakland, CA: The Independent Institute, 2004.

Hodge, Charles. *Systematic Theology*. Vol. 3. Peabody, MA: Hendrickson Publishers, [1871–72] 1999.

Hoppe, Hans-Hermann. *A Theory of Socialism and Capitalism*. Boston, MA: Kluwer Academic Publishers, 1989.

Huerta de Soto, Jesus. *Money, Bank Credit, and Economic Cycles*. Auburn, AL: The Ludwig von Mises Institute, 2006.

Jaki, Stanley L. *The Savior of Science*. Washington, DC: Regnery Gateway, 1998.

Krech, Shepard. *The Ecological Indian: Myth and History*. New York: W. W. Norton & Company, 1999.

Larpenteur, Charles. *Forty Years A Fur Trader on the Upper Missouri: The Personal Narrative of Charles Larpenteur, 1833–72*. Lincoln, NE: The University of Nebraska Press, [1933] 1989.

Lenin, V. I. *The State and Revolution*. Peking, China: Foreign Languages Press, [1917] 1970. Accessed August 6, 2008. Online: http://ia351424.us.archive.org/1/items/TheStateAndRevolution/MicrosoftWord-Document1.pdf.

Lomborg, Bjørn. *The Skeptical Environmentalist: Measuring the Real State of the World*. Cambridge, UK: Cambridge University Press, 2001.

Marx, Karl. *Capital: A Critical Analysis of Capitalist Production*, London: William Glaisher, Limited, 1912.

Mazour, Anatole G. *Soviet Economic Development: Operation Outstrip: 1921–1965*, Princeton, NJ: D. Van Nostrand Company, Inc., 1967.

Menger, Carl. *Principles of Economics*. Grove City, PA: Libertarian Press, Inc. [1871] 1994.

Mises, Ludwig von. "Foreign Investment." In *Economic Policy: Thoughts for Today and Tomorrow*, 75–91. Washington, DC: Regnery Gateway, 1989.

———. *Human Action*, Scholars ed. Auburn, AL: The Ludwig von Mises Institute, [1949] 1998.

———. "Lord Keynes and Say's Law." In *Planning for Freedom*, 64–71. South Holland, IL: Libertarian Press, 1980.

———. "Middle-of-the-Road Policy Leads to Socialism." In *Planning for Freedom*, 18–35. South Holland, IL: Libertarian Press, 1980.

———. "Monetary Stabilization and Cyclical Policy." In *The Causes of the Economic Crisis and Other Essays Before and After the Great Depression*, 53–153. Auburn, AL: The Ludwig von Mises Institute, 2006.

———. "Profit and Loss." In *Planning for Freedom*, 108–50. South Holland, IL: Libertarian Press, 1980.

———. *Socialism*. Indianapolis, IN: Liberty Classics, [1922] 1981.

———. "Stones into Bread, the Keynesian Miracle." In *Planning for Freedom*, 50–63. South Holland, IL: Libertarian Press, 1980.

———. *The Theory of Money and Credit*. Indianapolis, IN: Liberty Classics, [1912] 1981.

Murray, John. *The Epistle to the Romans*. Vol. II., Grand Rapids, MI: Wm. B. Eerdmans Publishing Company, 1968.

The New Oxford Annotated Bible, ed. by Herbert G. May and Bruce M. Metzger, New York: Oxford University Press, 1962.

North, Gary. "A Christian View of Labor Unions," *Biblical Economics Today*, 1 (2) (1978) 1–2.

———. *The Dominion Covenant*, rev. ed. Tyler, TX: Institute for Christian Economics, 1987.

———. *Honest Money: Biblical Principles of Money and Banking*. Biblical Blueprints Series. Nashville, TN: Thomas Nelson Publishers, 1986.

Osterfeld, David. *Prosperity Versus Planning: How Government Stifles Economic Growth*. New York: Oxford University Press, 1992.

Rector, Robert E., and Kirk A. Johnson. "Understanding Poverty in America," *Backgrounder* 1713, January 5, 2004.

Ritenour, Shawn. "What You Should Know About the Minimum Wage." No pages. Accessed August 6, 2008. Online: http://mises.org/story/1603

Rose, Tom. *Economics: Principles and Policy from a Christian Perspective*. Mercer, PA: American Enterprise Publications, 1986.

Rothbard, Murray N. *America's Great Depression*, Auburn, AL: The Ludwig von Mises Institute, 2000.

———. "Freedom, Inequality, Primitivism, and the Division of Labor." In *The Logic of Action Two: Applications and Criticism from the Austrian School*, 3–35. Cheltenham, UK: Edward Elgar, 1997.

———. "Karl Marx: Communist as Religious Eschatologist." *Review of Austrian Economics* 4 (1990) 123–79.

———. *Man, Economy, and State with Power and Market*. Scholar's ed. Auburn, AL: The Ludwig von Mises Institute, [1962] 2004.

———. "The Mantle of Science." In *Logic of Action One*, 3–23. Cheltenham, UK: Edward Elgar, 1997.

———. *The Mystery of Banking*. 2nd ed. Auburn, AL: The Ludwig von Mises Institute, 2008.

————. *What Has Government Done to Our Money?* Auburn, AL: The Ludwig von Mises Institute, 1990.

Schlossberg, Herbert. *Idols for Destruction: The Conflict of Christian Faith and American Culture*, Wheaton, IL: Crossway Books, 1990.

Shapiro, Milton, M. *Foundations of the Market-Price System*. Lanham, MD: University Press of America, Inc., 1985.

Smith, Adam. *An Inquiry into the Nature and Causes of the Wealth of Nations*. New York: The Modern Library, [1776] 1965.

Strong, James. *A Greek Dictionary of the New Testament* in *Strong's Exhaustive Concordance*. Gordonsville, TN: Dugan Publishers, Inc.

United States Department of Agriculture, "Farm Machinery and Technology," in *A History of American Agriculture 1776–1990*. No pages. Accessed August 5, 2008. Online: http://www.agclassroom.org/gan/timeline/farm_tech.htm.

Vedder, Richard K., and Lowell E. Gallaway. *Out of Work: Unemployment and Government in Twentieth-Century America*. New York: Holmes and Meier, 1993.

Watts, Isaac. *Logic*. Morgan, PA: Soli Deo Gloria Publications, 1996.

Wayland, Francis. *The Elements of Moral Science*, Boston, MA: Gould and Lincoln, 1856.

————. *The Elements of Political Economy*. Boston, MA: Gould, Kendall, and Lincoln, 1843.

Westminster Confession of Faith

The World Bank. *Doing Business in 2004*. Washington, DC: The World Bank, 2004.

Subject/Name Index